ATOMIC PROCESSES IN PLASMAS

Tenth Topical Conference

ATOMIC PROCESSES IN PLASMAS

Tenth Topical Conference

San Francisco, CA   January 1996

EDITORS
Albert L. Osterheld
William H. Goldstein
Lawrence Livermore National Laboratory

AIP CONFERENCE
PROCEEDINGS 381

American Institute of Physics   Woodbury, New York

Authorization to photocopy items for internal or personal use, beyond the free copying permitted under the 1978 U.S. Copyright Law (see statement below), is granted by the American Institute of Physics for users registered with the Copyright Clearance Center (CCC) Transactional Reporting Service, provided that the base fee of $6.00 per copy is paid directly to CCC, 222 Rosewood Drive, Danvers, MA 01923. For those organizations that have been granted a photocopy license by CCC, a separate system of payment has been arranged. The fee code for users of the Transactional Reporting Service is: 1-56396-552-6/ 96 /$6.00.

© 1996 American Institute of Physics

Individual readers of this volume and nonprofit libraries, acting for them, are permitted to make fair use of the material in it, such as copying an article for use in teaching or research. Permission is granted to quote from this volume in scientific work with the customary acknowledgment of the source. To reprint a figure, table, or other excerpt requires the consent of one of the original authors and notification to AIP. Republication or systematic or multiple reproduction of any material in this volume is permitted only under license from AIP. Address inquiries to Office of Rights and Permissions, 500 Sunnyside Boulevard, Woodbury, NY 11797-2999; phone: 516-576-2268; fax: 516-576-2499; e-mail: rights@aip.org.

L.C. Catalog Card No. 96-86304
ISBN 1-56396-552-6
DOE CONF- 960155

Printed in the United States of America

CONTENTS

Preface ix

ATOMIC PHYSICS IN TOKAMAK PLASMAS

Recent Developments in the Role of Atomic Processes in Future Tokamaks ... 3
    D. Post

$2l-nl'$ X-ray Transitions from Neonlike Charge States of the Row 5 Metals with $39 \leq Z \leq 46$ 11
    J. E. Rice, K. B. Fournier, J. L. Terry, M. Finkenthal, E. S. Marmar,
    W. H. Goldstein, and U. I. Safronova

The Contribution of Spectroscopy on the Texas Experimental Tokamak (TEXT) to Atomic Physics and Plasma Physics: A Final Report 21
    W. L. Rowan

ATOMIC PHYSICS IN ASTROPHYSICAL PLASMAS

EUV Spectroscopy of Stellar Coronae 31
    N. S. Brickhouse

Laboratory Astrophysics: Measurements of $n=n'$ to $n=2$ Line Emission in Fe^{16+} to Fe^{23+} 39
    D. W. Savin, P. Beiersdorfer, G. V. Brown, J. Crespo López-Urrutia,
    V. Decaux, S. M. Kahn, D. A. Liedahl, K. J. Reed, and K. Widmann

Atomic Physics for Modeling of Astrophysical Plasmas 47
    J. M. Shull

X-RAY SOURCES AND X-RAY LASERS

Demonstration and Study of a Discharge-Pumped, Table-Top Soft-X-ray Laser 59
    J. J. Rocca, F. G. Tomasel, V. N. Shlyaptsev, J. L. A. Chilla, D. Clark,
    and M. C. Marconi

Observation of Multiphoton-Induced Multiple Vacancy Production in Xe(L) Emission from Plasma Channels 71
    A. McPherson, A. B. Borisov, K. Boyer, and C. K. Rhodes

Short Wavelength Generation from Atomic Clusters 75
    T. D. Donnelly, T. Ditmire, M. D. Perry, and R. W. Falcone

ATOMIC PHYSICS IN COOL PLASMAS

Electron-Impact Ionization of Molecules of Interest to Processing Plasmas 85
    K. H. Becker and V. Tarnovsky

New Model for Electron-Impact Ionization Cross Sections
of Atoms and Molecules... 93
    Y.-K. Kim, W. Hwang, and M. E. Rudd

SPECTROSCOPY AND DIAGNOSTICS

Critical Tests of Line Broadening Theories by Precision Measurements....... 109
    S. H. Glenzer
Spectroscopic Temperature Measurements of Non-equilibrium Plasmas....... 123
    C. A. Back, S. H. Glenzer, R. W. Lee, B. J. MacGowan, J. C. Moreno,
    J. K. Nash, L. V. Powers, and T. D. Shepard
Various Applications of Atomic Physics and Kinetics Codes
to Plasma Modeling.. 131
    J. Abdallah, Jr., R. E. H. Clark, D. P. Kilcrease, G. Csanak, and C. J. Fontes
The Motional Stark Effect Diagnostic on TFTR 143
    F. M. Levinton
Polarization Spectroscopy of Ionized Gases 151
    S. A. Kazantsev
Line Shape Measurements of Visible Light Emission from the Alcator
C-Mod Tokamak... 159
    B. L. Welch, H. R. Griem, J. L. Weaver, J. L. Terry, R. L. Boivin,
    B. Lipschultz, D. Lumma, E. S. Marmar, G. McCracken, and J. C. Rost

ATOMIC DATA AND DATABASES

Ion-Atom Collisions Relevant to Fusion Plasmas 169
    H. B. Gilbody
The NIST Spectroscopic Database and Two New Critical Data Tables 177
    W. L. Wiese
Electron-Ion Collision Processes.. 187
    G. H. Dunn
On-line Atomic Data Access 197
    D. R. Schultz and J. K. Nash

ULTRA-SHORT PULSE LASER PLASMAS

Picosecond Time-Resolved Spectroscopy of a Controlled Preformed
Plasma Heated by an Intense Sub-ps Laser Pulse........................ 207
    C. Y. Côté, Z. Jiang, J. C. Kieffer, A. Ikhlef, H. Pépin, and O. Peyrusse

ATOMIC PHYSICS IN DENSE PLASMAS

Population Kinetics in Dense Plasmas.................................... 215
    M. Schlanges, Th. Bornath, R. Prenzel, and D. Kremp

De-excitation and Recombination Processes Involving Doubly Excited Levels in High-Density Plasma... 223
    T. Kawachi

Plasma Density Effects on Ionization 231
    M. S. Murillo

ATOMIC PHYSICS IN STRONG FIELDS

Atomic Emission Spectroscopy in High Electric Fields 245
    J. E. Bailey, A. B. Filuk, A. L. Carlson, D. J. Johnson, P. Lake,
    E. J. McGuire, T. A. Mehlhorn, T. D. Pointon, T. J. Renk, W. A. Stygar,
    Y. Maron, and E. Stambulchik

Laser Beam Propagation, Filamentation and Channel Formation in Laser-Produced Plasmas... 259
    P. E. Young, S. C. Wilks, W. L. Kruer, J. H. Hammer, G. Guethlein,
    and M. E. Foord

RADIATIVE OPACITY

A Radiation-Dependent Ionization Model for Laser-Produced Plasmas 271
    M. Busquet

New Methods for Probing the Opacity and Optical Properties of Dense Low-Temperature Plasmas 279
    A. N. Mostovych, L. Y. Chan, and K. J. Kearney

The Rosseland Mean Opacity of a Composite Material at High Temperatures.. 287
    T. J. Orzechowski, M. D. Rosen, H. N. Kornblum, J. L. Porter, L. J. Suter,
    A. R. Thiessen, and R. Wallace

Author Index... 295

PREFACE

This volume contains papers based on 30 of the 37 invited talks presented at the Tenth American Physical Society Topical Conference on Atomic Processes in Plasmas, held in San Francisco, California, on January 14–18, 1996. This biennial conference series provides a forum for those whose research overlaps atomic physics and plasma physics. The interaction between atomic and plasma physicists is essential for applied research in inertial fusion energy, magnetic fusion energy, astrophysics, X-ray lasers, and materials processing. The previous conference in this series was held in San Antonio, Texas, on September 19–23, 1993 (AIP Conference Proceedings No. 322).

The volume begins with a section on atomic physics in tokamak plasmas. These papers discuss important uses of radiation cooling in the design of future tokamak reactors, applications of spectroscopy to diagnostics of magnetic fusion energy plasmas, and the use of tokamaks to study atomic structure and electron-ion collision processes, and to perform a very detailed study of L-shell X-ray transitions.

Astrophysics and the study of atomic processes in plasmas have had a long and productive association. The section on atomic processes in astrophysics begins with an exposition of extreme ultraviolet spectroscopy and its impact on the current understanding of solar and stellar coronae. Another paper details experiment work testing the atomic physics underlying recently developed iron L-shell emission models. The final contribution reviews atomic physics models for ionization equilibrium and emission of hot astrophysical plasmas, and the use of absorption lines to derive elemental abundances.

There is a section devoted to the development of novel X-ray sources and short wavelength lasers. The first paper describes the recent development of a table-top soft X-ray laser using a capillary discharge. The other two papers describe the generation of high-energy radiation from vapors of atomic clusters and plasma channels driven by high-intensity lasers.

Plasma processing applications have led to an interest in atomic processes in cool plasmas. The two papers in this section discuss recent experimental and theoretical studies of electron impact ionization cross sections of atoms, molecules, and free radicals of interest in these applications.

Emission spectra provide a wealth of information on the conditions in a plasma. The section on spectral diagnostics discusses the physics underlying these diagnostics, as well as applications to a variety of plasmas. The section begins with a paper on critical tests of the line broadening theories. Other papers discuss spectroscopic diagnostics for inertial confinement fusion plasmas, applications of the motional Stark effect in a tokamak plasma, and line shape measurements of optical transitions to characterize the magnetic field, ion temperature, and electron density in the edge and divertor regions of a tokamak. An additional paper describes a suite of atomic physics and kinetics codes used for several plasma modeling applications. The final paper in this section reviews polarization spectroscopy of ionized gases, emphasizing applications to remote sensing.

Four papers deal with atomic data and database projects developed to facilitate the access to atomic data for applications. These contributions discuss ion-atom collisions relevant to tokamak plasmas, recent developments in the spectroscopic databases and critical data tables maintained by the National Institute of Standards and Technology, direct and indirect electron-ion collisional excitation, ionization and recombination processes, and the rapidly developing field of on-line access to atomic data.

Three sections deal with atomic physics in unusual plasma environments. One paper studies the transient and non-Maxwellian effects in a pre-formed plasma heated by an intense sub-picosecond laser pulse. Three papers deal with modifications of atomic processes in high-density plasmas. The first uses quantum kinetic equations to develop a description of many-body effects in population kinetics of very dense plasmas. The following paper discusses enhancements of de-excitation and recombination caused by multi-step collision processes involving multiply excited levels. The final paper in this section treats plasma density effects on collisional ionization, including the effects of dynamical screening. There are two papers on atomic physics in strong fields. The first discusses atomic emission spectroscopy in the high electric fields present in ion diodes used to accelerate ion beams for inertial confinement fusion applications. The other paper describes experimental studies of filamentation, channel formation and propagation of a high-intensity laser beam through an under-dense plasma.

Radiative opacity is one of the most important manifestations of atomic processes in plasmas, and plays a crucial role in many fields, including astrophysics and inertial confinement fusion. This section begins with a discussion of a radiation-dependent ionization model used for rapid computations of opacity for plasma simulation codes. The following paper presents recent measurements of opacities in strongly coupled plasmas. The final paper describes experimental measurements of the Rosseland mean opacity of composite materials used to increase the wall opacity of laser-heated hohlraums for laser fusion research.

We are grateful to the Office of Fusion Energy and the Office of Energy Research of the U.S. Department of Energy, and the Physics and Space Technology Directorate of Lawrence Livermore National Laboratory for providing partial financial support for this Conference and the publication of these proceedings.

Albert L. Osterheld
William H. Goldstein
Livermore, California
June 1996

ATOMIC PHYSICS
IN TOKAMAK PLASMAS

Recent Developments in the role of atomic processes in future tokamaks

D. Post

ITER Joint Central Team, San Diego, California

Abstract. Since the beginning of magnetic fusion research, reducing the impurity level in experiments has been strongly correlated with successful achievement of high performance plasmas. One of the most important examples of this was the recognition that the use of tungsten as a plasma facing material and the associated high radiative losses were responsible for the poor performance of the ORMAK and PLT tokamaks. Tungsten was replaced with graphite and the central plasma temperature in PLT increased a factor of ten. The magnetic fusion program is now planning on constructing an ignited fusion experiment. One of the major design issues is the reduction of the peak heat loads on the plasma facing components. It appears that the carefully controlled introduction of impurities can lead to a solution of the problem.

INTRODUCTION

Over the past 30 years, extensive efforts have been devoted to reducing the role of impurities in tokamaks. Experimental success in achieving ever higher degrees of plasma performance has generally gone hand in hand with success in lowering the impurity content. The understanding of impurity radiation losses has benefited greatly from progress in the atomic physics of highly charged ions. This progress has led to the identification of the adverse effects of radiation from high Z impurities and the deleterious effects of high levels of low Z impurities on plasma performance[1,2]. Impurity atomic physics calculations have advanced from "average-ion" models in the mid-1970's[3] to very sophisticated collisional radiative models with detailed treatments of the atomic structure and meta-stables[4]. Measurements have advanced from the early theta pinch rate experiments to detailed cross section measurements for electron impact ionization and excitation, dielectronic recombination and ion-atom charge transfer collisions[5].

Simple estimates for power exhaust requirements for the next generation of fusion experiments such as ITER indicate that the peak power loads will approximately 30 MW/m^2 or higher, too high to develop an acceptable engineering design[6]. It is therefore planned to introduce impurities such as Neon, Argon and Krypton to increase the radiation losses and spread the heating power over the first wall and divertor chamber walls to reduce the power on the divertor plates[7]. The impurity level must be high enough in the plasma edge that there is adequate impurity radiation from the divertor and plasma edge but not so high that the impurity contamination of the plasma core adversely affects the plasma performance. This type of divertor operation has been observed on ASDEX

Upgrade, DIII-D, Alcator C-Mod, JET and JT-60 and on TEXTOR, TORE-Supra and TFTR with limiters. However, these experiments have power levels below that of fusion experiments such as ITER, so that extrapolations from these experiments using computational models will be necessary. These models need to be able to calculate the impurity radiation levels very accurately and thus need high quality atomic data. The required accuracy for ionization, recombination and excitation rates is 30% or better. There is only a relatively narrow window in impurity concentrations between the impurity levels in the plasma edge needed for radiation losses and the impurity levels that lead to unacceptable contamination of the plasma center. Since the plasma edge is relatively cool (5 to 500 eV), accurate rates are needed at lower temperatures than for the plasma core where the temperature is in the multi-keV range.

THE ROLE OF TUNGSTEN IN THE 1970'S

Tokamaks in the 1960's and early 1970's used refractory limiters to define the plasma edge to avoid plasma contact with the entire wall area. Without such limiters, large quantities of loosely bound impurities from the wall such as oxygen and carbon contaminated the plasma and led to large energy losses due to radiation. The limiters were made of metals with high melting temperatures such as tungsten and molybdenum which were able to survive high peak loads during disruptions and other off-normal events. With modest heating powers, central electron temperatures of 500 to 1000 eV were obtained. In the mid-1970's new, larger experiments such as the Princeton Large Torus (PLT) and Oak Ridge Tokamak (ORMAK) with higher currents (and better confinement) and with high power neutral beam heating came into operation. It was expected that these experiments would reach temperatures in the 3 to 5 keV range. However, with the auxiliary heating, the temperatures in these experiments either increased very little or decreased. Indeed, the PLT tokamak had hollow temperature

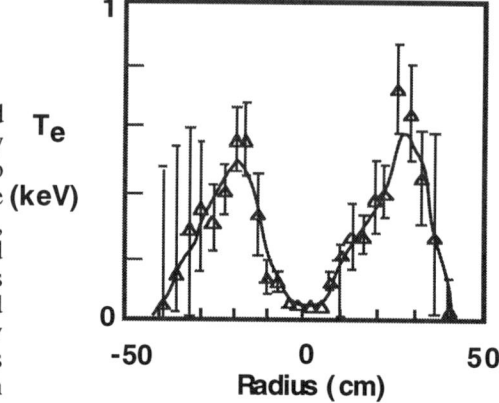

Fig. 1 Electron temperature profile in PLT with a tungsten limiter

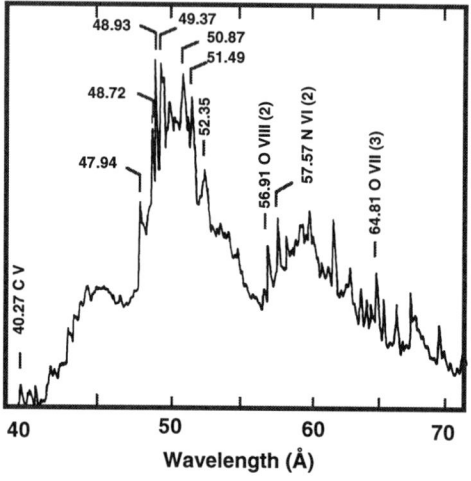

Fig. 2 Measured ORMAK spectrum

profiles (Fig. 1). Similar results were reported on the ORMAK experiment.

The reason for these profiles was eventually recognized to be energy losses from the center of the tokamak due to line radiation from tungsten. This was not immediately apparent because the tungsten emission spectrum was not known and the radiation emission rates were not known. Tungsten has many electrons (74) and detailed calculations of such a complicated system were at the frontier of atomic physics at that time. At the same time the experimentalists were inadvertently operating the world's largest tungsten lamps, two efforts came to fruition which identified the source of energy losses. One was an effort by R. Isler and R. Neideigh to measure the spectrum of the line emission from ORMAK which had a tungsten limiter and a gold liner (Fig. 2).

Together with calculations by R. Cowan of $\Delta n=0$ lines from WXXXI-WXXXV (Fig. 3), Isler and Neidigh were able to identify the major energy loss in ORMAK as due to tungsten[1]. Simultaneously, a collaborative effort between a Princeton group and a group at the Lawrence Livermore National Laboratory using an "average ion" model led to the calculation of the radiation rates from tungsten (Fig. 4)[2,3]. Together with bolometer measurements of the central radiation losses and confirmatory measurements on PLT, these measurements and calculations led to the recognition that a low Z refractory material such as graphite would be better than tungsten for use as a limiter. Graphite limiters were installed on PLT and almost immediately central temperatures of 4 keV were obtained (Fig. 5).

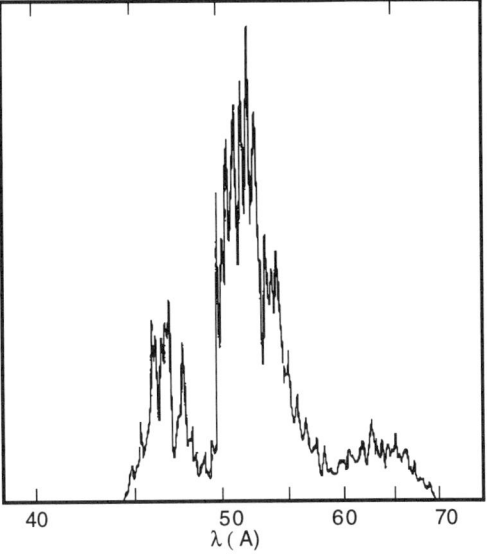

Fig. 3 Cowan calculated W spectrum

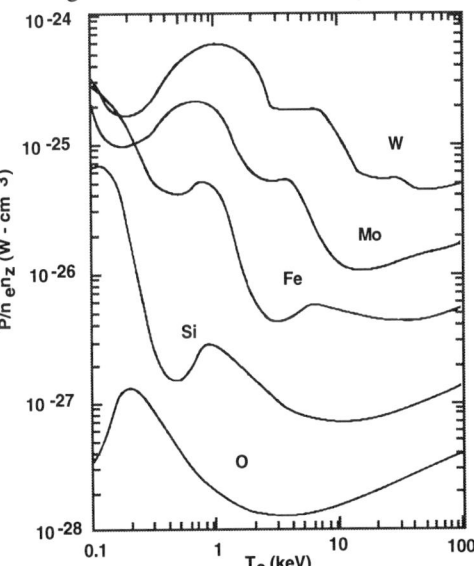

Fig. 4 Average ion emission rates

REDUCING THE PEAK HEAT LOADS IN DIVERTORS WITH IMPURITIES

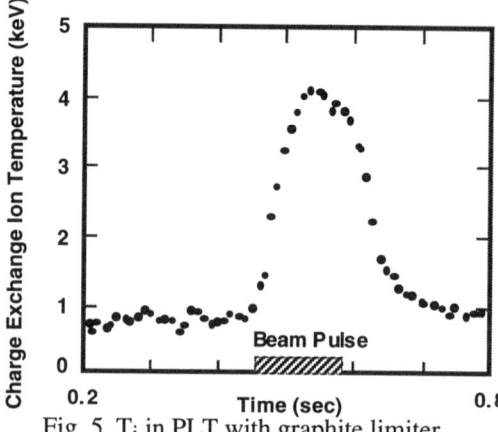

Fig. 5. T_i in PLT with graphite limiter

Power exhaust will be a key issue for the next generation of fusion experiments such as ITER[8]. The alpha heating power in ITER of about 300 MW will give a peak heat load on the divertor plates of about 30 MW/m^2 or larger. To maintain the surface temperature of the divertor plates at an acceptable level requires that the peak heat loads be reduced to ~ 5 MW/m^2. This can be accomplished by using impurity radiation, enhanced by the injection of impurities such as Ne or Ar, to spread out the plasma energy on the first wall and divertor chamber walls. An analysis of the various channels for energy losses from the plasma for the plasma core, the plasma edge inside the separatrix (mantle), the scrape-off layer and divertor plasma, and the divertor plates indicates that it should be possible to radiate about 250 MW of power to the walls before the energy reaches the plate (Fig. 6 and Table 1).

Table 1. Power flow in ITER

Bremsstrahlung	100 MW
Edge impurity radiation (inside separatrix)	50 MW
SOL/divertor impurity radiation (outside separatrix) + H and He radiation and charge exchange losses	100 (up to 250) MW
Power on target plates	50 MW
total heating power	300 (up to 400) MW

DETAILED POWER LOSSES IN ITER

Bremsstrahlung losses will be larger in ITER than in present experiments due to the higher electron density, temperature and energy confinement time. The ratio of the Bremsstrahlung emission P_{Brem} to the alpha heating, P_α, is independent of size and density (Eq. 1), where C_B is 2×10^{-40} Zeff MW/(m^{6}eV$^{0.5}$), n_e and T_e are the volume averaged density and temperature, $f_{DT} = n_{DT}/n_e$, E_α = 3.5 MeV, and $<\sigma v>_{DT}$ is the fusion reaction rate. For ITER, ($<T>$~10 keV, f_α~0.1, f_{Be}~0.01, f_{Ne}~0.003 and $P_{Brem}/P_\alpha \approx 0.33$ or about 100 MW in total . P_{Brem}/P_α decreases with T_e as $T_e^{-1.5}$, and increases with impurity level where $f_Z = n_{impurity}/n_e$.

There will be impurity radiation from the main plasma edge (mantle) just inside the separatrix to the first wall due to line emission from intrinsic impurities and injected impurities such as Neon or Argon. This radiation can be calculated from the radial heat conduction equation and the radiation emission rates for candidate injected impurities (Eq. 2). If one assumes that the density and the electron transport coefficients are roughly constant, as is approximately true for H-mode plasmas[9], the radiation losses are proportional to an integral of the radiation emission rate[10,11] (Fig. 7). The effect of heat conduction on the radiation losses and the different scaling

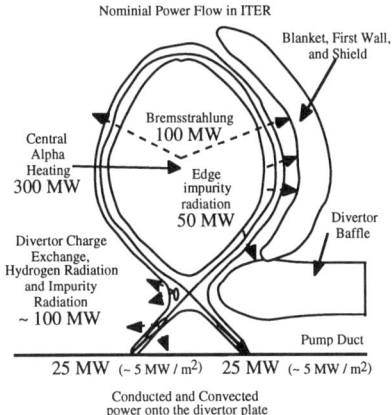

Fig. 6. Power flow in ITER.

with density and impurity fraction is significant. With conduction, $Q_\perp \approx n_e \sqrt{f_z}$, which is much weaker than the volume loss rate $\approx n_e^2 f_z$.

$$\frac{P_{Brem}}{P_\alpha} = \frac{C_B n_e^2 Z_{eff} T^{1/2}}{n_e^2/4 \, f_{DT}^2 \langle \sigma v \rangle_{DT} E_\alpha} = \frac{4 C_B Z_{eff} T^{1/2}}{f_{DT}^2 \langle \sigma v \rangle_{DT} E_\alpha} : \frac{Z_{eff}}{f_{DT}^2} \approx \frac{1 + 2 f_\alpha + \sum Z(Z-1) f_z}{1 - 2 f_\alpha - \sum Z f_z} \quad (1)$$

$$Q_\perp = \kappa \frac{\partial T}{\partial r} : \frac{\partial Q_\perp}{\partial r} = -n_e n_z L_z(T_e) \Rightarrow Q_\perp \frac{\partial Q_\perp}{\partial r} = -n_e^2 f_z L_z(T_e) \kappa \frac{\partial T}{\partial r} \quad (2)$$

$$\Rightarrow Q_{\perp center}^2 - Q_{\perp separatrix}^2 \approx 2 \int_{T_{separatrix}}^{T_{center}} n_e^2 f_z L_z(T_e) \kappa \, dT_e \approx 2 n_e^2 f_z \kappa \int_{T_{separatrix}}^{T_{center}} L_z(T_e) \, dT_e$$

For the range of expected conditions for ITER for $Z_{eff} \approx 1.6$ with $f_{He} = 0.1$, $f_{Be} = 0.01$, 8×10^{19} m^{-3} $\leq n_{edge} \leq 10^{20}$ m^{-3}, 0.125×10^{20} m^{-1} s^{-1} $\leq \kappa \leq 2 \times 10^{20}$ m^{-1} s^{-1}, Neon, Argon and Krypton can provide radiation losses of about 50 MW from the plasma edge ($Q_\perp \sim 0.05$ MW/m^2, Table 2). These analytic estimates are consistent with detailed calculations using the WHIST 1 1/2 tokamak transport code with a multi-species impurity transport package with the constraint that $\Delta P_\alpha / P_\alpha \leq 5\%$ and $\tau_E \approx 2 \times \tau_E^{ITERP-89}$ (Fig. 8).

A similar model can be used to estimate the radiation losses from the scrape-off layer and divertor plasma [11,12] (eq. 3). The parallel heat conduction is classical and the pressure is constant along the field lines. The radiation efficiency can be computed as a function of the upstream temperature (Fig. 9). The upstream temperature can be calculated by integrating the heat flux eq. from the midplane to the radiation region. The impurity radiation losses can be computed for candidate ITER upstream densities (Table 3). For heating powers of 100 to 200 MW, $Q_\| \approx$ 0.3 to 0.6 GW/m^2 for each divertor leg. Neon and higher Z impurities can radiate 100 MW or more from the divertor and SOL plasma. These losses will be supplemented by H, He and Be radiation and charge exchange losses. These results are consistent with detailed 2-D B2 code divertor modelling calculations using a multi-species transport model for Neon or Argon impurities (Table 4).

Table 2 Radial Impurity Radiation losses for Be, C, Ne, Ar and Kr for ITER

	Be	C	Ne	Ar	Kr
Assumed fraction ($Z_{eff} \leq 1.6$)	0.01	0.01	0.0033	0.001	.00025
$Q_\perp/(n_e (\kappa_\perp f_z)^{0.5})$ MW m*	0.1	0.4	1.0	3.0	10
$Q_\perp$ (MW/m^2) for $5 \times 10^{19} \leq n_e < 10^{20}$; $0.125 \leq \kappa \leq 2$	0.002 to 0.014	0.007 to 0.06	0.01 to 0.08	0.02 to 0.13	0.03 to 0.22

*$\kappa_\perp$ in 10^{20} m^{-1} s^{-1}, n_e in 10^{20} m^{-3}

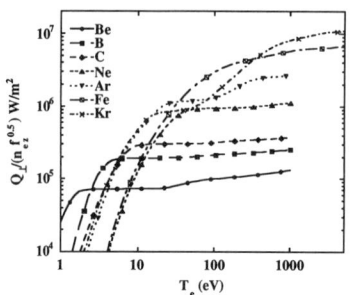

Fig. 7. Normalized radiation losses

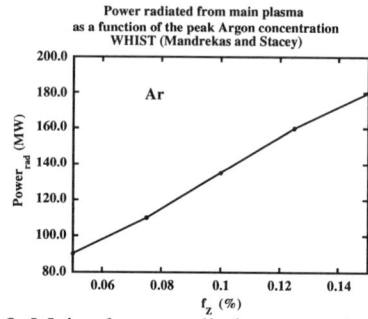

Fig. 8. Main plasma radiation vs n_{Ar}/n_e.

$$Q_\perp \Big/ n_e f_z^{0.5} \kappa^{0.5} = \sqrt{2 \int_0^T L_z(T_e)\, dT_e}\ .$$

$$\frac{\partial Q_\parallel}{\partial x} = -n_e n_z L_z(T_e) \ ;\ \ Q_\parallel = -\kappa_o T_e^{2.5} \frac{\partial T_e}{\partial x}\ ;\ \ p_e = n_e T_e \ \Rightarrow$$

$$\frac{1}{2}\frac{\partial Q_\parallel^2}{\partial x} \approx \frac{p_e^2}{T_e^2} \kappa_o f_z L_z T_e^{2.5} \frac{\partial T_e}{\partial x} \approx p_e^2 \kappa_o f_z L_z T_e^{0.5} \frac{\partial T_e}{\partial x} \Rightarrow \frac{1}{2} dQ_\parallel^2 \approx p_{es}^2 \kappa_o f_z L_z T_e^{0.5} dT_e \quad (3)$$

$$\Rightarrow \frac{\Delta Q_\parallel}{n_{es}\sqrt{f_z/Z_{eff}}} \approx \sqrt{2\bar{\kappa}_o T_{es}^2 \int_0^{T_{es}} L_z(T_e) T_e^{0.5} dT_e}\ \text{with}\ \kappa_o = \bar{\kappa}_o \frac{T_e^{2.5} Z_{eff}}{\ln \Lambda}$$

Table 3 Divertor radiation efficiencies for DIII-D and ITER.

Element	Be	C	Ne	Ar	Kr
f_z(%) for $Z_{eff} \sim 1.6$ with f_{He} = 0.1, $f_{Be} \sim 0.01$	0.01	0.01	.0033	.001	.00025
$Q_{\parallel ITER}/\sqrt{(f_z(\%)/Z_{eff})}$ at 250 eV (GW/m^2)	0.08	0.2	0.7	2	4
$Q_{\parallel ITER}$ (coronal equilibrium) for $n_e \approx 10^{20}$ m^{-3}	0.065	0.16	0.33	0.52	0.52

These radiation levels can be enhanced by charge exchange recombination and rapid recycling of the impurities[13-15] and the radiation losses calculated as before(Fig. 10). To radiate the required 100 MW in the divertor plasma takes relatively modest levels of recycling and neutral densities (Table 5).

The power balance in Table 1 thus appears to be achievable for operation with 300 MW of ignited operation. Driven operation with 400 MW of total heating will require larger radiation levels from the divertor.

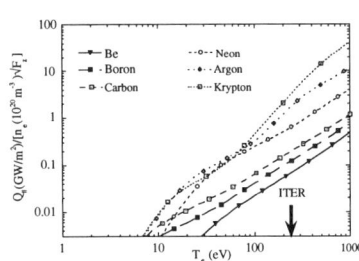

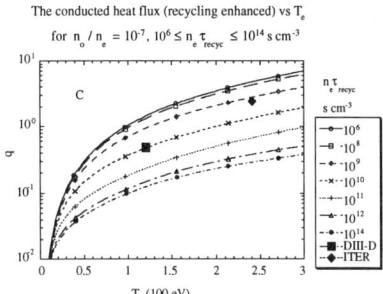

Fig. 9. Parallel radiation integral.

Fig. 10. Recycling enhanced radiation integral for carbon.

Table 4 B2 calculations of the radiated power in the ITER divertor for Neon

n_e^{sep} (10^{19} m^{-3})	f_{Ne} (%)	P_{rad} (MW)
4	0.5	70
4	0.25	40
6	0.5	100
6	0.25	60

Table 5 Enhanced Be, C, Ne, and Ar divertor radiation efficiencies for ITER.

Element	Be	C	Ne	Ar
f_Z(%) for Z_{eff}~1.6 : f_{He}=0.1, f_{Be}~0.01	1	1	0.33	0.1
$Q_{\|\|ITER}/\sqrt{(f_Z(\%)/Z_{eff})}$ at 250 eV required to radiate 100 MW	0.8	0.8	1.4	2.5
n_o/n_e (required to radiate 100 MW)	6×10^{-3}	10^{-3}	3×10^{-4}	10^{-3}
$n_e\tau_{recycle}$ (s cm^{-3}), for 100MW radiation	2×10^{10}	8×10^{10}	2×10^{11}	10^{10}

CONCLUSIONS AND SUMMARY

Impurities, particularly high Z impurities, were a severe problem in tokamaks in the early 1970's. The replacement of high Z materials by low Z materials for the plasma facing components allowed the achievement of the high temperatures and good energy confinement found on the present generation of tokamaks such as TFTR, JET and JT-60/U. This high level of plasma performance has lead to very demanding requirements for power exhaust on the next generation of fusion experiments such as ITER. To reduce the peak heat loads, impurity radiation from the divertor plasma due to the carefully controlled injection of impurities is to be used to spread out the power and reduce the peak power loads.

ACKNOWLEDGMENTS

The author is grateful for discussions with and assistance from R. Isler, R. Clark, B. Braams and J. Mandrekas.

REFERENCES

[1] R. C. Isler, R. V. Neidigh, and R. D. Cowan, Phys. Let. **63A** (3), 295-297 (1977).

[2] R. V. Jensen, D. E. Post, W. H. Grasberger *et al.*, Nucl. Fusion **17**, 1187-1196 (1977).

[3] D. E. Post, R. V. Jensen, C. B. Tarter *et al.*, Atomic Data and Nuclear Data Tables **20**, 397-439 (1977).

[4] R. Clark, J. Abdalleh, and D. Post, Journal of Nuclear Materials **220-222**, 1016-1120 (1995).

[5] R. K. Janev and H. W. Drawin, "Atomic and Plasma-Material Interaction Processes in Controlled Thermonuclear Fusion," (Elsevier, Amsterdam, 1993), pp. 484.

[6] K. J. Dietz, G. Janeschitz, Y. Igitkhanov *et al.*, "The ITER Divertor Concept," presented at the 15th International Conference on Plasma Physics and Controlled Nuclear Fusion Research, Seville, Spain, Sept. 26—Oct. 1, 1994 .

[7] D. E. Post, B. Braams, J. Mandrekas *et al.*, Contributions to Plasma Physics **(to appear)** (1996).

[8] K. Borrass G. Janeschitz, G. Federici, Y. Igitkhanov, A. Kukushkin, H.D. Pacher, G.W. Pacher and M. Sugihara, Journal of Nuclear Materials **220-222**, 73-88 (1995).

[9] D. Stork and JET Team, Plasma Physics and Controlled Fusion **36**, A23—A38 (1994).

[10] P. H. Rebut and B. J. Green, Plasma Physics and Controlled Nuclear Fusion Research 1976 (International Atomic Energy Agency, Vienna) **Vol. 2**, 3-17 (1977).

[11] L. Lengyel, Report No. 1/191, Max Planck Institute For Plasma Physics, Garching, 1981.

[12] M. Shimada, M. Nagami, K. Ioki *et al.*, Nuclear Fusion **22** (5), 643-655 (1982).

[13] R. A. Hulse, D. E. Post, and D. R. Mikkelsen, J. Phys. B **13**, 3895-3907 (1980).

[14] P. G. Carolan and V. A. Piotrowicz, Plasma Physics **25** (10), 1065—1086 (1983).

[15] S. Allen, M. Rensink, D. Hill *et al.*, Journal of Nuclear Materials **196-198**, 804 (1992).

$2l$ - nl' X-ray Transitions from Neonlike Charge States of the Row 5 Metals with $39 \leq Z \leq 46$

J. E. Rice, K. B. Fournier[+], J. L. Terry, M. Finkenthal[!],
E. S. Marmar, W. H. Goldstein[+] and U.I. Safronova[*]
Plasma Fusion Center, MIT, Cambridge, MA 02139-43070

[+] *Lawrence Livermore National Laboratory, Livermore, CA 94550*
[!] *Racah Institute of Physics, The Hebrew University, Jerusalem, Israel, 91904*
[*] *Institute for Spectroscopy, Russian Academy of Sciences, Troitsk, 142092, Russia*

Abstract

X-ray spectra of $2l$ - nl' transitions with $3 \leq n \leq 12$ in the row five transition metals zirconium (Z=40), niobium (Z=41), molybdenum (Z=42) and palladium (Z=46) from charge states around neonlike have been observed from Alcator C-Mod plasmas. Accurate wavelengths ($\pm$ .2 mÅ) have been determined by comparison with neighboring argon, chlorine and sulphur lines with well known wavelengths. Line identifications have been made by comparison to *ab initio* atomic structure calculations, using a fully relativistic, parametric potential code. For neonlike ions, calculated wavelengths and oscillator strengths are tabulated for 2p-nd transitions in Y (Z=39), Tc (Z=43), Ru (Z=44) and Rh (Z=45) with n = 6 and 7. The magnitude of the configuration interaction between the $(2p^5)_{\frac{1}{2}} 6d_{\frac{3}{2}}$ $J = 1$ level and the $(2p^5)_{\frac{3}{2}} 7d_{\frac{5}{2}}$ $J = 1$ levels is demonstrated as a function of atomic number for successive neonlike ions. Measured spectra of selected transitions in the aluminum-, magnesium-, sodium- and fluorinelike isosequences are also shown.

Introduction

Recently there has been considerable interest in x-ray transitions in high Z ions with charge states around the neonlike isosequence[1-11]. X-ray lasing[12,13] has been demonstrated in neonlike ions, and a need to understand the kinetics of this system has motivated development of very precise collisional-radiative modelling tools[14]. The identifications of many x-ray lines from neonlike ions allow high resolution experimental data to be used for benchmarking multi-electron atomic structure calculations[15-19]. Most of the work which has been done in the past has been limited to 3-3, 2-3 and 2-4 transitions in the NeI isosequence and adjacent charge states. The high temperature, optically thin tokamak plasmas enable the measurement of many lines originating in transitions from levels having n ≥ 5; in fact, all of the transitions in the 2p-nd series in Mo^{32+} lying under the ionization potential have been measured[10,11]. The availability of a large number of transitions in several adjacent elements provides the opportunity to study the systematics of configuration interaction effects.

In this paper are presented spectra of selected 2p-nd transitions with n between 3 and 7 in near-neonlike Zr, Nb, Mo and Pd, obtained from Alcator C-Mod plasmas[20]. A comprehensive study of numerous transitions in these elements is given in Ref. (11). The 2p-nd transitions considered here are strongly split by the j-value (in jj-coupling) of the 2p hole in the ionic core. The splitting is very apparent in the neonlike ions, where the resonance transitions with upper states containing a $2p_{\frac{1}{2}}$ hole are at much shorter wavelengths than the corresponding transitions with a $2p_{\frac{3}{2}}$ hole. This splitting can lead to significant configuration interaction when a $(2p^5)_{\frac{1}{2}}$nd orbital is close in energy to a $(2p^5)_{\frac{3}{2}}$n'd (n'>n) orbital. Interaction between the orbitals will perturb transition wavelengths and re-distribute oscillator strength within a class of transitions[21]. The magnitude of the configuration interaction between the $(2p^5)_{\frac{1}{2}}6d_{\frac{3}{2}}$ $J=1$ level and the $(2p^5)_{\frac{3}{2}}7d_{\frac{5}{2}}$ $J=1$ levels has been measured as a function of energy level spacing for the neonlike ions of Zr, Nb and Mo[11]. In this paper calculations are presented for these interacting levels in neonlike ions with 39 ≤ Z ≤ 46.

The x-ray observations described here were obtained from the Alcator C-Mod[20] tokamak, a compact high field device with all molybdenum plasma facing components. For these measurements, the plasma parameters were in the range of 7.7 x 10^{13}/cm^3 ≤ n_{e0} ≤ 2.0 x 10^{14}/cm^3 and 1500 eV ≤ T_{e0} ≤ 3400 eV. A laser blow-off impurity injection system[22], which has been used to study impurity transport, was used to inject niobium, palladium and zirconium into Alcator C-Mod plasmas. The spectra presented here were recorded by a five chord, independently spatially scannable, high resolution x-ray spectrometer array[23]. Wavelength calibration[2,3] has been achieved by determining the instrumental dispersions in reference to H- and He-like argon, chlorine and sulphur lines and previously measured molybdenum[10] lines.

Ab initio atomic structure calculations for the aluminum- through fluorine-like isosequences (ground states $2p^63s^23p$ to $2s^22p^4$, respectively) have been performed using the RELAC code[24,25], which solves the Dirac equation by optimizing a parametric potental. RELAC has been used to calculate the full

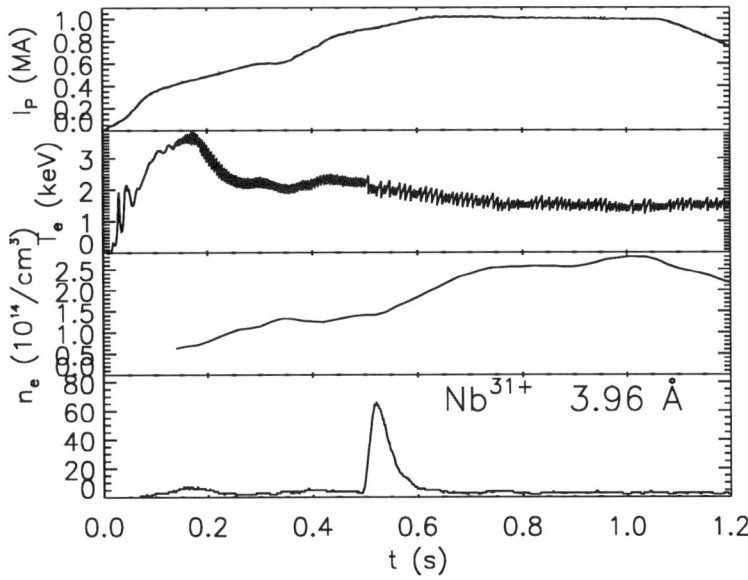

Figure 1: Plasma current, electron temperature, electron density and Nb x-ray (3.96 Å) brightness time histories in a discharge with a niobium injection at .5 sec.

multi-configuration transition wavelengths and oscillator strengths for all lines observed in this paper.

X-ray Transitions

Shown in Fig.1 are the time histories of several quantities of interest for a typical Alcator C-Mod 5.3 T, deuterium discharge. There was a niobium injection into this particular discharge at 0.5 seconds, when the plasma current was 0.9 MA, the central electron temperature was 2200 eV and the central electron density was 1.3 x 10^{14}/cm^3. The niobium stayed in the plasma for about 100 ms, as shown by the bottom frame of the figure, indicative of anomalously fast impurity transport[22]. In Fig.2 is shown an x-ray spectrum taken during an injection which demonstrates the strongest niobium line which falls within the wavelength range of the spectrometer, the $2p^6$ - $(2p^5)_{\frac{3}{2}}4d_{\frac{5}{2}}$ transition in neon-like Nb^{31+} at 3957.3 mÅ. Also apparent in this spectrum are some sodiumlike Nb^{30+} 2p-4d lines at 4008.4 and 4011.3 mÅ, and some weaker lines from the magnesiumlike and fluorinelike charge states. A synthetic spectrum, generated

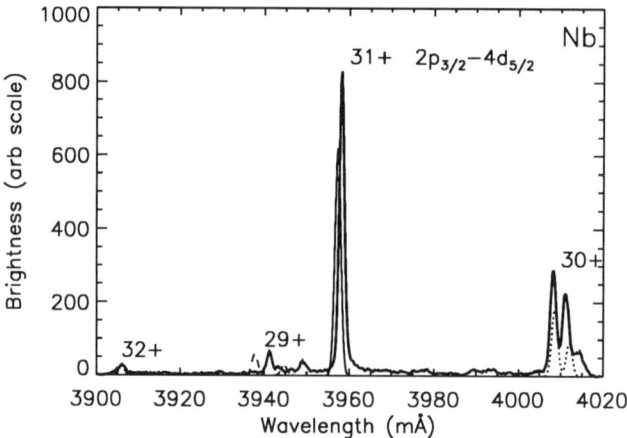

Figure 2: 2p-4d transitions in Nb^{29+} - Nb^{32+}. Theoretical lines for neonlike Nb^{31+} (solid), Nb^{30+} (dotted), Nb^{29+} (dashed) and Nb^{32+} (dash-dot-dash) are shown at the bottom, where the relative intensities within a given charge state are proportional to the oscillator strengths of each transition.

using calculated wavelengths[11], typical instrumental and Doppler line widths, and line amplitudes proportional to the oscillator strengths[11] within a given charge state, is shown at the bottom of the figure. The observed niobium lines are within 1 mÅ of the calculated wavelengths[11]. Wavelength calibration was obtained from several nearby Ar^{16+} lines[26,27], S^{14+} lines[26] and S^{15+} lines[28]. A calibration spectrum showing several sulphur lines is shown in Fig.3. This spectrum includes the (unresolved) Lyman β doublet and the 1s^2-1snp series in S^{14+} with n between 5 and 13. Calculated wavelengths and radiative transition probabilities for this series are given in Table I.

Another strong neonlike Nb^{31+} line, the 2p^6 - (2p^5)$_{\frac{1}{2}}$4d$_{\frac{3}{2}}$ transition at 3843.8 mÅ, is shown in Fig.4a. Also prominent in the figure are the 2p-4d lines at 3892.8 and 3822.9 mÅ, from sodium- and fluorinelike niobium, respectively. The corresponding 2p-5d transitions are shown in Fig.4b. At the bottom of each figure is a synthetic spectrum and the wavelength agreement is very good.

For higher n transitions in neonlike systems, the upper levels of certain lines in the 2p^6 - (2p^5)$_{\frac{3}{2}}$nd$_{\frac{5}{2}}$ series and the 2p^6 - (2p^5)$_{\frac{1}{2}}$nd$_{\frac{3}{2}}$ series can have nearly identical energies, giving rise to significant configuration interaction. In particular, the effect is seen in the enhancement of the intensity of the 2p^6

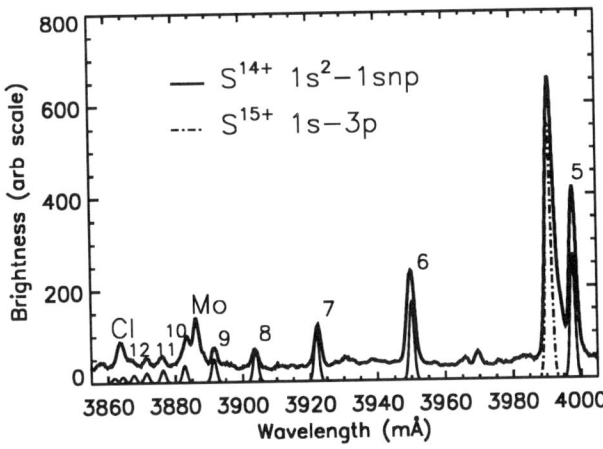

Figure 3: $1s^2$-$1snp$ transitions in S^{14+} with n between 5 and 12. Also apparent is the Lyman β doublet of S^{15+}, a 2p-4s line of Mo^{32+} at 3886.3 mÅ, and a satellite of Cl^{15+} K_β at 3863.7 mÅ.

Table 1: Calculated $1s^2$-$1snp$ transition wavelengths and radiative transition probabilities for heliumlike S^{14+}.

Transition	wavelength (mÅ)	A (10^{12}/s)
$1s^2\ ^1S_0$ - $1s4p\ ^1P_1$	4088.46	7.28
$1s^2\ ^1S_0$ - $1s5p\ ^1P_1$	3997.73	3.63
$1s^2\ ^1S_0$ - $1s6p\ ^1P_1$	3950.10	2.07
$1s^2\ ^1S_0$ - $1s7p\ ^1P_1$	3921.92	1.29
$1s^2\ ^1S_0$ - $1s8p\ ^1P_1$	3903.84	.860
$1s^2\ ^1S_0$ - $1s9p\ ^1P_1$	3891.54	.602
$1s^2\ ^1S_0$ - $1s10p\ ^1P_1$	3882.79	.438
$1s^2\ ^1S_0$ - $1s11p\ ^1P_1$	3876.34	.328
$1s^2\ ^1S_0$ - $1s12p\ ^1P_1$	3871.45	.252
$1s^2\ ^1S_0$ - $1s13p\ ^1P_1$	3867.66	.198

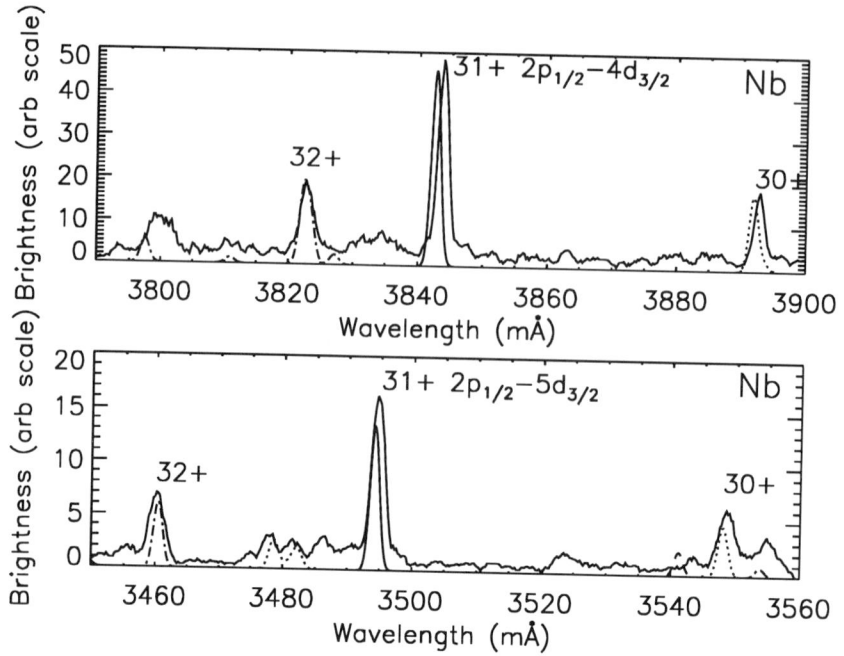

Figure 4: (a) 2p-4d and (b) 2p-5d transitions in neonlike Nb^{31+} (solid), Nb^{30+} (dotted) and Nb^{32+} (dash-dot dash).

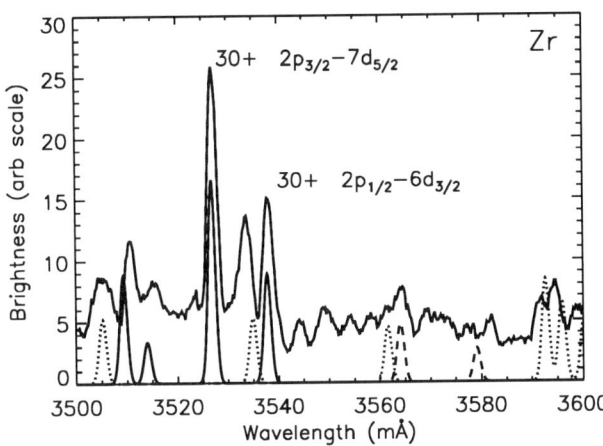

Figure 5: $2p_{\frac{3}{2}}$ - $7d_{\frac{5}{2}}$ and $2p_{\frac{1}{2}}$ - $6d_{\frac{3}{2}}$ transitions in neonlike Zr^{30+}, and the calculated neonlike (solid), sodiumlike (dotted) and fluorinelike (dashed) lines. Also shown are 2s-5p transitions in Zr^{30+} (solid) at 3510.3 and 3515.1 mÅ.

- $(2p^5)_{\frac{3}{2}}7d_{\frac{5}{2}}$ transition at the expense of the $2p^6$ - $(2p^5)_{\frac{1}{2}}6d_{\frac{3}{2}}$ transition[10,11] in Mo^{32+}, where the difference in the upper state energy levels is 3.5 eV, and the interaction is quite large. A spectrum of these two lines in Zr^{30+} is shown in Fig.5. In the case of zirconium, the separation is 11 eV (11 mÅ) and there is little configuration interaction at all (the two lines are within a factor of 2 in intensity). This effect is summarized in Fig.6a where the calculated oscillator strengths of the $2p^6$ - $(2p^5)_{\frac{3}{2}}6d_{\frac{5}{2}}$, the $2p^6$ - $(2p^5)_{\frac{1}{2}}6d_{\frac{3}{2}}$, the $2p^6$ - $(2p^5)_{\frac{3}{2}}7d_{\frac{5}{2}}$ and the $2p^6$ - $(2p^5)_{\frac{1}{2}}7d_{\frac{3}{2}}$ lines are plotted as a function of atomic number. The oscillator strengths of the $2p^6$ - $(2p^5)_{\frac{3}{2}}6d_{\frac{5}{2}}$ lines and the $2p^6$ - $(2p^5)_{\frac{1}{2}}7d_{\frac{3}{2}}$ lines are relatively insensitive to atomic number. The magnitude of the configuration interaction between the $2p^6$ - $(2p^5)_{\frac{1}{2}}6d_{\frac{3}{2}}$ level and the $2p^6$ - $(2p^5)_{\frac{3}{2}}7d_{\frac{5}{2}}$ level is quite apparent; as the atomic number increases from Y to Mo, the g*f value of the $7d_{\frac{5}{2}}$ line increases while the value of the $6d_{\frac{3}{2}}$ line decreases. At technetium (Z=43), this effect dramatically switches; for Tc and above, the $2p^6$ - $(2p^5)_{\frac{1}{2}}6d_{\frac{3}{2}}$ line is at shorter wavelength and the $2p^6$ - $(2p^5)_{\frac{3}{2}}7d_{\frac{5}{2}}$ line is the *weaker* of the two. The wavelength differences between the two levels is shown in Fig.6b.

For the electron temperatures of Alcator C-Mod, palladium can just reach the neonlike state, and the lower charge states are present in abundance.

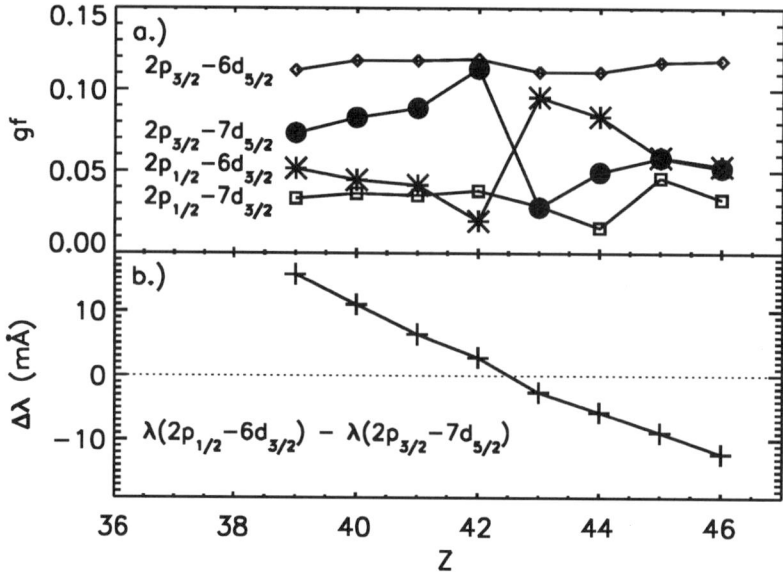

Figure 6: Calculated oscillator strengths (a) of 2p-6d and 2p-7d transitions and calculated wavelength differences (b) between $2p_{\frac{1}{2}}$ - $6d_{\frac{3}{2}}$ transitions and $2p_{\frac{3}{2}}$ - $7d_{\frac{5}{2}}$ transitions as a function of atomic number Z.

Table 2: Calculated neonlike 2p-nd E1 transition wavelengths and oscillator strengths for n = 6 and 7 in Y^{29+}, Tc^{33+}, Ru^{34+} and Rh^{35+}. The upper level designations in the first column are indicated by three jj-coupled orbitals where '-' indicates $l - s$ coupling and '+' indicates $l + s$ coupling: the first two orbitals show the occupancy of the $2p_{\frac{1}{2}}$ and $2p_{\frac{3}{2}}$ subshells, respectively, and the third orbital is where the 2p-electron has been promoted.

Upper level	Y^{29+}		Tc^{33+}		Ru^{34+}		Rh^{35+}	
	λ_T (mÅ)	g*f	λ_T (mÅ)	g*f	λ_T (mÅ)	g*f	λ_T (mÅ)	g*f
$(2p_-)^2(2p_+)^3 6d_+$ J=1	3856.8	.112	3055.4	.111	32894.1	.111	2744.3	.117
$(2p_-)(2p_+)^4 6d_-$ J=1	3765.6	.0516	2967.1	.0955	2806.7	.0836	2657.7	.0587
$(2p_-)^2(2p_+)^3 7d_+$ J=1	3749.9	.0733	2969.7	.0280	2812.4	.0494	2666.6	.0583
$(2p_-)(2p_+)^4 7d_-$ J=1	3663.0	.0334	2885.7	.0289	2729.4	.0155	2584.3	.0460

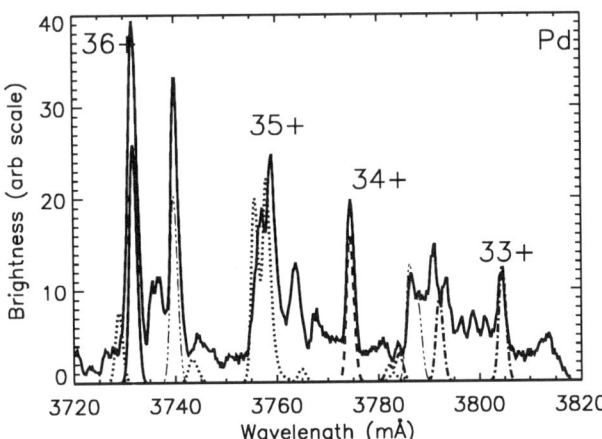

Figure 7: 2p-3d transitions in Pd^{33+} - Pd^{36+}. Theoretical lines for neonlike Pd^{36+} (solid), Pd^{35+} (dotted), Pd^{34+} (dashed) and Pd^{33+} (dash-dot-dash) are shown at the bottom, where the relative intensities within a given charge state are proportional to the oscillator strengths of each transition. Molybdenum transitions are shown by the thin dash-dot-dot-dot-dash lines.

Shown in Fig.7 is a spectrum of Pd 2-3 transitions in the vicinity of the $2p^6$ - $(2p^5)_{\frac{1}{2}}3d_{\frac{3}{2}}$ line at 3731.7 mÅ. The corresponding lines in Pd^{35+}, Pd^{34+} and aluminumlike Pd^{33+} are clearly identified.

Conclusions

X-ray transitions in the magnesiumlike through fluorinelike charge states in zirconium, niobium, molybdenum and palladium have been observed from Alcator C-Mod plasmas. Line identifications have been made by comparison to the results of *ab initio* calculations and overall wavelength agreement is very good. The magnitude of the configuration interaction between the $(2p^5)_{\frac{1}{2}}6d_{\frac{3}{2}}$ level and the $(2p^5)_{\frac{3}{2}}7d_{\frac{5}{2}}$ level has been calculated as a function of atomic number for $39 \leq Z \leq 46$. 2p-3d transitions in aluminumlike Pd^{33+} have also been identified.

Acknowledgements

The authors would like to thank F. Bombarda for assistance with the spectrometer system, J. Irby for electron density measurements, A. Hubbard for electron temperature measurements and the Alcator C-Mod operations group for expert running of the tokamak. Work supported at MIT by DoE Contract No. DE-AC02-78ET51013 and at LLNL by DoE Contract No. W-7405-ENG-48.

References

[1] E.Källne, J.Källne and R.D.Cowan, Phys. Rev. A **27**, 2682 (1983).
[2] P.Beiersdorfer et al., Phys. Rev. A **34**, 1297 (1986).
[3] P.Beiersdorfer et al., Phys. Rev. A **37**, 4153 (1988).
[4] E.V.Aglitskii et al., Physica Scripta **40**, 601 (1989).
[5] P.Beiersdorfer et al., Phys. Rev. Lett. **65**, 1995 (1990).
[6] R.Hutton et al., Phys. Rev. A **44**, 1836 (1991).
[7] M.B.Schneider et al., Phys. Rev. A **45**, R1291 (1992).
[8] Steven Elliott et al., Phys. Rev. A **47**, 1403 (1993).
[9] P.Beiersdorfer et al., Physica Scripta **51**, 322 (1995).
[10] J.E.Rice et al., Phys Rev A **51**, 3551 (1995).
[11] J.E.Rice et al., accepted for publication in Phys. Rev. A (1995).
[12] D.L.Matthews et al., Phys. Rev. Lett., **54**, 110 (1985).
[13] M.D.Rosen et al., Phys. Rev. Lett., **54**, 106 (1985).
[14] A.L. Osterheld et al., J. Quant. Spectrosc. Radiat. Transfer, **51**, No. 1/2, 263 (1994).
[15] P.Beiersdorfer, M.H.Chen, R.E.Marrs and M.Levine, Phys. Rev. A **41**, 3453 (1990).
[16] G.A.Chandler, M.H.Chen, D.D.Dietrich, P.O.Egan, K.P.Ziock, P.H.Mokler, S.Reusch and D.H.H.Hoffmann, Phys. Rev. A, **39**, 565 (1989).
[17] D.D.Dietrich, G.A.Chandler, P.O.Egan, K.P.Ziock, P.H.Mokler, S.Reusch and D.H.H.Hoffmann, Nucl. Instrum. Methods B, **24/25**, 301 (1987).
[18] E. Avgoustoglou, W.R. Johnson, Z.W. Liu and J. Sapirstein, Phys. Rev. A, **51**, 1196 (1995).
[19] W.R. Johnson, J. Sapirstein and K.T. Cheng, Phys. Rev. A, **51**, 297 (1995).
[20] I.H.Hutchinson et al., Phys. Plasmas **1**, 1511 (1994).
[21] R.D.Cowan, The Theory of Atomic Structure and Spectra, University of California Press, pp.433-434 (1981).
[22] M.A.Graf et al., Rev. Sci. Instrum. **66**, 636 (1995).
[23] J.E.Rice and E.S.Marmar, Rev. Sci. Instrum. **61**, 2753 (1990).
[24] M.Klapisch, Comput. Phys. Commun. **2**, 269 (1971).
[25] M.Klapisch, J.L.Schwob, B.S.Fraenkel and J.Oreg, J. Opt. Soc. Am. **67**, 148 (1977).
[26] L.A.Vainshtein and U.I.Safronova, Physica Scripta **31**, 519 (1985).
[27] J.E.Rice, E.S.Marmar, E.Källne and J.Källne, Phys. Rev. A **35**, 3033 (1987).
[28] G.W.Erickson, J. Phys. Chem. Ref. Data **6**, 831 (1977).

The Contribution of Spectroscopy on the Texas Experimental Tokamak (TEXT) to Atomic Physics and Plasma Physics: A Final Report

William L. Rowan

Fusion Research Center
The University of Texas at Austin, Austin, Texas 78712

Abstract. Research on TEXT from inception of operation in 1980 to its end in 1995 resulted in major contributions to the general advance of both plasma physics and atomic physics that has been made using tokamaks. In some experiments, spectroscopy was applied to understand the physics of transport. These experiments identified the role of turbulence in particle transport and made significant progress toward identifying the source of the turbulence. TEXT was also an excellent platform for the generation of atomic data for investigation of atomic structure and electron-ion collisions. The atomic physics applications have increased the availability of atomic data and the accuracy of the atomic database.

INTRODUCTION

Spectroscopy on the Texas Experimental Tokamak (TEXT) has contributed to the general advance of plasma physics and atomic physics that has been made using tokamaks. The recent decision to close the TEXT facility motivated this review of that part of the experimental work that depended heavily on spectroscopic diagnostics. This review may be useful for guiding future research, and it may provide some insight for other modest-sized research programs.

PLASMA PHYSICS EXPERIMENTS

The plasma physics studies centered on the phenomenology and physics of particle transport. The study of particle transport contributes to understanding of plasma physics. Progress in understanding the physics of transport will contribute to the attainment of controlled thermonuclear fusion. The research on TEXT began with attempts to characterize impurity transport in various discharge regimes and to compare measurement with the prevailing neoclassical theory. Subsequently, the research turned to studies of the effect of turbulence on transport and then to a search for the driving mechanisms for turbulence.

Impurity Transport

Two components of impurity transport were investigated, the transport transverse to the confining magnetic field and parallel flow of impurities along the confining field. Both components were compared to neoclassical prediction.

In one experiment, a measured O^{+8} distribution was compared to a neoclassical prediction.(1, 2) The measurement was derived from spectroscopic observations of the interaction of a high velocity beam of neutral hydrogen with fully stripped oxygen. The unusual feature of this discharge in which the measurements were made was that it was fueled by injection of pellets of frozen hydrogen rather than in the more common fashion by gas puffing. After the injection of the pellet, the oxygen was seen to evolve to a distribution that agreed well with the prediction of neoclassical theory, but the apparent good agreement between measurement and experiment did not hold up under more detailed examination. Though the neoclassical theory would predict 100 ms for the characteristic profile evolution time, the actual characteristic time was 20 ms.

The conclusions of the previous experiment hinged on the experimentalists being rather clever in extracting time dependent information to avoid a specious conclusion. In this second example,(3) a similar conclusion is reached, but the experiment is done in a manner that allows simpler resolution of the critical time dependence issues. As illustrated in Fig. 1, the impurity of interest is injected into the plasma in small quantity using laser injection(4) so that it little affects the plasma parameters. The impurity remains in the plasma only transiently. The impurity flux Γ is modeled by a quantity that is the sum of two parts, one proportional to the impurity gradient and another part proportional to the impurity density n_z, $\Gamma = -D \frac{\partial n_z}{\partial r} + n_z V$. With this transport flux, a simulation of the experiment can determine D, the diffusion coefficient, and V, the convective velocity. These are also shown in the Fig. 1. Neoclassical theory would predict D = 0.0120 $m^2 s^{-1}$, and hence does not adequately describe the experiment. In fact, for this experiment, it was concluded that a diffusion coefficient based on drift wave turbulence provided a better representation.

From the results of these two experiments, it appears that cross field transport cannot be described by neoclassical theory. In spite of that, parallel impurity flow is consistent with the predictions of neoclassical transport. In the experiments leading to this conclusion, the Doppler shifts of spectra emitted by C^{+4} were used to infer the flow velocity of the impurity along the magnetic field.

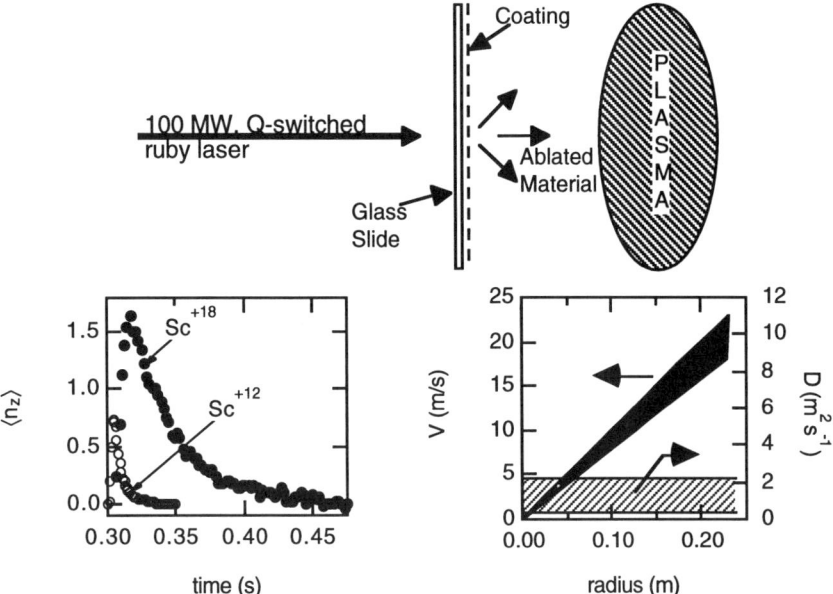

FIGURE 1. In laser ablation (upper diagram), the material to be injected is coated onto a glass slide. It is ablated from the slide by a high power laser pulse and then flows into the plasma. The temporal evolution of Sc injected into a He plasma is indicated at lower left (intermediate ionization stages are omitted for clarity). Transport coefficients for Sc determined in this experiment are shown at lower right.

In this experiment,(5-7) it was found that the flow of C^{+4} in the edge of a helium plasma was consistent with well-known neoclassical theory.(8) When this same theory was applied to describe impurity flow in a hydrogen plasma, theory agreed with experiment only when theory was corrected(9) to account for momentum loss due ion-neutral charge exchange ($H^0 + H^+ \rightarrow H^+ + H^0$). The correction accounted for an obvious atomic process: charged particles are converted to unconfined neutrals in the plasma edge, and then lost to the plasma along with the momentum that they carry.

Particle Transport in the Plasma Edge

Though neoclassical theory omits turbulence, the plasma is turbulent. To determine the effect of turbulence on transport, the experimental plan was to measure the part of the transport induced by turbulence and compare that to measurements of the total particle transport.(10, 11) Thus, a standard measurement of particle confinement was needed, and a spectroscopic measurement filled that need.

The confinement of particles in a tokamak is described by the particle flux Γ which is related to the production rate for charged particles S by the continuity equation $\nabla \cdot \Gamma = S$. In a plasma with a limited impurity content, S is just the rate of ionization of the fueling gas, in this case, hydrogen. A collisional radiative model for the excited states of H can be used to predict S in terms of the emission of spectral lines. (See reference (12), for example.) Measurement of the spatial distribution of the spectra of neutral H then leads in a straightforward way to determination of Γ.

In Fig. 2, the particle flux from the spectroscopic measurement is compared with that from the turbulence measurement.(10) The electrostatic turbulence flux is based on detailed measurements from Langmuir probes in the plasma edge. In this one case, there is good agreement at the plasma edge, a critically important region for the measurement in terms of interpretation and availability of cross checks. From extensive results similar to this one example, there is excellent reason to believe that turbulence causes transport. Thus, the problem is now to find the cause for the turbulence.

Spectroscopic measurements were used in experiments which sought to complete the picture of transport in tokamaks by identifying the source of the turbulence. In this series of experiments, several different suggestions were tested. These included turbulence induced by atomic processes, by the conductivity of plasma-facing surfaces, and finally by the curvature of magnetic field lines. Of these, only the last has so far proved to be interesting.

The idea of turbulence being driven by the magnetic curvature is straightforward. In a tokamak, fluid modes may be unstable for the appropriate average curvature of the magnetic field lines with respect to the pressure gradient.(13) Consequently, curvature is destabilizing on the low field side of the poloidal cross section of a tokamak and stabilizing on the high field side. In common magnetic configurations, magnetic field lines run through both regions, but it is possible to create a magnetic configuration in some tokamaks that isolates the two regimes from each other. A double null diverted configuration does just that. Hence, if curvature generates turbulence, then an experiment might expect to find different levels of turbulence in the two isolated regions of the double null configuration.

New spectroscopic techniques(14, 15) were used to look for these characteristic spatial asymmetries in the turbulence. Normally, the Langmuir probe would be the diagnostic of choice, but it is exceptionally difficult to insert common Langmuir probes on the high field side of a tokamak because of spatial constraints. Consequently, the high frequency emission from neutral H was compared for the two regions of the plasma. Intuitively, this does not seem to be a viable measurement because of the familiar chord averaging effect that is common to virtually every spectroscopic measurement. Yet, by imaging the light from either side on a multichannel detector and then considering only the cross

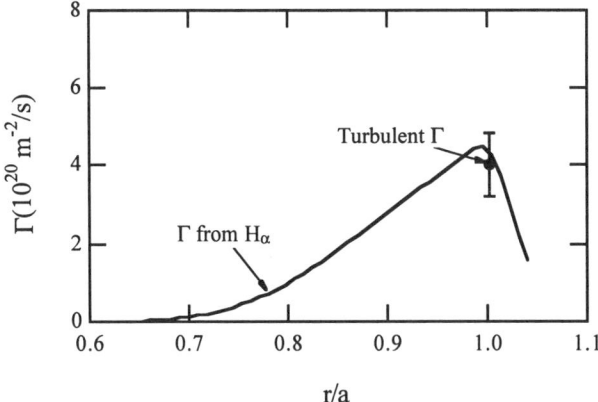

FIGURE 2. The total particle flux determined from measurements of the spectrum of neutral hydrogen is compared to the flux determined from detailed measurements of plasma turbulence. The edge of this hydrogen discharge is at r/a = 1. This and similar measurements demonstrate that particle flux is driven by electrostatic turbulence in the plasma edge.

correlated signal between neighboring channels, it is possible to discriminate against emission from the unfocused side of plasma and select out the part of the signal related to plasma turbulence.(14, 15) The spectroscopic measurements confirmed the existence of curvature drive for turbulence in the plasma.(16) That result is consistent with those on other devices.(17)

ATOMIC PHYSICS EXPERIMENTS

The significance of the atomic physics studies was equal to that of the plasma physics studies, but the atomic physics work grew from the main plasma physics studies. To facilitate the impurity transport work, techniques were developed to inject impurities into the plasma, and spectroscopic instrumentation was acquired for spectral observations from the near infrared to the soft x-ray. The opportunities for new atomic studies attracted experimentalists with interests in the spectra of highly ionized species and in the rates for atomic collision processes. Their work led principally to an increase in the availability of atomic data and to an improvement in the accuracy of the atomic database for ions and charge states not readily attainable with other experimental apparatus.

The TEXT plasma was been used for investigations of atomic spectra and structure (18-49) and of atomic collision processes.(50-59) The elements that were the object of study were C, N, O, F, Ne, Mg, Al, Si, S, Cl, Ar, Ca, Sc, Ti, V, Cr, Fe, Ni, Cu, Zn, Ga, Ge, As, Se, Br, Kr, Y, Zr, Nb, Mo, Ru, Rh, Pd, Ag, Cd, In, Sn, Sb, I, Xe, Cs, Hf, W, Pt, Au, Pb, Bi, Pr, Nd, Sm, Eu, Dy, Er, Yb, and U.

Charge states as high as +32 were studied. Clearly, laser injection techniques virtually removed limitations to the elements that could be studied. The Ohmic power deposition in this medium scale device was sufficient to avoid radiative collapse even for injection of useful levels of heavy impurities. Exceptional engineering design assured the high discharge repetition rate that made experiments fast and the high reproducibility that allowed accurate results. Transport is sufficiently slow in tokamaks to allow the excited densities in the ions to reach an equilibrium, and thus allows the complete knowledge of the profiles of the plasma temperature and density to be effectively applied to interpretation of the spectra.

The spectra are characterized by narrow line widths determined solely by thermal effects and are completely free of optical depth problems. Shifts in the spectra are induced by Doppler effects, but these are well known from the plasma physics studies of impurity flow so these were well diagnosed and could be demonstrated to be ignorable with proper experimental design.

CONCLUSION

In the plasma physics experiments, some limits for the applicability of neoclassical transport were identified, but turbulence was shown to play the key role in transport. Though the source of the turbulence remains uncertain, some turbulence drive mechanisms were shown to be negligible and circumstantial evidence was found for magnetic field line curvature as a drive mechanism. The atomic physics studies arose as opportunities made available by appropriate diagnostics and device. These studies have materially increased the atomic database and its accuracy.

ACKNOWLEDGMENTS

This work was supported by the USDOE under grant DE-FG03-95ER54296.

REFERENCES

1. Synakowski, E. J. *Light Impurity Studies on TEXT Using Charge-Exchange Recombination Spectroscopy*, Ph. D. Thesis, The University of Texas at Austin (1988).
2. Synakowski, E. J., Bengtson, R. D., Ouroua, A., Wootton, A. J. and Kim, S. K. *Nuclear Fusion* **29**, 311-315 (1989).
3. Horton, W. and Rowan, W. *Physics of Plamas* **1**, 901-908 (1994).
4. Leung, W. K. *Investigation of Impurity Transport in the Texas Experimental Tokamak*, Ph. D. Thesis, The University of Texas at Austin (1984).
5. Rowan, W. L., Meigs, A. G., Solano, E. R. and Valanju, P. M. *Physics of Fluids B* **5**, 2485 (1993).

6. Rowan, W. L., Meigs, A. G., Hickok, R. L., Schoch, P. M., *et al. Physics of Fluids B* **4**, 917 (1992).
7. Meigs, A. G. and Rowan, W. L. *Physics of Plasmas* **1**, 960 (1994).
8. Hazeltine, R. D. *Physics of Fluids* **17**, 961-968 (1974).
9. Valanju, P. M., Calvin, M. D., Hazeltine, R. D. and Solano, E. R. *Physics of Fluids B* **4**, 2675-2679 (1992).
10. Rowan, W. L., Klepper, C. C., Ritz, C. P., Bengtson, R. D., *et al. Nuclear Fusion* **27**, 1105 (1987).
11. Klepper, C. C. *An Investigation of Particle Transport through the Measurement of the Electron Source in the Texas Experimental Tokamak*, Thesis, The University of Texas at Austin (1985).
12. Johnson, L. C. and Hinnov, E. R. *Journal of Quantitative Spectroscopy and Radiative Transfer* **13**, 333 (1973).
13. Garbet, X., Laurent, L., Rubin, J. P. and Samain, A. *Nuclear Fusion* **31**, 967-972 (1991).
14. Hurwitz, P. D., Hall, B. F. and William L. Rowan. *Review of Scientific Instruments* **63**, 4614 (1992).
15. Hurwitz, P. D. and Rowan, W. L. *Review of Scientific Instruments* **66**, 441 (1995).
16. Hurwitz, P. D. *Optical Imaging of Turbulence Fluctuations in the Texas Experimental Tokamak-Upgrade*, Ph. D. Thesis, The University of Texas at Austin (1995).
17. Endler, M., Niedermeyer, H., Giannone, L., Holzhauer, E., *et al. Nuclear Fusion* **35**, 1307-1339 (1995).
18. Datla, R. U., Roberts, J. R. and Rowan, W. L. *Journal of the Optical Society of America B: Optical Physics* **4**, 428-429 (1987).
19. Datla, R. U., Roberts, J. R., Woodward, N., Lippman, S. and Rowan, W. L. *Physical Review A: Atomic, Molecular, and Optical Physics* **40**, 1484-1487 (1989).
20. Finkenthal, M., Lippmann, A. S., Huang, L. K., Yu, T. L., *et al. Journal of Applied Physics* **59**, 3644-3649 (1986).
21. Finkenthal, M., Lippmann, S., Moos, H. W., Mandelbaum, P. and TEXT Group, *Physical Review A: Atomic, Molecular, and Optical Physics* **39**, 3717-3720 (1989).
22. Finkenthal, M., Lippmann, S., Huang, L. K., Zwicker, A., *et al. Physical Review A: Atomic, Molecular, and Optical Physics* **45**, 5846-5853 (1992).
23. Fournier, K. B., Goldstein, W. H., Osterheld, A., Finkenthal, M., *et al. Physical Review A: Atomic, Molecular, and Optical Physics* **50**, 2248-2256 (1994).
24. Fournier, K. B., Finkenthal, M., Lippmann, S., Holmes, C. P., *et al. Physical Review A: Atomic, Molecular, and Optical Physics* **50**, 3727-3733 (1994).
25. Lippmann, S., Finkenthal, M., Huang, L. K., Moos, H. W., *et al. Astrophysical Journal* **316**, 819-825 (1987).
26. Reader, J., Kaufman, V., Sugar, J., Ekberg, J. O., *et al. Journal of the Optical Society of America B: Optical Physics* **4**, 1821-1828 (1987).
27. Reader, J., Sugar, J., Acquista, N. and Bahr, R. *Journal of the Optical Society of America B: Optical Physics* **11**, 1930-1934 (1994).
28. Roberts, J. R., Pittman, T. L., Sugar, J., Kaufman, V. and Rowan, W. L. *Physical Review A: Atomic, Molecular, and Optical Physics* **35**, 2591-2595 (1987).
29. Sugar, J., Kaufman, V. and Rowan, W. L. *Journal of the Optical Society of America B: Optical Physics* **4**, 1927-1930 (1987).
30. Sugar, J., Kaufman, V. and Rowan, W. L. *Journal of the Optical Society of America B: Optical Physics* **5**, 2183-2189 (1988).
31. Sugar, J., Kaufman, V. and Rowan, W. L. *Journal of the Optical Society of America B: Optical Physics* **5**, 236-242 (1988).
32. Sugar, J., Kaufman, V., Indelicato, P. and Rowan, W. L. *Journal of the Optical Society of America B: Optical Physics* **6**, 1437-1443 (1989).
33. Sugar, J., Kaufman, V. and Rowan, W. L. *Journal of the Optical Society of America B: Optical Physics* **7**, 152-158 (1990).
34. Sugar, J., Kaufman, V., Baik, D. H., Kim, Y.-K. and Rowan, W. L. *Journal of the Optical Society of America B: Optical Physics* **8**, 1795-1798 (1991).

35. Sugar, J., Kaufman, V. and Rowan, W. L. *Journal of the Optical Society of America B: Optical Physics* **8**, 913-916 (1991).
36. Sugar, J., Kaufman, V. and Rowan, W. L. *Journal of the Optical Society of America B: Optical Physics* **8**, 22-26 (1991).
37. Sugar, J., Kaufman, V. and Rowan, W. L. *Journal of the Optical Society of America B: Optical Physics* **8**, 2026-2027 (1991).
38. Sugar, J., Kaufman, V. and Rowan, W. L. *Journal of the Optical Society of America B: Optical Physics* **9**, 1959-1961 (1992).
39. Sugar, J., Kaufman, V. and Rowan, W. L. *Journal of the Optical Society of America B: Optical Physics* **9**, 344-346 (1992).
40. Sugar, J., Kaufman, V. and Rowan, W. L. *Journal of the Optical Society of America B: Optical Physics* **10**, 799-801 (1993).
41. Sugar, J., Kaufman, V. and Rowan, W. L. *Journal of the Optical Society of America B: Optical Physics* **10**, 1977-1979 (1993).
42. Sugar, J., Kaufman, V. and Rowan, W. L. *Journal of the Optical Society of America B: Optical Physics* **10**, 1321-1325 (1993).
43. Sugar, J., Kaufman, V. and Rowan, W. L. *Journal of the Optical Society of America B: Optical Physics* **10**, 13-15 (1993).
44. Sugar, J., Reader, J. and Rowan, W. L. *Physical Review A: Atomic, Molecular, and Optical Physics* **51**, 835-837 (1995).
45. Sugar, J. and Rowan, W. L. *Journal of the Optical Society of America B: Optical Physics* **12**, 1403-1405 (1995).
46. Kaufman, V., Sugar, J. and Rowan, W. L. *Journal of the Optical Society of America B: Optical Physics* **5**, 1273-1274 (1988).
47. Kaufman, V., Sugar, J. and Rowan, W. L. *Journal of the Optical Society of America B: Optical Physics* **6**, 142-145 (1989).
48. Kaufman, V., Sugar, J. and Rowan, W. L. *Journal of the Optical Society of America B: Optical Physics* **6**, 1444-1446 (1989).
49. Kaufman, V., Sugar, J. and Rowan, W. L. *Journal of the Optical Society of America B: Optical Physics* **7**, 1169-1175 (1990).
50. Datla, R. U., Roberts, J. R., Rowan, W. L. and Mann, J. B. *Physical Review A: Atomic, Molecular, and Optical Physics* **34**, 4751-4756 (1986).
51. Datla, R. U., Roberts, J. R., Durst, R. D., Hodge, W. L., et al. *Physical Review A: Atomic, Molecular, and Optical Physics* **36**, 5448-5450 (1987).
52. Datla, R. U., Roberts, J. R. and Bhatia, A. K. *Physical Review A: Atomic, Molecular, and Optical Physics* **43**, 1110-1113 (1991).
53. Finkenthal, M., Yu, T. L., Lippmann, S., Huang, L. K., et al. *Astrophysical Journal* **313**, 920-927 (1987).
54. Finkenthal, M., Moos, H. W., Bar-Shalom, A., Spector, N., et al. *Physical Review A: Atomic, Molecular, and Optical Physics* **38**, 288-295 (1988).
55. Huang, L. K., Lippmann, S., Yu, T. L., Stratton, B. C., et al. *Physical Review A: Atomic, Molecular, and Optical Physics* **35**, 2919-2927 (1987).
56. Huang, L. K., Lippmann, S., Stratton, B. C., Moos, H. W. and Finkenthal, M. *Physical Review A: Atomic, Molecular, and Optical Physics* **37**, 3927-3934 (1988).
57. Wang, J.-S., Griem, H. R., Hess, R., Kochanski, T. P. and Rowan, W. L. *Physical Review A* **33**, 4293 (1986).
58. Wang, J.-S., Griem, H. R. and Rowan, W. L. *Physical Review A* **36**, 951 (1987).
59. Wang, J.-S., Griem, H. R., Hess, R. and Rowan, W. L. *Physical Review A* **38**, 4761 (1988).

ATOMIC PHYSICS
IN ASTROPHYSICAL PLASMAS

EUV Spectroscopy of Stellar Coronae

Nancy S. Brickhouse

*Smithsonian Astrophysical Observatory
Cambridge, Massachusetts 02138*

Abstract. The EUV spectral region contains a wealth of plasma diagnostics for stellar coronae ($T_e \sim 5 \times 10^5$ to 2×10^7 K). Of particular importance for understanding coronal structure are the observable emission lines of highly ionized iron (Fe VIII — XXIV), which allow the determination of electron temperatures (and the detailed temperature distributions) and electron densities. Comparison of continuum emission and lines from other elements with the iron lines provides diagnostics for relative abundances in the stellar atmospheres.

Recent work in both solar and stellar coronal physics has greatly changed our picture of the corona, with EUV spectroscopy providing critical pieces of the puzzle. Here we discuss some important new spectral diagnostic results, examining in particular the quality of the theoretical atomic physics used in the data interpretation.

INTRODUCTION

Basic coronal processes such as coronal heating mechanisms, flare and mass ejection events, and wind acceleration are not well understood despite decades of effort. The proximity of the Sun facilitates the construction of detailed physical models of different coronal structures. These structures may be closed field line regions, such as the magnetic loops that confine hot material above active regions, or coronal holes, which are the open field line regions at the source of the fast solar wind. Studies of cool stars beyond the Sun provide critical information on the scaling of coronal parameters with intrinsic properties of the star (e.g. gravity, age, rotation rate). Both solar and stellar studies require us to determine the physical conditions of the corona — the temperatures, densities, abundances, and their distributions — primarily by spectroscopic measurements of the hot emitting plasma.

The Sun is our only laboratory for detailed studies of coronal phenomena, as well as being a unique light source for benchmarking our spectroscopic models. While solar observations cannot replace systematic laboratory measurements of the collisional and radiative rates and cross sections, the dearth of laboratory measurements to confirm the vast amount of theoretical atomic data leaves us highly dependent on theory. Benchmarking the coronal models with solar observations at least gives us some idea as to the overall accuracy of our analysis.

High resolution EUV spectra of the Sun recently obtained with the Solar EUV Rocket Telescope and Spectrograph (*SERTS*) have spectral resolution $\lambda/\triangle\lambda >$

10,000 over the range 170 – 450 Å (1). The *SERTS* active region catalogue contains over 250 emission lines from highly ionized species, including some newly identified emission lines. The *SERTS* spectrograph has sufficient calibration accuracy for detailed comparisons with atomic models. EUV spectroscopy of the Sun will continue with the spectral instruments onboard the Solar and Heliospheric Observatory (*SOHO*), launched in late 1995 (see Ref. 2 for a review of spectroscopic diagnostics useful for specific missions).

The Extreme Ultraviolet Explorer satellite (*EUVE*), launched in 1992, has obtained EUV spectra of dozens of cool stars in the wavelength range 70 – 700 Å. Stellar sources observed by *EUVE* have coronal temperatures ranging from solar (peaking at $T_e \sim 10^6$ K) to active binary (peaking at $T_e > 10^7$ K) temperatures (3). Active binaries such as the RS CVn systems exhibit brighter and more variable X-ray emission, as well as more frequent and more energetic flaring, than the Sun. For active binaries and for some single stars, the spectra obtained by *EUVE* contain iron emission lines from Fe XVIII to XXIII. These lines have been previously studied in tokamak plasmas with measured N_e and T_e profiles (4), as well as in solar flare events (5). Some of these iron ions now have highly accurate atomic models (6, and references therein). Iron is an astrophysically abundant metal, and the hot, quiescent (i.e. nonflaring) stellar plasmas provide detailed diagnostics of coronal structure through the individual emission lines of the different iron ions. While X-ray observations of essentially the same stellar coronal plasma component (e.g. the $\triangle n > 0$ transitions of highly ionized iron around 1.0 keV) encompass a larger sample of sources than *EUVE*, the X-ray instruments do not have sufficient spectral resolution ($\lambda/\triangle\lambda < 50$) to determine the detailed temperature structure. Spectroscopic studies of the physical conditions in these stars have important implications for our understanding of coronal processes, as we discuss below.

Given N_e, T_e, and a set of elemental abundances, spectral emission codes produce model spectra to be compared with observations. Raymond and Brickhouse (7) review the most important radiative and collisional processes under coronal conditions, as well as a number of other processes that can be important in special situations. For solar and stellar coronae, optical depth effects in strong resonance lines (8) and time-dependent processes (9) are potentially important, but the accuracy of the atomic models and the applicability of simplifying assumptions (e.g. single temperature source models) also need to be considered.

We present examples that demonstrate the importance of accurate atomic data for the interpretation of the new coronal observations. Brickhouse, Raymond, and Smith (6) describe the iron spectral models and the sources of atomic data used in the examples I present here. For iron, some of the high ionization stages now have collision strengths as accurate as 10%, although many of the intermediate ionization stages of iron may have collision data accurate only to factors of two. Collisional excitation rates have been recently reviewed for *SOHO* and recommendations for electron collisional data are compiled in Lang (10); some ions are currently being recalculated (11). Ionization and recombination rates for iron have been compiled and applied to low density, coronal equilibrium conditions (12). While large un-

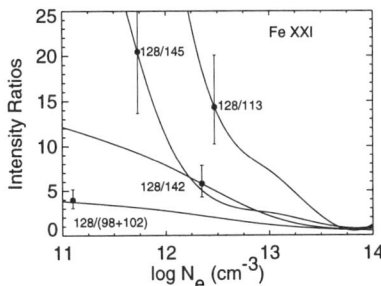 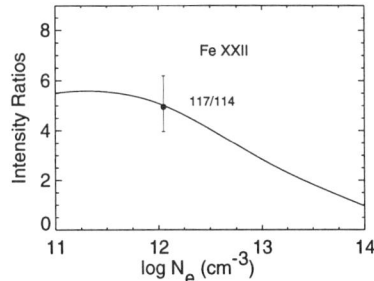

FIGURE 1. Densities derived here from different line ratios are more discrepant with each other than can be accounted for by statistical and atomic rate uncertainties. Curves are theoretical emissivity ratios for EUV emission lines (identified by their wavelengths in Å) of Fe XXI (*left*) and Fe XXII (*right*) vs N_e. The observed ratios from the summed (280 ksec) *EUVE* spectrum of Capella are overplotted as dots (1-σ errors) so that N_e may be determined from each line ratio.

certainties remain for iron, ionization and recombination rate coefficients for other astrophysically abundant elements such as Si, S, and Ni are in need of review.

THE ELECTRON DENSITY

The early *EUVE* calibration data of the bright active binary system Capella indicated the possibility of high density ($N_e \sim 10^{12}$ cm^{-3}), but density derivations from different line ratios were found to be inconsistent with each other (13). Subsequent observations confirm the large spread in derived densities (Fig. 1), suggesting the possibility that multiple densities exist (14). Recently the Hubble Space Telescope observed the Fe XXI λ1354.1 forbidden line on Capella with the Goddard High Resolution Spectrograph. Line profile measurements show roughly equal contributions from the two stars of the binary system (15), lending credence to our suggestion.

Densities derived from the *SERTS* solar active region catalogue show nearly two orders of magnitude spread (Fig. 2). While multiple densities may account for some of the large spread, the inaccuracy of the theoretical atomic data contributes to the spread as well, rendering the density determination highly inaccurate.

THE EMISSION MEASURE DISTRIBUTION

The emission measure $\int N_e N_H dV$, taken over a specified temperature interval, describes the amount of emitting material necessary to produce an observed line intensity. We construct a continuous emission measure distribution for the entire temperature range, over which all line emissivities are integrated, and then compare the predictions with a set of observed line intensities. The emission measure

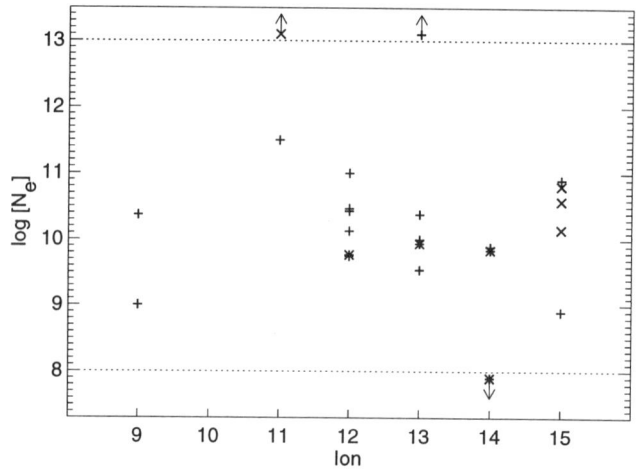

FIGURE 2. Electron densities derived for the *SERTS* solar active region spectrum, using different density-sensitive line ratios, show a large spread. The derived densities for each line ratio (both + and x) are plotted vs. their relative ionization stages (e.g. 9 = Fe IX). Derived densities represented by + use line ratios listed in Ref. 16, if observed; derived densities represented by x use line ratios listed from Ref. 2. The large spread in derived densities shown here is consistent with other criteria for the selection of line ratios, such as those from Ref. 6.

distribution model provides information on the components of the energy balance. The radiative loss term is calculated directly from the emission measure distribution. From the emission measures in conjunction with a measurement of the electron density, the thermal conduction (and mass motion) term may be calculated. These loss terms in turn balance the heating term.

The study of the temperature structure of cool stars in active binary systems with *EUVE* has led to surprising results. Whereas one might expect a constant emission measure gradient above the transition region (i.e. in the optically thin corona), as observed in the Sun, many stellar sources show an enhancement of the emission measure over a narrow temperature range, as shown in Figure 3 for Capella. The Capella emission measure distribution "bump" is constrained primarily by Fe XVIII and Fe XIX line intensities, which have remained constant to within 30% over two years of monitoring, despite factors of four variation in lines from higher ionization stages. Thus this nearly isothermal enhancement is remarkably constant over time. While the Capella "bump" is the most pronounced emission measure distribution feature we have found, many other *EUVE* cool star sources show similar enhancements (18).

Determining how isothermal these features are is severely hampered by uncertainties in the ionization balance for highly ionized iron. As Figure 3 shows, different choices of ionization and recombination rates from the literature over the

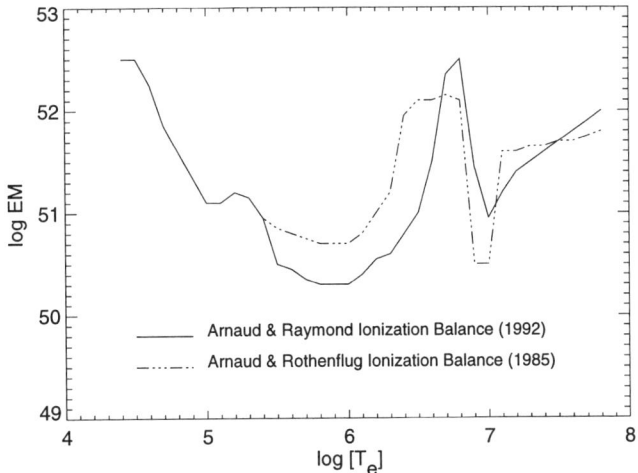

FIGURE 3. Models of the emission measure distribution for the Capella EUVE data. The solid line is the model based on the ionization equilibrium of Arnaud and Raymond (12); the dashed line is the model based on Arnaud and Rothenflug (17). Both models assume $N_e = 10^{11}$ cm^{-3}. For $T_e <\sim 3 \times 10^5$ K, ultraviolet emission lines are used (13, 14).

past decade lead to quite different models for Capella. The dominant remaining uncertainties here are the dielectronic recombination rates for Fe XVI to XIX.

Figure 4 shows the emission measure distribution for 44*i* Boo, a W UMa type contact binary which undergoes a partial eclipse. This source has the highest N_e observed so far with *EUVE* (19). The good agreement among the electron density line ratio diagnostics in the temperature range of the "bump" suggests that the emission measure enhancements are due to density rather than volume effects.

ELEMENTAL ABUNDANCES

The study of elemental abundances with EUV spectroscopy has led to other surprising results. The first ionization (FIP) effect, first observed with in situ particle detectors measuring the solar wind, has been corroborated with EUV spectroscopy of the full solar disk (20). This effect is an enhancement in the corona of elements with low FIP relative to those with high FIP, compared with their photospheric values. New solar observations indicate that this effect depends on the type of coronal structure and origin of material. Some cool stars observed with *EUVE*, but not all, exhibit a FIP effect, as determined by comparing emission lines from elements with different FIP (3).

For the cool stars with the hottest coronae, a different type of abundance anomaly has been observed from both X-ray spectra (21, and references therein) and *EUVE* spectra (22). This effect takes the form of reduced "metallicities," i.e. the line-to-

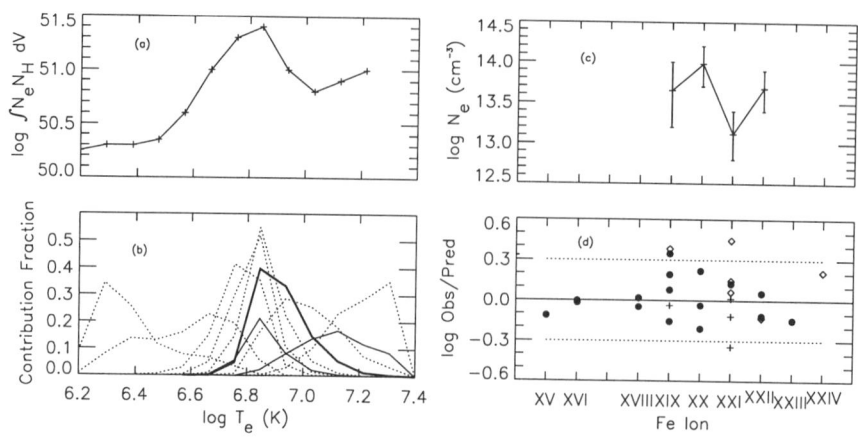

FIGURE 4. *(a)*: Emission measure distribution of 44i Boo derived from Fe line intensities. The highest temperatures are not well constrained since Fe XXIV is the highest ionization stage observable with EUVE. *(b)*: Model contribution fractions for strong Fe lines used in determining the emission measure distribution. For each value of $log T_e$ the contribution to a given line intensity is the line emissivity times the emission measure distribution. The solid thin line shows both Fe XX and Fe XXIII contributions to the strong, blended line $\lambda 132.85$. The solid thick line highlights the Fe XXI contribution. *(c)*: N_e derived from line ratios of different ions. *(d)*: Agreement of the observed to predicted line intensities. • represents strong (S/N > 3.0), unblended lines. + represents upper limits for weak (S/N < 3.0) or unobserved lines. ◊ represents strong, but blended lines, as determined either from model predictions or from the observed profile.

continuum ratio is observed to be lower than predicted by plasma emission models which assume photospheric abundances.

For Capella, the line-to-continuum ratio is consistent with predictions based on a photospheric abundance (14). The analysis includes line emission primarily coming from the "bump." We have shown that modeling the continuum emission correctly around 100 Å requires an accurate emission measure distribution over several orders of magnitude in temperature, and thus is susceptible to the errors in the ionization balance for iron as mentioned above.

DISCUSSION

We summarize important new EUV results for solar and stellar coronae:

(1). Electron densities in some stellar systems exceed solar active region densities by three to four orders of magnitude. The pressures involved are up to five orders of magnitude times the pressures typical of solar active regions. These high plasma pressures imply confining magnetic fields in the corona of order 100 to 1000 G (still higher in the photosphere). At high densities and temperatures, the dominant loss mechanism (for photospheric abundances) is radiation rather than conduction.

(2). The emission measure distribution for systems with hot coronae ($T_e \sim 10^7$ K) often has an enhancement that is inconsistent with scaling the emission measure gradient from solar active regions. The most likely explanation seems to be that these enhancements represent emission from a very high density region.

(3). Preliminary results indicate that only systems with a rapidly rotating component show an enhancement in the emission measure distribution (18). Doppler imaging and Zeeman Doppler imaging studies of photospheric lines for a few active stars show long-lived polar active regions, as well as shorter-lived equatorial active regions as found on the Sun (23, 24). We have suggested that the emission measure enhancements are the *coronal* counterpart of the polar active regions observed in the photosphere. A plausible picture is that the polar open field lines overlay small, dense, hot magnetically confined plasma loops.

(4). Anomalous stellar coronal abundances have drastic implications for physical modeling. One immediate impact is on the radiative cooling curve (25), since the radiative power loss at high temperature is dominated by iron emission lines. Radiative cooling in turn affects the thermal stability of the plasma. Element differentiation in the corona may also be a key to distinguishing between acoustic and magnetic heating mechanisms (26, and references therein).

ACKNOWLEDGMENTS

The author sincerely appreciates the contributions of her collaborators Andrea Dupree and John Raymond, and stimulating discussions with Steve Vogt and Aad van Ballegooijen. This work has been supported by NAGW-528 and NAG5-2330 to the Smithsonian Institution.

REFERENCES

1. Thomas, R. J., and Neupert, W. M., *Astrophys. J. Supp.*, **91**, pp. 461–482 (1994).
2. Mason, H. E., and Monsignori-Fossi, B. C., *Astron. & Astrophys. Rev.*, **6**, pp. 123–185 (1994).
3. Drake, J. J., "Stellar Spectroscopy with EUVE," to appear in *Ninth Cambridge Workshop on Cool Stars, Stellar Systems and the Sun*, R. Pallavacini and A. K. Dupree, eds., 1996.
4. Stratton, B. C., Moos, H. W., Suckewer, S., Feldman, U., Seely, J. F., and Bhatia, A. K., *Phys. Rev. A*, **31**, pp. 2534–2547 (1985).
5. Mason, H. E., Bhatia, A. K., Kastner, S. O., Neupert, W. M., and Swartz, M., *Sol. Phys.*, **92**, pp. 199–216 (1984).
6. Brickhouse, N. S., Raymond, J. C., and Smith, B. W., *Astrophys. J. Supp.*, **97**, pp. 551–570 (1995).
7. Raymond, J. C., and Brickhouse., N. S., "Atomic Processes in Astrophysics," in *Laboratory, Solar and Stellar, Diffuse Astrophysical and Extragalactic Plasmas*, Dordrecht: Kluwer Academic Publishers, 1996, in press.
8. Waljeski, K., Moses, D., Dere, K. P., Saba, J. L. R., Strong, K. T., Webb, D. F., and Zarro, D. M., *Astrophys. J.*, **429**, pp. 909–923 (1994).

9. Raymond, J. C., *Astrophys. J.*, **365**, pp. 387–390 (1990).

10. Lang, J., ed., *At. Data Nucl. Data Tbls.*, **57**, nos. 1/2 (1994).

11. Mason, H. E., private communication.

12. Arnaud, M. and Raymond, J. C., *Astrophys. J.*, **398**, pp. 394–406 (1992).

13. Dupree, A. K., Brickhouse, N. S., Doschek, G. A., Green, J. C., and Raymond, J. C., *Astrophys. J.*, **418**, pp. L41–L44 (1993).

14. Brickhouse, N. S., "Dissecting the EUV Spectrum of Capella," in *IAU Colloquium No. 152 on Astrophysics in the Extreme Ultraviolet*, Dordrecht: Kluwer Academic Publishers, 1996, pp. 105–112.

15. Linsky, J. L., private communication.

16. Harrison, R. A., ed., "The Coronal Diagnostic Spectrometer for SOHO: Scientific Report", Rutherford Appleton Laboratory report, SC-CDS-RAL-RP-89-0002 (1989).

17. Arnaud, M. and Rothenflug, R., *Astron. & Astrophys. Supp.*, **60**, pp. 425–457 (1985).

18. Dupree, A. K., "EUVE Spectroscopy of Active Binaries," to appear in *Ninth Cambridge workshop on Cool Stars, Stellar Systems and the Sun*, R. Pallavacini and A. K. Dupree, eds., 1996.

19. Brickhouse, N. S., and Dupree, A. K., in progress.

20. Laming, J. M., Drake, J. J., and Widing, K. G., *Astrophys. J.*, **443**, pp. 416-422 (1995).

21. Singh, K. P., White, N. E., and Drake, S. A., *Astrophys. J.*, **456**, pp. 766–776 (1996).

22. Stern, R. A., Lemen, J. R., Schmitt, J. H. M. M., and Pye, J. P., *Astrophys. J.*, **444**, pp. L45–L48 (1995).

23. Vogt, S. S., and Hatzes, A. P., "Doppler Images of HR 1099 from 1981–1993," to appear in *IAU Symposium No. 176 on Stellar Surface Structure*, 1996.

24. Donati, J.-F., Brown, S.F., Semel, M., Rees, D.E., Dempsey, R.C., Matthews, J.M., Henry, G.W., & Hall, D.S., *Astron. & Astrophys.*, **265**, pp. 682–700 (1992).

25. Cook, J.W., Cheng, C.-C., Jacobs, V.L., & Antiochos, S.K., *Astrophys. J.* **338**, pp. 1176–1183 (1989).

26. Drake, J. J., Laming, J. M., and Widing, K. G., *Astrophys. J.*, **443**, pp. 393–415 (1995).

Laboratory Astrophysics: Measurements of $n=n'$ to $n=2$ Line Emission in Fe^{16+} to Fe^{23+}

D. W. Savin*, P. Beiersdorfer[†], G. V. Brown[‡], J. Crespo López-Urrutia[†], V. Decaux[†], S. M. Kahn[§], D. A. Liedahl[†], K. J. Reed[†], and K. Widmann[†]

*Space Sciences Laboratory, 366 LeConte, University of California, Berkeley, CA 94720
[†]Department of Physics and Space Technology, Lawrence Livermore National Laboratory, University of California, Livermore, CA 94550
[‡]Department of Physics, Auburn University, Auburn, AL 36849
[§]Department of Physics, Columbia University, New York, NY 10027

Abstract. One of the dominant forms of astronomical line emission in the 6 Å to 18 Å spectral region is line emission produced by $n=n'$ to $n=2$ transitions in Fe^{16+} to Fe^{23+} (i.e., Fe L-shell n-2 line emission). Using the Lawrence Livermore National Laboratory electron beam ion trap (EBIT) facility, we have carried out a number of measurements designed to address astrophysical issues concerning Fe L-shell line emission. Desired ions are produced and trapped using the nearly monoenergetic electron beam of EBIT. Trapped ions are collisionally excited and the resulting X-ray line emission detected using Bragg crystal spectrometers. We have recently completed a line survey of Fe L-shell 3-2 line emission. The line survey will allow a more reliable accounting of line blending in astronomical spectra. We have now begun a series of broadband, high resolution line ratio measurements. These measurements are designed to benchmark atomic calculations used in astronomical plasma emission codes and also for comparison with X-ray spectral observations of astronomical objects. Initial measurements have been carried out in Fe^{23+}. Preliminary results agree with distorted wave calculations to within 20 percent and better.

INTRODUCTION

In astronomical plasmas, transitions in the iron L-shell ions (Fe^{16+} to Fe^{23+}) of the type $n=n'$ to $n=2$ (hereafter, n'-2) are one of the dominant forms of line

© 1996 American Institute of Physics

emission in the 6 Å to 18 Å spectral region. Iron L-shell line emission (*i.e.*, n-2) is seen in spectra from the sun (1,2), other stars (3,4), supernova remnants (5), cataclysmic variables (6), X-ray binaries (7), active galactic nuclei (8), and clusters of galaxies (9). Astrophysical interpretation of these spectra, however, are compromised by fundamental uncertainties that still remain in our understanding of the basic atomic physics processes that generate the observed X-ray spectra. Two examples serve to demonstrate this point.

The first example involves a long standing problem in solar observations. L-shell lines from neonlike Fe^{16+} are among the most intense X-ray lines emitted from the solar corona. These line are observed both in the presence and in the absence of flares (2). The observed Fe^{16+} line ratios, however, do not agree with theoretical predictions for inferred coronal temperatures. A number of different solar physics and atomic physics explanations have been put forward in an attempt to explain the apparent discrepancies. Possible solar physics explanations include resonant scattering of the strongest Fe^{16+} lines (2) or explosive events in the corona producing ionizing plasmas (10). Possible atomic physics explanations have been addressed by carrying out new, more detailed atomic calculations (11-13). The resolution of these discrepancies may have major implications for solar physics. If the explanation is resonant scattering, then that may offer the possibility of a new density diagnostic (14,15). And if explosive events are the explanation, then that would have implications for the heating of the solar corona (10). Laboratory benchmark measurements can aid greatly in resolving the apparent Fe^{16+} line ratio discrepancies and addressing the associated solar physics issues.

The second example involves a recent observation by the *Advanced Satellite for Cosmology and Astrophysics* (*ASCA*) of cooling flows in the Centaurus cluster of galaxies (9). The launching of *ASCA* in 1993 of the has opened a new era in high energy astrophysics. The large collecting area and moderate spectral resolving power of the spectrometers on *ASCA* are yielding high signal-to-noise X-ray spectra of many extrasolar astronomical objects (16). These spectra challenge the quality of the atomic data used in existing standard plasma emission codes. For example, in the 8 Å to 18 Å spectral band, significant discrepancies exist between *ASCA* observations of cooling flows in the Centaurus cluster of galaxies and model spectra from plasma emission codes (9). These discrepancies cannot be eliminated by invoking multi-temperature, multi-density distributions, nor by varying relative elemental abundances. Because cooling flows are believed to be optically thin, quasi-static plasmas (17), other astrophysical effects are expected to be insignificant. Hence, the discrepancies indicate the existence of errors in the atomic data used in the plasma codes, and have been attributed to errors in the electron impact excitation (EIE) rate coefficients for producing n-2 line emission in Fe^{22+} and Fe^{23+} (18).

The importance of accurate atomic data for n-2 transitions in iron L-shell ions is not limited just to *ASCA* or solar spectra. The planned launches later this decade of the *Advanced X-Ray Astrophysics Facility* (*AXAF*, 19), the *X-Ray Multimirror Mission* (*XMM*, 21) and *ASTRO-E* (22) will put in orbit spectrometers which will collect extrasolar spectra in the 6 Å to 18 Å spectral band with resolving powers, $\lambda/\Delta\lambda$, over an order of magnitude greater than those of *ASCA* spectrometers. Interpreting *AXAF*, *XMM*, and *ASTRO-E* spectra will, therefore, require atomic data with an accuracy and completeness significantly greater than exists at the present time.

For several years, we have been carrying out experimental measurements of n-2 line emission from the iron L-shell ions in order to address these present and future needs of high energy astrophysics. Measurements are carried out using the Lawrence Livermore National Laboratory electron beam ion trap (EBIT) facility (22,23). Iron ions of interest are produced, trapped, and excited by collisions with the nearly monoenergetic electron beam in EBIT. The emitted L-shell X-ray lines are recorded and analyzed with high resolution Bragg crystal spectrometers (24).

ELECTRON BEAM ION TRAP MEASUREMENTS

Line Surveys

Iron L-shell line emission is comprised of numerous strong transitions which, depending on the temperature of the emitting source, result typically from transitions in Fe^{14+} to Fe^{23+}. The analysis of iron L-shell spectra thus requires consideration of a large number of ionization stages and a large number of possible transitions. A reliable analysis of astronomical spectra must account for all lines and needs to employ accurate line position information to properly model line blending. To address these issues, we have made a survey of the iron L-shell 3-2 transitions in the 10 Å to 17.5 Å spectral range. A representative Fe^{16+} L-shell spectrum is shown in Fig. 1 (24). Spectra were collected from Fe^{16+} through to Fe^{23+} (25). Charge balances consisting primarily of a single ionization stage of iron were produced in EBIT by setting the electron beam energy to just below the ionization threshold for the charge state of interest. This allowed observed spectral lines to be unambiguously assigned to a given ionization stage of iron. The wavelength scale was determined by observing K-shell (*i.e.*, n-1) transitions in hydrogenlike and heliumlike oxygen, fluorine, neon, and magnesium. An accuracy of about 0.1-0.3 eV was achieved in the determination of the transition energies. For example, the measured wavelength of 3E, the weakest Fe^{16+} line in Fig. 1, was 15.456±0.004 Å while that of 3C, the strongest line, was 15.009±0.001 Å (24). Identifications of lines in the higher charge states of iron that have not been identified in earlier measurements were made by comparing observed intensities and measured wavelength with calculations from a collisional-radiative model which uses the Hebrew University-Lawrence Livermore set of atomic codes (HULLAC, 26,27).

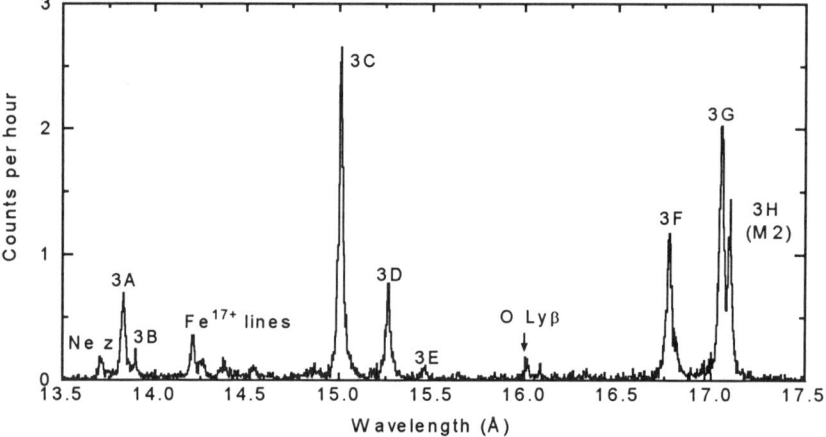

Figure 1. Fe^{16+} L-shell spectrum recorded using a thallium hydrogen phthalate crystal in first order. The electron beam energy was 1300 eV. The spectrum is a composite of data from three different crystal settings. Fe^{16+} lines are labeled using the notation of Ref. 1.

Line Ratio Measurements

X-ray spectral observations of astronomical objects are typically broadband. To be useful, laboratory measurements must, accordingly, be broadband. And for comparison with atomic calculations it is necessary to collect spectra at a resolving power high enough that individual observed spectral lines are fully resolved. These two requirements, broadband spectra with high resolving power, have been met by using two crystal spectrometers to observe simultaneously line emission from EBIT. Crystal spectrometers, however, are inherently narrowband devices with X-ray detection efficiencies which over a broad band may change dramatically. Broadband line ratio measurements are carried out by using the two spectrometers to observe simultaneously different narrowband regions (each containing lines of interest).

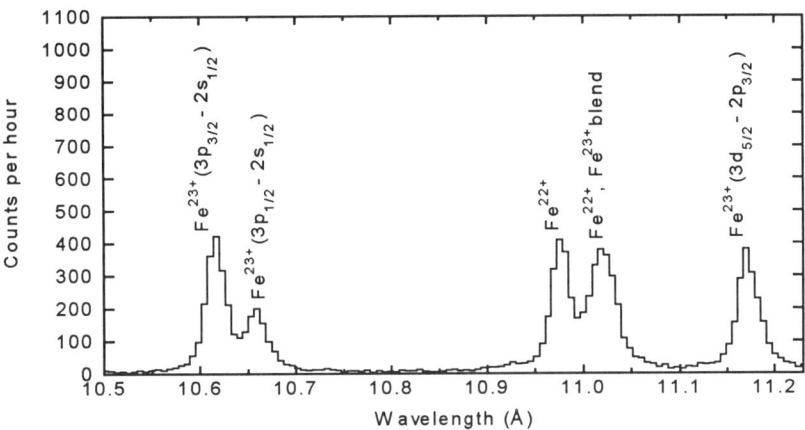

Figure 2. Iron *L*-shell 3-2 spectrum recorded using a thallium hydrogen phthalate crystal in first order. The electron beam energy was 4500 eV.

The ion selected for our initial broadband line ratio measurements was Fe^{23+} This ion was chosen for a number of reasons. Fe^{23+} is lithiumlike which makes it a relatively simple ion to work with both experimentally and theoretically. This makes Fe^{23+} a good initial test-case for carrying out quantifiable, broadband line ratio measurements. And astrophysically, Fe^{23+} was chosen to provide benchmark measurements for the HULLAC calculations (18) which appear largely to remove the reported (8) discrepancies between *ASCA* observations of the Centaurus cluster of galaxies and predictions from plasma emission codes.

For our initial experiments we have measured the EIE-generated Fe^{23+} $(3p_{3/2}$-$2s_{1/2})/(3p_{1/2}$-$2s_{1/2})$, $(3d_{5/2}$-$2p_{3/2})/(3p_{1/2}$-$2s_{1/2})$, and $(4p_{3/2,1/2}$-$2s_{1/2})/(3p_{1/2}$-$2s_{1/2})$ line ratios (28). Representative Fe^{23+} spectra are shown in Figs. 2 and 3. Measurements have been carried out at electron beam energies of 2.5 and 4.5 keV. These energies are above the ionization potential for Fe^{23+}, which ensures that the measured line ratios are due solely to EIE and are free of any contributions from dielectronic recombination. At these beam energies, significant amounts of Fe^{24+} exist in EBIT. Recombination-cascade contributions due to radiative recombination of Fe^{24+} with beam electrons and from charge transfer of Fe^{24+} with background gas in EBIT were determined to have an insignificant effect on the measured line intensities. EBIT uses a directed beam of electrons which can produce anisotropically emitted, linear polarized radiation (29). These effects on the measured line intensities were accounted for using the formalism of Ref. (30) combined with distorted wave calculations using the code of Ref. (31).

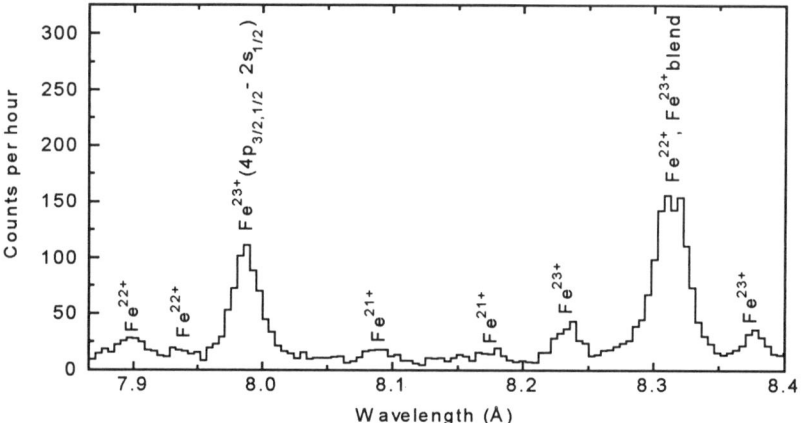

Figure 3. Iron L-shell 4-2 spectrum recorded using a thallium hydrogen phthalate crystal in first order. The electron beam energy was 4500 eV.

The X-ray detection efficiency for one of the crystal spectrometers was established by calibrating the thin foil windows and thallium acid phthalate crystal used for the present measurements. X-rays were detected in the spectrometer using a flowing gas proportional counter (FGPC) with an absorbing gas of 90% Ar and 10% CH_4 (24). X-ray detection efficiencies were calculated from the measured depth of the FGPC gas volume and using photoabsorption cross sections which are well known experimentally and theoretically (32). The narrow band

($3p_{3/2}$-$2s_{1/2}$)/($3p_{1/2}$-$2s_{1/2}$) and ($3d_{5/2}$-$2p_{3/2}$)/($3p_{1/2}$-$2s_{1/2}$) line ratios were measured using the calibrated spectrometer. The broad band ($4p_{3/2,1/2}$-$2s_{1/2}$)/($3p_{1/2}$-$2s_{1/2}$) line ratio was measured using both spectrometers. The uncalibrated spectrometer was cross-calibrated with the calibrated spectrometer for detection of the $3p_{1/2}$-$2s_{1/2}$ line by simultaneously observing this line with both spectrometers. The ($4p_{3/2,1/2}$-$2s_{1/2}$)/($3p_{1/2}$-$2s_{1/2}$) line ratio was measured using the calibrated spectrometer to observe the $4p_{3/2,1/2}$-$2s_{1/2}$ line and the cross-calibrated spectrometer to observe the $3p_{1/2}$-$2s_{1/2}$ line.

A preliminary data analysis indicates that the measured line ratios and predictions by HULLAC and by the distorted wave code of Ref. (31) all agree to within 20 percent and better. For example, the measured ($4p_{3/2,1/2}$-$2s_{1/2}$)/($3p_{1/2}$-$2s_{1/2}$) line ratio at 4.5 keV is $0.492 \pm 0.074(1\sigma)$ while the ratio predicted by HULLAC is 0.409 and by the code of Ref (31) is 0.386. The uncertainty quoted here represents the total experimental uncertainty in the measured line ratio. The accuracies of the present line ratio measurements are limited by the resolving powers of the crystal spectrometers and the resulting uncertainty in determining the background level in collected spectra. The spectra shown in Figs. 2 and 3 were collected at a resolving power of 500. Future measurements will be carried out using spectrometers with resolving powers in excess of 1000.

CONCLUSIONS

Using EBIT we have carried out a number of measurements of iron L-shell line emission designed to address astrophysical issues. Our recently completed line survey will allow a more reliable accounting of line blending in astronomical spectra. More recently, we have demonstrated that broadband, high resolution, iron L-shell line ratio measurements can be carried out using EBIT. These iron L-shell line ratio measurements are the beginning of a series of measurements which will benchmark atomic calculations used in plasma emission codes and also be used for comparison with X-ray spectral observations of astronomical objects.

ACKNOWLEDGEMENTS

This work was performed under the auspices of the U. S. Department of Energy by the Lawrence Livermore National Laboratory under contract number

W-7405-ENG-48 and supported by NASA High Energy Astrophysics X-Ray Astronomy Research and Analysis grant NAGW-4185.

REFERENCES

1. Parkinson, J. H., *Astron. Astrophys.* **24**, 215-218 (1973).
2. Rugge, H. R. and McKenzie, D. L., *Astrophys. J.* **297**, 338-346 (1985).
3. Vedder, P. W. and Canizares, C. R., *Astrophys. J.* **270**, 666-670 (1983).
4. Agrawal, P. C. et al., *Mon. Not. R. Astron. Soc.* **213**, 761-771 (1985).
5. Winkler, P. F. et al., *Astrophys. J.* **246**, L27-L31 (1981).
6. Mauche, C. W. and Raymond, J. C., in *New Horizon of X-Ray Astronomy*, edited by F. Kamino and T. Ohashi, Tokyo: Univeral Academy Press, Inc., 1994, pp 399-400.
7. Kahn, S. M., Seward, F. D, and Chlebowski, T., *Astrophys. J.* **283**, 286-294 (1984).
8. Iwasawa, I. et al., *Publ. Astron. Soc. Japan* **46**, L167-L171 (1994).
9. Fabian, A. C. et al., *Astrophys. J.* **436**, L63-L66 (1994).
10. Feldman, U., *Comm. At. Mol. Phys.*, **31**, 11-20 (1995).
11. Raymond, J. C. and Smith, B. W., *Astrophys. J.* **306**, 762-766 (1986).
12. Chen, M. H. and Reed, K. J., *Phys. Rev. A* **40**, 2292-2300 (1989).
13. Goldstein, W. H. et al., *Astrophys. J.* **344**, L37-L40 (1989).
14. Schmelz, J. T., Saba, J. L. R., and Strong, K. T., *Astrophys. J.* **398**, L115-L118 (1992).
15. Waljeski, K. et al., *Astrophys. J.* **429**, 909-923 (1994).
16. Tanaka, Y. Inoue, H, and Holt, S. M., *Publ. Astron. Soc. Japan* **46**, L37-L41 (1994).
17. Fabian, A. C., *Ann. Rev. Astron. Astrophys.* **32**, 277-318 (1994).
18. Liedahl, D. A., Osterheld, A. L., and Goldstein, W. H., *Astrophys. J.* **438**, L115-L118 (1995).
19. Markert, T. H., in *UV and X-ray Spectroscopy of Laboratory and Astrophysical Plasmas*, by E. Silver and S. M. Kahn, eds., Cambridge: Cambridge University Press, 1993, pp. 459-468.
20. Brinkman, A. C., in *UV and X-ray Spectroscopy of Laboratory and Astrophysical Plasmas*, E. Silver and S. M. Kahn, eds., Cambridge: Cambridge University Press, 1993, pp. 469-482.
21. http://www.astro.isas.ac.jp/xray/mission/astroe/astroeE.html.
22. Levine, M. A. et al., *Physica Scripta* **T22**, 157-163 (1988).
23. Levine, M. A. et al., *Nucl. Instrum. Methods* **B43**, 431-440 (1989).
24. Beiersdorfer, P. and Wargelin, B. J., *Rev. Sci. Instrum.* **65**, 13-17 (1994).
25. Brown, G. V. et al. (in preparation).
26. Klapisch, M. et al., *J. Opt. Soc. Am.* **67**, 148-155, (1977).
27. Bar-Shalom, A., Klapisch, M., and Oreg, J., *Phys. Rev. A* **38**, 1773-1784 (1988).
28. Savin, D. W. et al. (in preparation).
29. Percival, I. C. and Seaton, M. J., *Phil. Trans. Roy. Soc. London Ser. A* **251**, 113-138.
30. Steffen, R. M. and Alder, K. in *The Electromagnetic Interaction in Nuclear Spectroscopy*, edited by W. D. Hamilton, Amsterdam, North-Holland Publishing Company, 1975, ch. 12.
31. Zhang, H. L., Sampson, D. H., and Clark, R. E. H., *Phys. Rev. A* **41**, 198-206 (1990).
32. Henke, B. L. Gullikson, E. M., and Davis, J. C., *At. Data Nucl. Data Tables*, **54**, 181-342 (1993).

Atomic Physics for Modeling of Astrophysical Plasmas

J. Michael Shull[*]

[*] Department of Astrophysical, Planetary, & Atmospheric Sciences; CASA and JILA, University of Colorado and National Institute of Standards and Technology, Campus Box 440, Boulder, CO 80309.

Abstract. This paper describes the various types of ionization equilibrium applicable to the modeling of astrophysical plasmas, particularly collisional ionization and photoionization equilibrium. Modern astrophysical studies rely heavily on atomic data for derivation of physical parameters from observations, as well as for theoretical modeling. I will describe the current state of atomic data for two areas: (1) Abundances and absorption-line oscillator strengths; (2) Hot plasma ionization and emission.

"THERE IS SOMETHING FASCINATING ABOUT SCIENCE. FROM SUCH A TRIFLING INVESTMENT OF FACT, ONE GETS SUCH WHOLESALE RETURNS OF CONJECTURE.
Mark Twain, *Life on the Mississippi*

Introduction

"Why model astrophysical plasmas?" What does one learns physically about hot gas in space through remote sensing of its continuum and line emission? The primary issues for the modeling of astrophysical plasmas involve finding appropriate spectrum diagnostics to determine the type of ionization equilibrium (or disequilibrium), infer the electron density and temperature, and to derive the abundances of key heavy elements relative to H and He. By measuring these simple properties of the plasma, astrophysicists are able to measure such macroscopic properties as the mass, volume, energy budget, and chemical history of the gas.

In the recent past, ionization modeling focussed on several simplified cases, such as collisional ionization equilibrium, photoionization equilibrium, or local thermodynamic equilibrium (LTE or Saha). These are limiting cases of more general situations. After reviewing these archetypes of ionization-state models, I then describe the current status of the required atomic data and modeling techniques. I conclude with a few specific examples from studies of the intergalactic medium (IGM) and the interstellar medium (ISM) of galaxies.

© 1996 American Institute of Physics

Basic Plasma Diagnostics

The most basic properties one needs to model an astrophysical plasma are the electron density (n_e), electron temperature (T_e), and extinction (A_λ or optical depth τ_λ) at wavelength λ. To infer these quantities, the standard technique begins by measuring the energy fluxes of two spectral lines (usually from the same atom or ion) and comparing their ratio to theoretically predicted values. These line ratios then yield preferred values of n_e and T_e. From these parameters, one then derives the volume emissivity of the plasma, the "emission measure", and finally the masses and elemental abundances.

For example, n_e is typically derived from two emission lines with very similar upper-state excitation energies, but different rates of (electron-impact) de-excitation. Thus, the "critical densities", $n_{\rm cr} = A_{21}/C_{21}$, for de-exciting the line result in a sensitivity to n_e of the line ratio for densities within about a factor of five on either side of $n_{\rm cr}$. In the optical, these density-sensitive line ratios include the forbidden lines from the [O II] doublet (at 3726 and 3729 Å) and the [S II] doublet (at 6716 and 6731 Å). For higher-temperature, more ionized plasmas, astronomers use other lines such as C III] 1907, 1909 Å and a variety of ultraviolet and soft X-ray emission lines from highly ionized Si, S, or Fe.

In contrast, T_e is typically derived from two emission lines with considerably different excitation energies. This allows one to sample the tail of the thermal distribution of particle velocities, since excitation rates and line ratios are sensitive to the Boltzmann factor, $\exp(-E_{\rm exc}/kT_e)$. In the optical, the usual temperature-sensitive line ratios are the two forbidden lines arising from the $^1D_2$ upper states of [O III] (lines at 5007 and 4959 Å) and of [Ne III] (lines at 3967 and 3869 Å). As is the case for n_e, the temperature of more ionized plasmas can be diagnosed by a variety of ultraviolet and soft X-ray emission lines from Si, S, or Fe.

After these basic diagnostics have been obtained, additional parameters can be obtained. The "emission measure" (EM) is defined as the volume integral,

$$\text{EM} = \int n_e n_i \, d^3 r \,, \tag{1}$$

measuring the product of the densities of the ion and electron collision partners (usually $i = $ H). This quantity is easily derived from an emission line flux, provided one knows the distance to the source and the volume emissivity (a function of T_e). From the EM, one can then derive the plasma mass,

$$M = \int n_H \, \mu \, d^3 r \,, \tag{2}$$

where n_H is the density of hydrogen in all forms and μ is the mean molecular weight of the plasma, typically 1.3–1.4m_H correcting for the helium abundance. For a parcel of plasma with density n_H and volume V, the EM $\propto n_H^2 V$ while the mass $M \propto n_H V$. Thus, $M \propto$ (EM)$/n_H$. Therefore, a mass measurement requires an accurate knowledge of the density, as well as T_e. The primary density diagnostics for n_e and T_e are particularly critical if the plasma contains fluctuations in these quantities, owing to the exponential dependence of emissivity on T_e.

Once the density, temperature, and mass of an astrophysical plasma have been ascertained, one can then turn to measurements of its elemental abundances and chemical (nucleosynthetic) history. As with the other parameters, the heavy element abundances of a plasma are usually derived from ratios of emission lines. The intensity of an emission line connecting two states (1,2) of ion i is proportional to

$$I_{21} \propto \int n_e n_i\, \Omega_{12}(T_e) \exp(-E_2/kT_e)\, d^3r \,, \qquad (3)$$

where $\Omega_{12}(T_e)$ is the Maxwellian-averaged, electron-impact collision strength of the transition, related to the excitation cross section (1). The abundance of ion i relative to hydrogen can therefore be derived from the ratio, $I_{21}/$(EM), given sufficient knowledge of the ionization fraction of the element, which depends on the particular model applied (collisional ionization, photoionization, shocks, etc.).

Abundances may also be determined by measuring absorption lines toward sources with strong continuum emission. The optical depth of a line connecting two levels (1,2) of ion i is proportional to the integral of the ion density along the sightline to the source,

$$\tau_\nu \propto \int n_i f_{12}\, \phi_\nu\, dx \,, \qquad (4)$$

where f_{12} is the absorption oscillator strength and ϕ_ν is the frequency profile function of the line, related to the Doppler velocity distribution. Thus, one may obtain the "column density" of the ion, $N_i = \int n_i\, dx$, and compare it with similar measurements of the abundant species (e.g., hydrogen) to obtain a mean abundance. The crucial atomic data for these measurements are the set of oscillator strengths, particularly for the weak lines of atoms and ions. These weak (often semi-forbidden) transitions provide the best abundance diagnostics, since their absorption lines do not suffer the line saturation effects that plague measurements from the stronger lines of C, N, O, Si, S, or Fe.

I have described how line diagnostics provide basic and secondary plasma parameters, such as density, temperature, mass, and abundances. Armed with this information, astrophysicists can next derive the energy input rate to the plasma required to balance the radiative cooling rate,

$$\text{Cooling Rate} = \int n_e n_H \, \Lambda(T) \, d^3r \tag{5}$$

where $\Lambda(T)$ is a cooling function (ergs cm^3 s^{-1}) that depends on the mixture of heavy element abundances, the collisional excitation rates, and the distribution of ions with temperature. A recent determination (2) gives $\Lambda(T)$ for hot, collisionally ionized plasmas. In the next two sections, I will discuss abundances and ionization mechanisms in more depth.

Abundances

Among the primary observational parameters for studying the ISM and IGM are the abundances of perhaps 20 – 25 elements. In the diverse environment of the ISM, these elements are often spread over many ionization states. In the colder phases of the medium, these elements are partially contained in molecular form (e.g., CO, H$_2$O, SiO) or solid-state form (graphite and silicate grains, polycyclic aromatic hydrocarbons, etc.). The ISM in its variety of phases contains ranges of temperature between 10 K and 10^7 K, densities between 10^{-3} and 10^6 cm^{-3}, and ultraviolet radiation fields fluctuating by several orders of magnitude about the mean value (3) at 912 Å of $J_o \approx 10^{-23}$ ergs cm^{-2} s^{-1} Hz^{-1} sr^{-1} or $\Phi_o = 3 \times 10^4$ photons cm^{-2} s^{-1}.

In the ISM, as in unevolved stellar atmospheres, the most abundant elements are hydrogen (H) and helium (He), whose abundances are set primarily by Big Bang nucleosynthesis (4). The primordial universe, ten minutes after the the Big Bang, contained mostly H and He, with trace amounts of deuterium ($^2_1$H) and lithium ($^7_3$Li). The heavier elements (e.g., C, N, O, Si, S, Fe) were produced much later through stellar nucleosynthesis. If X, Y, and Z denote the fractions, *by mass*, of H, He and all heavier trace elements, the current observations are consistent with a primordial helium abundance $Y = (0.239 \pm 0.015)$ and a primordial lithium abundance $Z \approx 10^{-9}$.

The present-day ISM has built up a heavy-element abundance $Z \approx 0.02$, owing to several generations of massive-star formation. The most massive stars (masses 8 – 100 times that of our Sun) produce strong stellar winds and supernova explosions, which expel the heavy elements synthesized in their interiors into the ISM. Lower-mass stars spend about 10% of their lives, during "old age", in the form of extended red giants and supergiants. These stars also shed their mass in the form of slow winds and planetary nebulae, which disperse the products of stellar nucleosynthesis back into the ISM. Since the

ISM is fairly well mixed, one expects interstellar abundances to agree with the "cosmic values" found in the Sun's atmosphere or in meteorites.

The oscillator strengths of major interstellar species were tabulated during the *Copernicus* satellite era (5) and updated (6) after the launch of the *Hubble Space Telescope*. I would note the need for more accurate oscillator strengths in Fe-group elements, such as Zn II, Ni II, Cr II, and Co II. With lower cosmic abundances, these elements have lines that are generally weaker than those of Si II and Fe II, and they offer a chance to obtain accurate abundances for nucleosynthetic products from stars other than those producing the CNO-group and Si/S-group. Recent experimental measurments have been made for lines of Cr II, Zn II, Si II, and Fe II (7,8,9).

Equally intriguing are the detections of these same Cr II, Zn II, Ni II, and Co II lines in high-redshift "damped Lyα" absorption systems towards quasi-stellar objects (10,11,12). These trace metals are used to determine the heavy-element abundance in the intervening galaxies responsible for this absorption. By measuring the metal abundances in a variety of absorbers, at different redshifts and therefore different "look-back times" over the history of the universe, astronomers can gauge the buildup of heavy elements through star formation, supernovae, and other nucleosynthetic processes.

Ionization Models for Hot Plasmas

A full discussion of the astrophysics of hot plasmas would take far too long for this article. I will confine my remarks to the major contributions in this field and the remaining atomic data needs. Astrophysical plasmas are generally heated by ionizing radiation (EUV, X-rays, γ-rays) or by collisions with particles (cosmic rays, hot electrons). Although in many cases both types of ionization occur, the extreme cases are distinguished as "photoionization equilibrium" and "coronal equilbrium" (low-density collisional ionization equilibrium). For a more complete discussion of the sorts of thermal and ionization phases produced by these processes see (13,14,15).

In the photoionzation (nebular) model, the ratio of ion densities in adjacent ionization states $(z, z+1)$ is determined by an equilibrium between photoionization $(z \to z+1)$ and recombination $(z+1 \to z)$. Thus, one has,

$$\frac{n_{z+1}}{n_z} \propto \frac{\text{Flux}}{n_e \alpha_z(T)} \propto U , \qquad (6)$$

where $\alpha_z(T)$ is the rate coefficient for radiative plus dielectronic recombination to ion z and where $U = n_\gamma/n_H$ is the "photoionization parameter", the density ratio of ionizing photons (ionizing flux) to baryons (hydrogen density). This dimensionless parameter determines the distribution of ionization states of the plasma, and secondarily determines the equilibrium temperature through the radiative cooling function. Thus, the ion distribution in a photoionized plasma depends on the ratio of the radiation field to density.

In contrast, for a collisionally ionized plasma ("coronal model") the ionization distribution is set by an equilibrium between the rate of electron-impact collisional ionization and recombination. Since electrons control both ionization and recombination processes in this model, the resulting ion fractions are independent of n_e,

$$\frac{n_{z+1}}{n_z} = \frac{C_z(T)}{\alpha_z(T)}, \qquad (7)$$

where $C_z(T)$ is the rate coefficient (cm^3 s^{-1}) for collisional ionization of ion stage z. Thus, the ionization fractions depend only on T_e. Charge exchange processes with H or He complicate these relations, as does inner-shell photoionizations by X-rays, which can produce one or more additional photoelectrons through the Auger effect, thereby coupling more than just adjacent ion stages.

In many astrophysical situations, hybrid models are required, since both collisional ionization and photoionization are important. For example, in radiative shock waves the hot post-shock gas is collisionally ionized, but ionizing photons produced by collisional excitation and recombination can photoionize the gas both upstream (radiative precursor) and downstream (plasma recombination zone). Both processes also appear to be important in "turbulent mixing layers" (16) and in the "galactic fountain" (17) produced by the stellar winds and supernovae from massive stars in the Galactic disk.

A further complexity in ionization modeling arises when the timescale t is insufficient for an equilibrium to be set up. The relevant processes for approaching equilibrium are the photoionization time, $t_{\rm ph} = \Gamma_z^{-1}$ (Γ_z is the photoionization rate per ion), the collisional ionization time, $t_{\rm ion} = [n_e C_z(T)]^{-1}$, and the recombination time, $t_{\rm rec} = [n_e \alpha_z(T)]^{-1}$. In situations when $t < t_{\rm ph}$ or $t < t_{\rm ion}$, the plasma can be under-ionized relative to equilibrium. If $t < t_{\rm rec}$, the plasma may be over-ionized. In both cases, a more complicated, time-dependent calculation is required to model the plasma.

In order to diagnose a hot plasma, astronomers use both emission and absorption lines, as well as UV/X-ray continuum radiation. For example, the hot gas at 10^{5-7} K in the Galactic halo has been studied in soft X-ray emission (18), through UV absorption lines (19) from highly-ionized gas, through optical emission lines (20), and through diffuse UV emission lines of C IV $\lambda 1549$, O III] $\lambda 1663$ (21) and O VI $\lambda 1035$ (22). These diagnostics are also applied to hot gas in external galaxies, including optical emission-line studies (23,24), soft X-ray emission (25), and optical and UV measurements of redshifted absorption-line systems in the spectra of QSOs (26).

Theoretical modeling of these processes has centered around understanding the temperatures and energy sources for hot gas. Models of the gaseous Galactic halo can be characterized by the heating mechanism invoked: external UV radiation (from O-stars, white dwarfs, quasars, or Seyfert galaxies), internal energy sources (cosmic rays or magnetic reconnection), or energy injection from the disk (supernova remnants, stellar winds, and superbubbles). Two currently popular models involve the "galactic fountain" (17,27) and "turbulent mixing layers" (16). For studies of QSO absorption-line systems, accurate photoionization codes (15,28) allow astronomers to derive the "photoionization parameter" (ratio of ionizing flux to hydrogen density) and heavy-element abundances of the gas. Generally, collisionally-ionized gas (the coronal model) attains equilibria much hotter (10^{5-7} K) than photoionized gas ($10^{4.0-4.3}$ K). However, nature is often complex, and many situations such as the fountains and mixing layers noted above involve both ionizing mechanisms.

In order to analyze these datasets or build models, astrophysicists require the following atomic data: collisional ionization rates, radiative and dielectronic recombination rates, photoionization cross sections, and collisional excitation cross sections (the collisional rates are dominated by electron-impact). In cases where the plasma contains a small ($H^o/H_{tot} \geq 10^{-3}$) neutral fraction, charge-transfer reactions with H^o and He^o are important. As a sidelight of these models, a "radiative cooling curve" can be derived which determines the energy budget, the power radiated (ergs cm^{-3} s^{-1}) from an optically-thin, low-density, metal-seeded plasma (2). For temperatures above $10^{7.5}$ K, thermal bremsstrahlung dominates the cooling. Collisionally-excited line radiation from ions of Fe, Si, Mg, Ne, C, N, and O dominates the cooling between 10^4 and 10^7 K. Between 10^6 K and 10^7 K, Fe ions provide between 50% and 75% of the cooling. These lines include both allowed and semi-forbidden transitions of elements of both high abundance *and* high atomic number. Between $10^{4.5}$ K and $10^{5.5}$ K, the Li-like (2s-2p) transitions of C IV, N V, and O VI are especially important coolants.

The atomic data needs in this area of astrophysical modeling are enormous, often involving H and He plus 5 – 10 abundant heavy elements (C, N, O, Ne, Mg, Si, S, Ar, Ca, Fe) to monitor the primary thermal mechanisms, and 5 or 10 more trace elements to model the spectral diagnostics. For each of the elements, between 6 and 27 ion stages may be involved. The collisional ionization rates are in fairly good shape; see the critically evaluated reviews (29,30). Photoionization cross sections have also been computed to a much higher degree of accuracy through the "US-UK Opacity Project" (31).

Probably the weakest links in the spectral modeling are the recombination rates and the excitation rates. An enormous amount of work has been done on the ionization and recombination rates of iron. However, the dielectronic recombination rates are still controversial and uncertain. A set of recent calculations and analytic fits to radiative recombination rates (32) is based on the Milne relation of detailed balance and Opacity Project photoionization cross sections. Similar programs to derive dielectronic rates from the resonances in photoionization cross sections (Pradhan, Nahar, Verner, Ferland) are now underway.

In addition, the electron-impact excitation cross sections for the major coolants are often uncertain, except for the high-accuracy experimental and theoretical work devoted to the important Li-like (2s-2p) transitions of C IV With the accuracy of today's theoretical codes and experimental cross-beam procedures, there is no reason why the modeling should use only a single-parameter fit (the "collision strength" Ω) to the Maxwellian-averaged excitation rate. The most accurate models now use multi-parameter fits to $\bar{\Omega}(T)$, including the effects of auto-ionizing resonances in the scattering cross section.

Beyond the two generic needs mentioned above, there is still an uneasiness that charge-transfer cross sections at low and moderate energies may be less accurate than needed, and that collisional ionization rates for some species (Si, S, Fe) at high temperatures need further work. However, despite these misgivings, the state of atomic physics needed for modeling hot plasmas is in a remarkably good state, thanks to the efforts of the atomic physics and plasma physics communities.

Acknowledgements

This research was supported by the NASA Astrophysical Theory Program for studies of the Interstellar and Intergalactic Medium (NAGW-766).

References

1. Osterbrock, D. E. 1989 Astrophysics of Gaseous Nebulae and Active Galactic Nuclei, (Mill Valley: University Science Books)

2. Sutherland, R. S., & Dopita, M. A. 1993, ApJS, 88, 253

3. Maloney, P. 1992, ApJ, 398, L89

4. Walker, T. P., Steigman, G., Schramm, D. N., Olive, K. A., & Kang, H.-S. 1991, ApJ, 376, 51

5. Morton, D. C., & Smith, W. H. 1973, ApJS, 26, 333

6. Morton, D. C. 1991, ApJS, 77, 119

7. Bergeson, S. D., & Lawler, J. E. 1993a, ApJ, 408, 382

8. Bergeson, S. D., & Lawler, J. E. 1993b, ApJ, 414, L137

9. Bergeson, S. D., Mullman, K. L., & Lawler, J. E. 1994, ApJ, 435, L157

10. Meyer, D. M., & York, D. G. 1987, ApJ, 319, L49

11. Pettini, M., Boksenberg, A., & Hunstead, R. W. 1990, ApJ, 378, 6

12. Wolfe, A. M., et al. 1994, ApJ, 435, L101

13. Lepp, S., McCray, R., Shull, J. M., Kallman, T., & Woods, D. T. 1985, ApJ, 288, 58

14. Kallman, T., & McCray, R. 1982, ApJS, 50, 263

15. Donahue, M., & Shull, J. M. 1991, ApJ, 383, 511

16. Slavin, J., Shull, J. M., & Begelman, M. C. 1993, ApJ, 407, 83

17. Shapiro, P. R., & Benjamin, R. A. 1993, in *Star-Forming Galaxies and their Interstellar Medium*, ed. J. J. Franco (Cambridge: Cambridge Univ. Press), 273

18. Snowden, S. L., et al. 1995, ApJ, 454, 643

19. Sembach, K. R., & Savage, B. D. 1992, ApJS, 83, 147

20. Reynolds, R. J 1990, ApJ, 349, L17

21. Martin, C., & Bowyer, S. 1990, ApJ, 350, 242

22. Dixon, W. V., Davidsen, A. F., & Ferguson, H. C. 1996, ApJ, in press

23. Rand, R., Kulkarni, S., & Hester, J. J. 1990, ApJ, 352, L1

24. Hunter, D. A., & Gallagher, J. S. 1990, ApJ, 363, 480

25. Wang, Q., & Helfand, D. J. 1991, ApJ, 370, 541

26. Weymann, R. J., Carswell, R. F., & Smith, M. G. 1981, Ann. Rev. Astr. Ap., 19, 41

27. Bregman, J. N. 1980, ApJ, 236, 577

28. Hammann, F., & Ferland, G. 1992, ApJ, 391, L53

29. Bell, K. L., Gilbody, H. B., Hughes, J. G., Kingston, A. E., & Smith, F. J. 1983, J. Phys. Chem Ref. Data, 12, 891

30. Lennon, M. A., Bell, K. L., Gilbody, H. B., Hughes, J. L., Kingston, A. E., Murray, M. J., & Smith, F. J. 1988, J. Phys. Chem. Ref. Data, 17, 1285

31. Seaton, M. J. et al. 1992, Rev. Mex. Astron. Astrofis., 23, 19

32. Verner, D. A., & Ferland, G. J. 1996, ApJ, in press

X-RAY SOURCES
AND X-RAY LASERS

Demonstration and Study of a Discharge-pumped, Table-top Soft-X-ray Laser

J.J. Rocca, F.G. Tomasel, V.N. Shlyaptsev, J.L.A. Chilla, D. Clark, and M.C. Marconi

Department of Electrical Engineering, Colorado State University, Fort Collins, CO 80523

Abstract. We review results of the first demonstration of large soft-x-ray amplification in a discharge-created plasma. A fast compressive capillary discharge was utilized to create elongated plasma columns with length-to-diameter ratios in excess of 500:1 having the required density and temperature for soft-x-ray amplification by collisional excitation of Ne-like ions. The most recent measurements, conducted in capillary discharges up to 20 cm in length, have yielded an amplification of ~exp(14) in the 46.9 nm line of Ne-like argon. The dependence of the laser line intensity with discharge parameters and the dynamics of the capillary discharge plasma column under lasing conditions are reported. Prospects for laser operation at shorter wavelengths are discussed.

I. INTRODUCTION

We have demonstrated a new type of soft-x-ray laser that uses direct excitation by a compact discharge. Since their demonstration in 1984 (1,2), successful x-ray lasers have been pumped by high power lasers. In spite of the important progress that the x-ray laser field experienced in the last decade, the widespread use of these lasers in a number of important applications has been hampered by their size, cost and complexity. Recently, researchers have sought means to implement more compact and efficient laser sources, often referred to as "table-top" soft-x-ray lasers, that could have extensive use in applications. With this objective several groups have explored amplification schemes that require smaller laser drivers (3-5) In particular, the use of high-power picosecond and femtosecond laser systems has led to important progress during the last two years (6-8).

Alternatively, it has been recognized that direct discharge excitation of the soft-x-ray lasing medium could result in increased efficiency and simplicity. However, despite the large success in the realization of discharge-pumped lasers at longer wavelengths, the problem of demonstrating large amplification at wavelengths below 100 nm had until recently remained unresolved. Figure 1 illustrates the rapid progress achieved in demonstrating lasers operating at increasingly shorter wavelengths during the first decade of discharge-pumped laser development, starting with the discovery

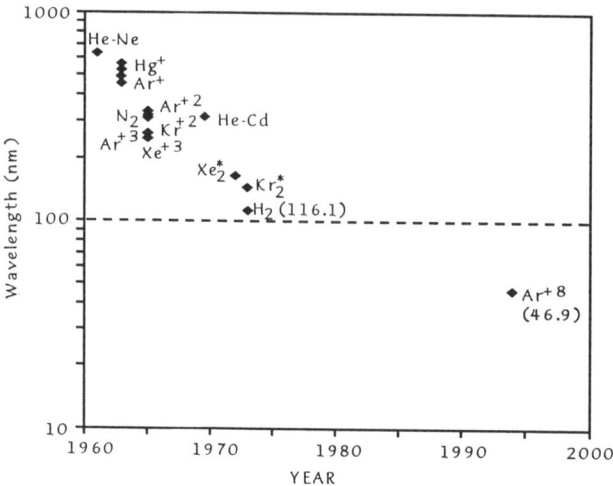

FIGURE 1. Progress in the development of discharge-pumped lasers at increasingly shorter wavelengths. To illustrate the rapid progress in the period that expands from the discovery of the HeNe laser in 1961 to the demonstration of the 116 nm molecular hydrogen laser in 1972, only several of the many visible, UV and VUV discharge-pumped lasers discovered in that period are shown.

of the helium-neon laser in 1961. However, after the demonstration in 1972 of lasing in hydrogen at 116 nm (9), further progress stalled for more than 20 years, until the experiments discussed herein. A major problem has resided in that shorter wavelength amplifiers require the creation of uniform plasma columns with increasingly hotter and denser plasma conditions. For example, while lasing in the blue at 488 nm can be obtained in singly ionized argon in a plasma with an electron temperature of about 5 eV and an electron density of 1×10^{14} cm^{-3}, lasing at 46.9 nm in the same element requires to ionize the atom eight times (to obtain neon-like argon), with more than an order of magnitude increase in the electron temperature, to about 80 eV, and nearly a five orders of magnitude increase in the electron density, to about 1×10^{19} cm^{-3}. Whereas the requirements of high temperature and high density can be relatively easy to meet by high power discharges, that of simultaneously maintaining high plasma uniformity is not. Axial inhomogeneities and severe distortions in the plasma columns, produced by non-symmetric compressions and instabilities, often develop destroying the amplification. To overcome these limitations we have proposed and implemented the excitation by fast capillary discharges (10). These discharges can produce hot plasma columns of small diameter by fast compressions starting from highly homogenous initial plasma conditions, results in more uniform plasma columns.

Utilizing a fast capillary discharge to create and excite neon-like argon ions, we have recently demonstrated for the first time large soft x-ray amplification in a discharge-created plasma, in the $J=0$-1 line of Ne-like argon at 46.9 nm (11).

The plasma columns generated by fast compressional capillary discharges are also of interest to produce soft-x-ray amplification by plasma recombination. Experiments conducted to explore amplification using this mechanism have produced encouraging results (12-14), however, further work is necessary in this case to demonstrate amplification of the laser line well above the intensity of strong, non-lasing plasma lines. Herein we focus our attention in discussing recent experiments conducted at Colorado State University, which have resulted in exponential gains up to approximately exp(14) in the 46.9 nm line of Ne-like argon.

The next section summarizes the measurement of the gain and of the variation of the laser intensity with discharge parameters. Section III discusses the dynamics of the plasma column. Section IV describes the observation of enhanced output characteristics under an externally applied magnetic field, and Section V discusses the prospects for extending the discharge excitation scheme to shorter wavelengths.

II. AMPLIFICATION IN NE-LIKE ARGON

The discharge set up utilized to obtain amplification in Ne-like Ar is illustrated in Fig. 2. The pulse generator consists of a 3 nF liquid dielectric capacitor that is pulse-charged by a Marx generator. To generate the plasma columns, the capacitor is rapidly discharged through the capillary channel by a spark-gap switch. The radiation from the plasma was analyzed with a 2.2 m grazing incidence spectrograph having a 1200 l/mm, gold-coated grating.

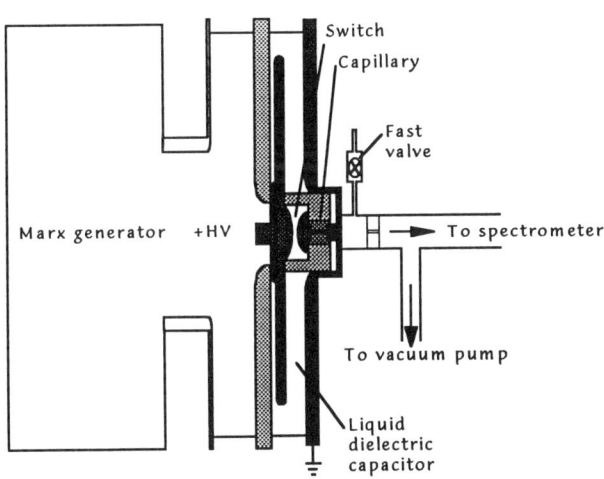

FIGURE 2. Schematic diagram of the pulse generator and capillary discharge

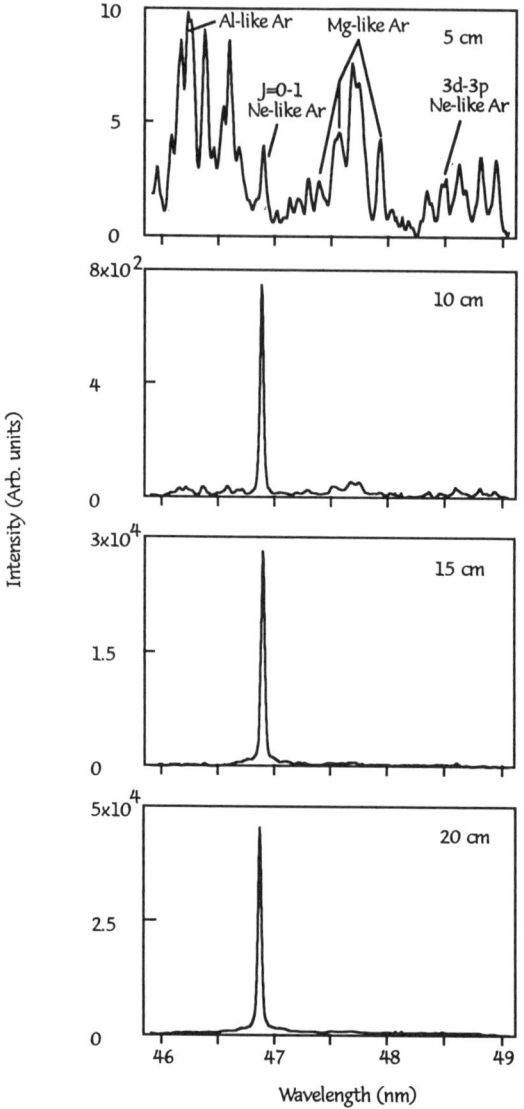

FIGURE 3. Spectra from the capillary discharge argon plasma showing the dramatic increase on the intensity of the 46.9 nm laser line as a function of plasma column length. The spectra correspond to a 4-mm-diameter capillary filled with approximately 720 mTorr of argon, excited by a 39 kA current pulse of 72 ns first half-cycle duration.

The observation of large amplification in the 46.9 nm ArIX line was first realized in 4-mm-diam capillaries up to 12 cm in length containing argon gas at a pressure of approximately 700 mTorr. The plasma columns were generated by a current pulse with a half-period of about 60 ns and a peak amplitude of near 40 kA. The fast current pulse rapidly compresses and heats up the plasma, to form a narrow column with measured length-to-diameter ratios exceeding 500:1. Toward the final stage of the compression, the plasma reaches nearly ideal conditions for soft-x-ray amplification by collisional excitation in Ne-like and Ni-like ions. The value of the gain was determined by measuring the increase of the laser line intensity as a function of the capillary length. The first experiments yielded a gain-length product of 7.2. The laser pulse, which is observed to occur about 6 ns before the time of maximum plasma compression, was measured to have a divergence slightly smaller than 6 mrad[15].

A more recent series of experiments, conducted at optimized plasma conditions in capillaries up to 20 cm in length, further increased the laser line intensity by more than two orders of magnitude[15]. Gain coefficients of up to 1.1 cm^{-1} have been measured. The spectra in Fig. 3 show the dramatic increase in the intensity of the laser line as a function of plasma column length. The laser line intensity is observed to increase by nearly four orders of magnitude when the plasma column length is increased from 5 to 15 cm. The variation of the integrated laser intensity as a function of capillary length is shown in Fig. 4.

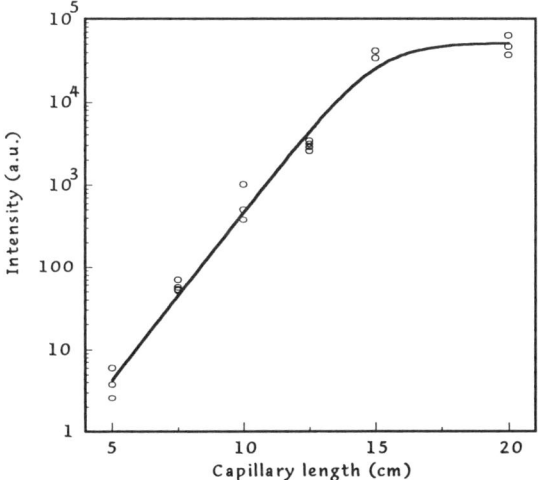

FIGURE 4. Integrated line intensity of the 46.9 nm line of ArIX as a function of plasma column length. The capillary diameter is 4 mm and the argon pressure is ~ 720 mTorr. The maximum gl value reached is 14.1 or 13.4, depending on two slightly different measurements of the attenuation of the filters used. The corresponding gain coefficients for lengths <15 cm determined by the Linford formula (16) are 0.94 cm^{-1} and 0.89 cm^{-1}, respectively.

The gain in the 15-cm-long columns was measured to reach ≈exp(14), the largest gain reported to date for a table-top soft-x-ray laser amplifier. For lengths greater than 15 cm, the laser line intensity is observed to saturate. The measured gain-length product approached the value at which gain saturation is expected to occur due to depletion of the population inversion by the intense stimulated emission. However, further experiments are needed to confirm that the roll-off of the intensity shown in Fig. 4 is due to the gain saturation.

III. DYNAMICS OF THE PLASMA COLUMN AND TEMPORAL CHARACTERISTICS OF THE LASER PULSE

To elucidate the dynamics of the capillary plasma column at the discharge conditions corresponding to amplification in the J=0-1 line of Ne-like Ar, time resolved sequences of end-on soft-x-ray images, such as that shown in Fig. 5, were obtained. The measurements were conducted with a x-ray pinhole camera, consisting of a 45 μm diameter pinhole placed at 36 cm from the end of the capillary and a gated multichannel plate (MCP) intensified coupled charge device (CCD) array detector. The camera has a calculated magnification of 3.14, and a spatial resolution limited, depending on the wavelength of the radiation, either geometrically by the size of the pinhole (to about 60 μm) or by diffraction. The temporal resolution is approximately 5 ns. Pinhole images of the plasma were obtained using a 1000 Å thick carbon foil filter, which limits the radiation observed to wavelengths below ~30 nm. Images were also obtained without a filter to allow for the observation of cooler plasma regions. In the latter case the wavelength of the radiation observed is limited to $\lambda \leq 1300$ Å by the spectral response of the MCP.

The study was conducted in a 4-mm-diam, 12-cm-long polyacetal capillary excited with a current pulse of 62 ns half period. The discharge conditions were near optimum for amplification: an Ar pressure of 700 mTorr and peak current of about 39 kA. The pinhole images clearly demonstrate that the plasma of the gas-filled fast capillary discharge rapidly contracts, heats up, and then expands, constituting a kind of wall influenced Z-pinch. Following the excitation of the pre-ionized plasma column with a fast voltage pulse, the current starts flowing near the capillary walls, in a region determined by the skin depth. During the first part of the current risetime the distribution of the current density remains localized near the capillary wall. The first pinhole image (A), acquired 26 ns after the initiation of the fast current pulse, shows that the majority of the soft-x-ray incoherent emission originates from a doughnut shaped region having an outer diameter of 3.5 mm. After this initial phase, the electromagnetic forces of the rapidly rising current pulse create a shock wave and compress the plasma. The soft-x-ray emitting region is rapidly compressed in about 10 ns to form a column 200-300 μm diameter (figures A-D). Hydrodynamic/atomic model calculations [17] show that during the compression, and in advance to the collapse of the plasma sheath on the capillary axis, the current flow distribution switches from the periphery to the central region of the capillary, heating and ionizing the plasma, and increasing the soft-x-ray emission near the axis. The emission reaches its maximum when the shock wave reaches the axis, causing the conversion of ion kinetic energy into

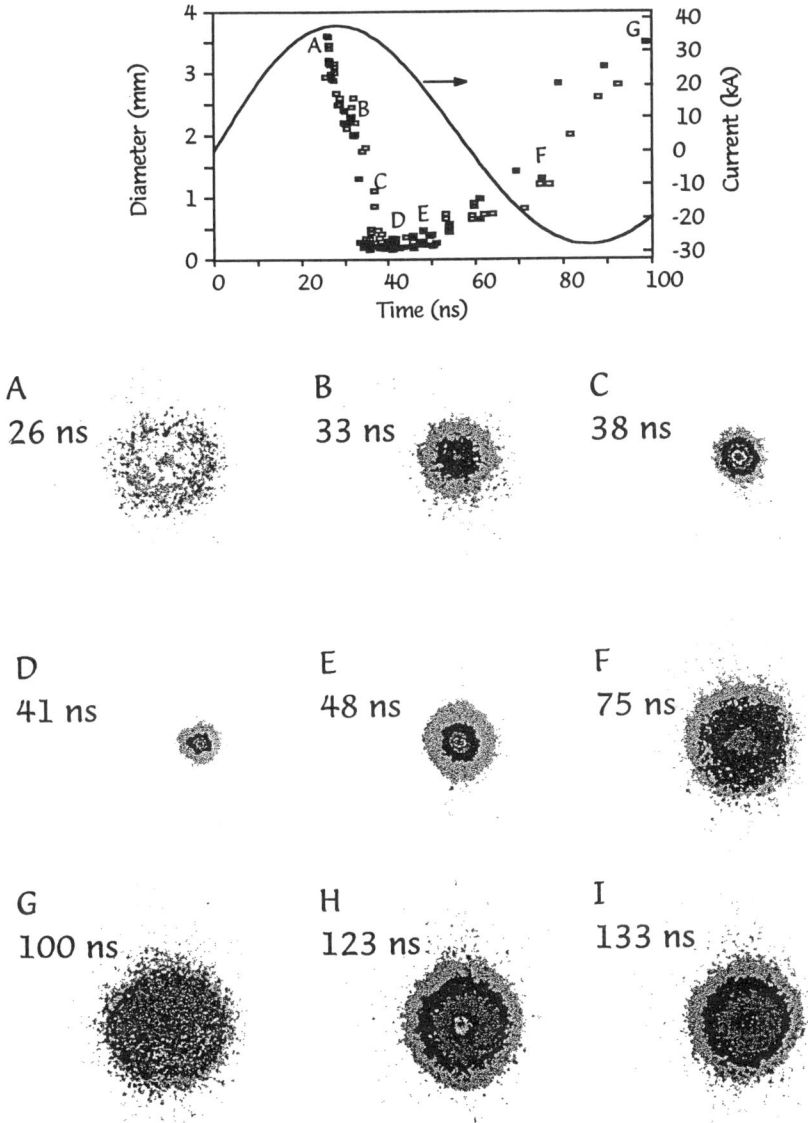

FIGURE 5. Series of time-resolved end-on images of the plasma column created in a 12-cm-long, 4-mm-diameter capillary channel filled with 700 mTorr of argon. The top portion of the figure illustrates the FWHM of the intensity of the XUV/soft-x-ray emitting region of the plasma as a function of time. The solid squares correspond to images obtained interposing a 1000 Å thick carbon foil. The timing with respect to the current pulse is shown.

thermal energy and a sharp increase in the electron density (fig. D). Lasing occurs during about 1 ns, several nanoseconds before the time of maximum compression and shortly after the current peak, when the electron density is about $0.3-1 \times 10^{19}$ cm^{-3} and the electron temperature reaches 60-90 eV. The pinhole images (E-G) show that subsequently the plasma column expands and cools. A second, less significant compression occurs near the end of the first cycle of the current pulse (fig. H). This time, a central column with a minimum diameter of about 1 mm and cooler temperatures, which are not of interest for the excitation of collisionally pumped soft-x-ray lasers, is observed to develop on top of a broad plasma background. The pinhole images acquired with ($\lambda<30$ nm) or without ($\lambda<130$ nm) the carbon filter are similar except near the time of plasma collapse, when the FWHM diameter of the radiating region are 200 μm and 300 μm, respectively.

OBSERVATION OF ENHANCED LASER INTENSITY AND BEAM PROFILE BY AN EXTERNALLY APPLIED MAGNETIC FIELD

An increase in the 46.9 nm line output intensity by a factor of 3-4 and a more uniform beam profile were obtained in 10-cm-long plasma columns by applying an axial magnetic field. Magnetic fields have been previously utilized to obtain more stable and reproducible compressions in Z-pinch discharges, which are known to present large inhomogeneities (18-21). We have previously studied the effects of axial magnetic fields of up to 7 T in the XUV emission of a capillary discharge (22). In the fast compressional capillary discharge of interest here, an externally applied magnetic field of 0.15 T reduces the plasma density gradient in the region of maximum gain, diminishing detrimental effects caused by large radial variations of the plasma density.

Experiments were conducted in capillaries 4 mm in diameter and 10 cm in length, surrounded by a solenoid for the generation of the axial magnetic field. The magnetic field was generated by discharging a capacitor that produced a current pulse having 100 μs half cycle through the solenoid. Figure 6 shows that the 46.9 nm laser intensity increases as a function of the externally applied axial magnetic field, to reach a maximum at approximately 0.15 T, value above which it decreases monotonically. The result of numerical simulations is also shown in the same Figure. Hydrodynamic calculations suggest that the enhancement of the laser characteristics is due to a reduction of the density gradients in the gain region. The computed density gradients decrease as a function of the increasing magnetic field, which in turn results in reduced refraction losses. At higher magnetic fields, the Zeeman effect is found to contribute to line broadening, decreasing the laser output power. In addition to Zeeman splitting, an externally applied magnetic field also lowers the gain by decreasing the plasma density, reducing transient effects associated with ionization and excitation, and increasing the optical depth. The larger optical depth at higher magnetic fields is a consequence of the reduction of the very important radial motional Doppler effect, which is in turn caused by the previously mentioned reduction of the density

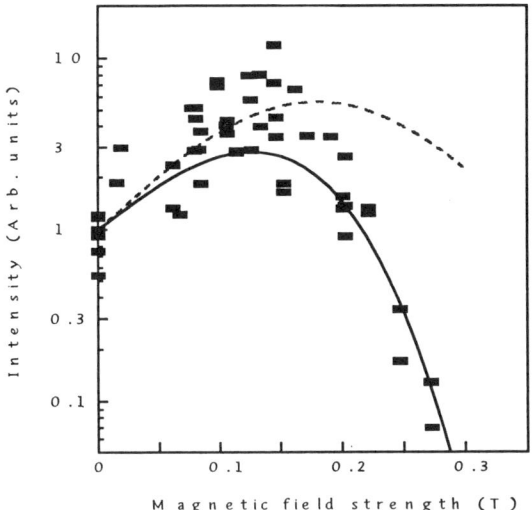

FIGURE 6. Variation of the integrated intensity of the 46.9 nm ArIX laser line as a function of the strength of the externally applied axial magnetic field. The solid markers correspond to data from discharges through 10 cm long, 4-mm-diam capillaries filled with approximately 600 mTorr of argon. The half cycle width of the current pulse is 64 ns. The solid and dashed curves are the result of computations with and without including Zeeman splitting, respectively.

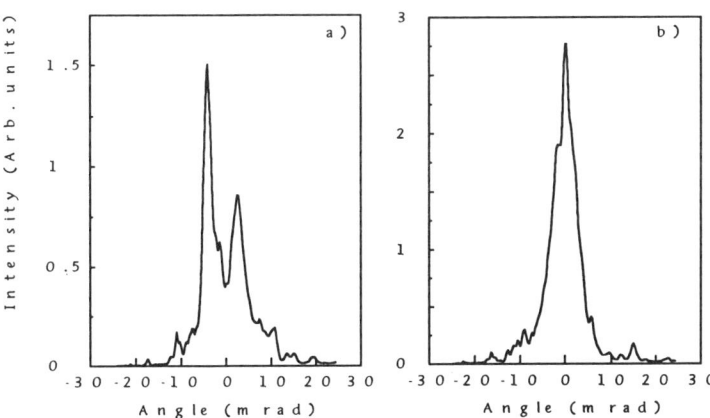

FIGURE 7. 46.9 nm laser beam profiles generated by a 39 kA discharge in a 4-mm-diam, 10-cm-long capillary filled with approximately 600 mTorr of argon. a) Without externally applied magnetic field. b) With an externally applied magnetic field of 0.15 T.

The effect of reduced refraction is observed in the recorded laser beam patterns. Figure 7 shows the laser beam profiles that resulted from discharges with and without the externally applied magnetic field, respectively, but with otherwise similar conditions. While the beam profile corresponding to the discharge without the magnetic field shows two dominant side lobes, which are caused by refraction, the profile corresponding to the 0.15 T case displays a single central peak.

PROSPECTS FOR LASING AT SHORTER WAVELENGTHS

The collisional excitation argon experiment discussed above can, in principle, be scaled in Z along the neon isoelectronic sequence to demonstrate lasing in a discharge-created plasma at shorter wavelengths. We have already reported[23] the observation of emission from the J=0-1 of CaXI, and possibly of the equivalent line in TiXIII, in capillary discharge plasmas created by discharge ablation of capillaries containing CaH_2 and TiH_2. Rapid excitation of 1.5 and 2.5 mm diameter capillaries made out of these materials with current pulses of less than 70 kA have produced Ca and Ti plasmas in which atoms are ionized up to the O-like and F-like ions, respectively.

To obtain lasing at wavelengths below 30 nm using the Ne-like sequence, excitation of higher-Z ions, such as Ne-like Cr (λ=28.5 nm) and Ne-like Fe (λ=25.5 nm) is required. It is foreseeable that scaling of the excitation to achieve 50-100 μm diameter plasma columns with electron densities near 2-5×10^{20} cm^{-3} and electron temperatures in the range of 500-700 eV, could result in lasing below 20 nm in Ne-like Kr, with gains of the order of 5-8 cm^{-1}. However, those plasma parameters require a significant increase in the discharge current and voltage. Alternatively, a more rapid scaling to shorter wavelengths could be achieved utilizing Ni-like ions[23,24]. In this case, amplification at wavelengths below 200 Å can be explored, for example, utilizing the $3d^9 4d^1S$ - $3d^9 4p\ ^1P$ transition of MoXIV (λ=18.92 nm). This transition has already been identified in a laser-created plasma, and gain has been measured in the equivalent line in Ni-like Nb (λ=20.42 nm) in a plasma created by a table-top laser[24]. Excitation of these Ni-like transitions in a fast compressive capillary discharge should still require modest currents, of the order of less than 100 kA. It is also possible that scaling of the excitation to generate plasma columns with electron densities near 2-5×10^{20} cm^{-3} and electron temperatures of about 450 eV could produce amplification with gain in the range of 4-6 cm^{-1} in Ni-like Xe in the vicinity of 10 nm.

ACKNOWLEDGEMENTS

We thank the contributions of O.D.Cortázar, D. Hartshorn, B.T. Szapiro and J.J. Gonzalez. We also acknowledge the technical assistance of B. Bach from Hyperfine, Inc. This work was supported by the National Science Foundation grants ECS-9401952, ECS-9412916 and ECS-9412106. Part of the diagnostics instrumentation was developed in collaboration with Hyperfine, Inc. (Boulder, CO), with the support of the Colorado

Advanced Technology Institute. Previous support from the Department of Energy and the U.S. National Research Council is also acknowledged.

REFERENCES

1. D.L. Matthews, P.L. Hagelstein, M.D. Rosen, M.J. Eckartr, N.M. Ceglio, A.N. Hazi, M. Medecki, B.J. MacGowan, J.E. Trebes, B.L. Whitten, E.M. Cambell, C.W. Hatcher, A.M. Hawryluk, R.L. Kaufman, L.P. Pleasance, G. Rambach, J.H. Scofield, G. Stone, and T.A. Weaver, Phys. Rev. Lett. **54**, 110 (1985).
2. S. Suckewer, C.H. Skinner, H. Milchberg, C. Keane, and D. Voorhees, Phys. Rev. Lett. **55**, 1753 (1985).
3. Proceedings of the Fourth International Colloquium on X-ray Lasers 1994, Williamsburg, VA. AIP Conf. Proc. 332. D.C. Eden and D.L.Matthews, Eds.
4. L.Y. Polonski, C.O. Park, K. Krushelnik, and S. Suckewer, SPIE J. **2012**, 75 (1993)
5. J. Goodberlet, S. Basu, M.H. Muendel, S. Kaushik, T. Savas, M. Fleury, and P.L. Hagelstein, J. Opt. Soc. Am. B **12**, 980 (1995).
6. Y. Nagata, K. Midorikawa, S. Kubodera, M. Obara, H. Tashiro, and K. Tokoda, Phys. Rev. Lett. **71**, 3774 (1993)
7. B. Lemoff, G.Y. Yin, C.L. Gordon III, C.P.J. Barty, and S.E. Harris, Phys. Rev. Lett. **74**, 1574 (1995).
8. P. V. Nickles, M. Schnuerer, M.P. Kalashnikov, I. Will, W. Sandner, and V.N. Shlyaptsev, SPIE J. **2520**, 373 (1995)
9. R.W. Waynant, Phys. Rev. Lett. **28**, 553 (1972).
10. J.J.Rocca, D.C.Beethe and M.C.Marconi, Opt. Lett, **13**, 565 (1988); J.J. Rocca, O.D. Cortázar, B. Szapiro, K. Floyd, and F.G. Tomasel, Phys. Rev. E **47**, 1299 (1993).
11. J.J.Rocca, V.N. Shlyaptsev, F.G. Tomasel, O.D. Cortázar, D. Hartshorn, and J.L.A. Chilla, Phys.Rev. Lett. **73**, 2192 (1994).
12. J.J. Rocca, M.C. Marconi, B.T. Szapiro, and J. Meyer, SPIE J. **1551**, 275 (1991).
13. C. Steden and H.-J. Kunze, Phys. Lett. A **151**, 534 (1990).
14. H.J. Shin, D.E. Kim, T.N. Lee, Phys. Rev. E **50**, 1376 (1994).
15. J. J. Rocca, M.C. Marconi, J.L.A. Chilla, D.P. Clark, F.G. Tomasel, and V.N. Shlyaptsev, IEEE J. of Selected Topics in Quantum Electronics **1**, 945 (1995).
16. G.J. Linford, E.R. Peressini, W.R. Sooy, and M.L. Spaeth, Appl. Opt. **13**, 379 (1974).
17. V.N.Shlyaptsev, J.J.Rocca, and A.L.Osterheld, SPIE J. **2520**,1995.
18. S. Glasstone and R.H. Lovberg, *Controlled thermonuclear reactions*, Princeton: D. van Nostrand, 1960.
19. F.S. Felberg, F.J. Wessel, N.C. Weld, H.U. Rahman, A. Fisher, C.M. Fowler, M.A. Liberman, and A.L. Velikovich, J. App. Phys. **64**, 3831 (1988).
20. S.A. Sorokin, A.V. Khachaturyan, and S.A. Chaikovskii, Sov. J. Plasma Phys. **17**, 841 (1992). [Fiz. Plasmii **17**, 1453 (1991)].
21. N.S. Edison, B. Etlecher, A.S. Chuvatin, S. Attelan, and R. Aliaga, Phys. Rev. E **48**, 3893 (1993).
22. M.C. Marconi, J.J. Rocca, J.F. Schmerge, M. Villagran, and F. Lehmann, IEEE J. Quantum Electron. **26**, 1809 (1990).
23. J.J. Rocca, O.D. Cortázar, F.G. Tomasel, and B.T. Szapiro, Phys. Rev. E **48**, R2378 (1993).
24. S. Basu, P.L. Hagelstein, and J.G. Goodberlet, Appl. Phys. B **57**, 303 (1993).

Observation of Multiphoton-Induced Multiple Vacancy Production in Xe(L) Emission from Plasma Channels

A. McPherson[1], A.B. Borisov[1],
K. Boyer[1], C.K. Rhodes[1,2]

[1]Department of Physics (M/C 273)
University of Illinois at Chicago
845 W. Taylor Street
Chicago, IL 60607-7059
USA

[2]TARA/Department of Applied Physics
University of Tsukuba
1-1-1 Tennohdai
Tsukuba, Ibaraki 305
JAPAN

ABSTRACT. The observation of multiple Xe 2p-vacancy production occurring in plasma channels generated with subpicosecond 248 nm radiation can be put in correspondence with multiple electron processes involved in ion-atom collisions.

Spatially resolved spectra of multiphoton-induced Xe(L) emission from self-trapped plasma channels [1] have enabled the intensity dependence [2] of the spectrum to be observed. These experimental data indicate the presence of a narrow threshold region which characterizes the onset of the production of multiple 2p vacancies. The comparison of the two spectra shown in Fig. (1) illustrates this transition.

The transition region is represented in Fig. (2). We observe that the multiple 2p hole production is correlated with a corresponding shift in the average charge state $\bar{q}$.

It has been possible to relate these observations to ion-atom collisional phenomena [2,5-7]. Specifically, a rather close correspondence can be made between the observed onset for multiphoton-induced multiple vacancy production [2] and collisional multi-electron processes [7]. The most interesting outcome of the comparison with the collisional processes is the indication that the interaction, under appropriate conditions, could be <u>dominated</u> by ordered multiple-electron motions.

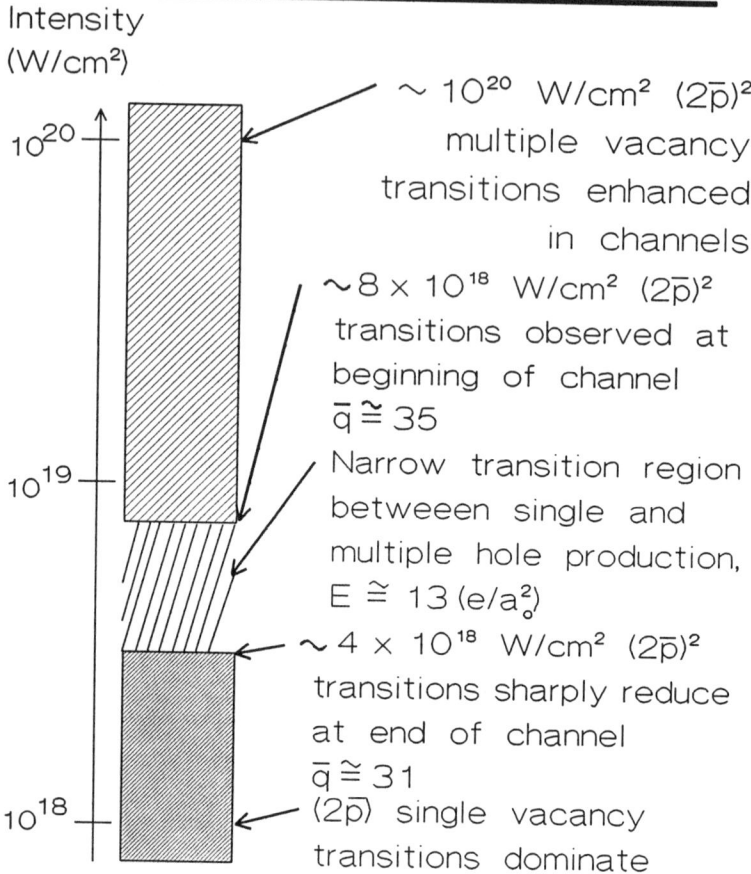

Figure 1: Spectral line-outs corresponding to spatial locations A and B as shown in the Z-λ image (Film #941014.4). The approximate intensity levels assigned to points A in the channel and B after the channel are $I_A \cong 8 \times 10^{18}$ W/cm² and $I_B \cong 4 \times 10^{18}$ W/cm², respectively. The regions corresponding to emission from the multiple vacancy species [$(2\bar{p})^2$ and $(2\bar{p})^3$] on $(2p^5 \leftarrow 2p^4 3d)$ transitions in the 2.625 - 2.655 Å and 2.60 - 2.625 Å ranges, respectively, are designated. The broad spectral features are characteristic of the emission from hollow atom configurations that have been previously described in Ref. [3]. The Xe^{q+} ionic charge states corresponding to the peaks of the observed emissions are marked. The calculational procedures used to locate the spectral positions of the ionic emissions are described in Ref. [4].

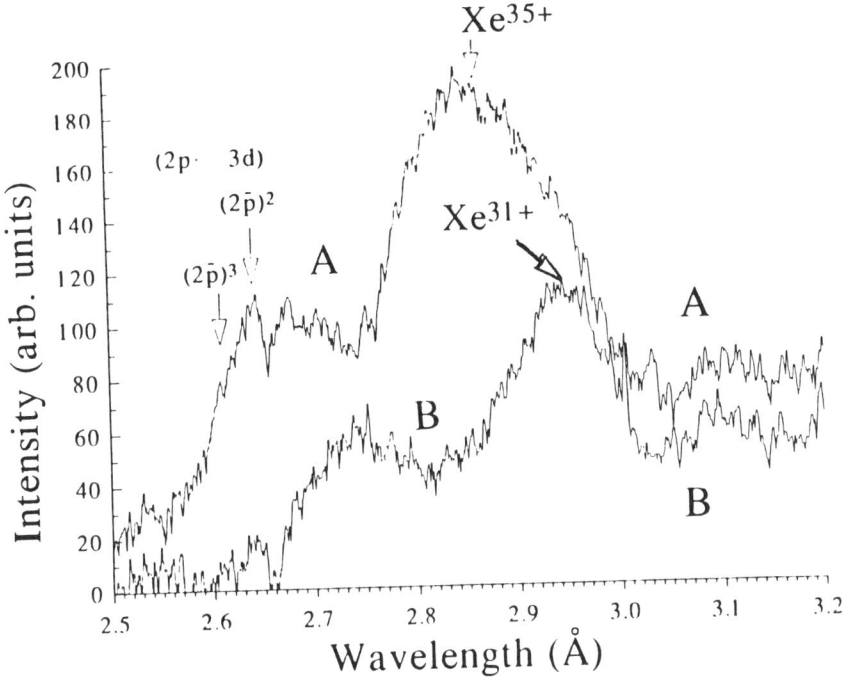

Figure 2: Intensity dependence of Xe(L) spectrum illustrating the transition region separating the multiple 2p vacancy zone from the region dominated by single 2p vacancy production. The transition is estimated [2] to occur between $\sim 4 \times 10^{18}$ W/cm^2 and $\sim 8 \times 10^{18}$ W/cm^2 for irradiation at 248 nm.

ACKNOWLEDGEMENTS

The authors respectfully acknowledge informative discussions with J. W. Longworth and the expert technical assistance of J. Wright and P. Noel. Support for this research was provided under contracts with SDI/NRL (N00014-94-K-2004), ARO (DAAH04-94-G-0089) and the University of California/Lawrence Livermore National Laboratory (B321297).

REFERENCES

1. Borisov, A. B., McPherson, A., Boyer, K., and Rhodes, C. K., *J. Phys.* **B29**, L113 (1996).
2. Borisov, A. B., McPherson, A., Boyer, K., and Rhodes, C. K., *J. Phys.* **B29**, L43 (1996).
3. McPherson, A., Thompson, B. D., Borisov, A. B., Boyer, K., and Rhodes, C. K., *Nature* **370**, 631 (1994).

4. The spectral positions of the ionic emission have been calculated with the procedures described in Cowan, R. D., *The Theory of Atomic Structure and Spectra*, Berkeley, CA: University of California Press, 1982.
5. Thompson, B. D., McPherson, A., Boyer, K., and Rhodes, C. K., *J. Phys.* **B27**, 4391 (1994).
6. Boyer, K., Gibson, G., Jara, H., Luk, T. S., McIntyre, I. A., McPherson, A., Rosman, R., Solem, J. C., and Rhodes, C. K., *IEEE Transactions on Plasma Science* **16**, 541 (1988).
7. Olsen, R. E., Ullrich, J., and Schmidt-Böcking, H., *J. Phys.* **B20**, L809 (1987).

Short Wavelength Generation from Atomic Clusters

T. D. Donnelly[1], T. Ditmire[2,3], M. D. Perry[2] and R. W. Falcone

Department of Physics, University of California at Berkeley, Berkeley, California 94720
[1]current address: Department of Physics, Swarthmore College, Swarthmore, PA 19081
[2]Lawrence Livermore National Laboratory, P.O. Box 808, L-443, Livermore, California, 94550
[3]current address: Imperial College of Science, Technology and Medicine, Prince Consort Road, London, SW7 2BZ

We report the generation of both incoherent and coherent short-wavelength radiation which results from the unique interaction of an intense laser pulse with atomic clusters. The atomic clusters are produced by a pulsed gas jet. In the higher intensity regime (10^{15}-10^{17} W/cm^2) the laser pulse ionizes and heats the atomic cluster resulting in a novel plasma with ~ 1 keV electron temperatures. This plasma strongly emits incoherent x-rays in the 20-500 Å wavelength range. Using laser pulses at intensities of 10^{14} W/cm^2, high order harmonics are created which, when compared to single atom gases, yield a higher appearance intensity for a given harmonic order, stronger non-linear dependence of harmonic signal on laser intensity, higher-order harmonics, and reduced saturation of the harmonic signal at high laser intensity.

Introduction

The interaction of high-intensity laser light with matter can be manipulated to produce both incoherent and coherent x-ray radiation. Atomic clusters, used as laser targets, are a unique form of matter which present the opportunity to increase the usefulness and production efficiency of such x-rays.

It is well known that x-rays will be produced when a high intensity laser is focused onto solid density material. It has been shown that microstructured, solid density targets increase the efficiency of x-ray production by a factor of 10^2 compared with conventional flat targets, for x-rays above 1 keV of energy [1]. These x-rays are broad band and incoherent, yet they can be produced with an energy efficiency of ~1%.

We have shown that "microstructured" gas phase targets (clusters) produced by a gas jet also show enhanced x-ray emission. Under appropriate conditions [2], gas jets flow will condense to form atomic clusters; the clusters' size range from 10-10^2 Å in the work described here. Compared to the microstructured targets in Ref. 1, these targets are easier to manufacture and to reproduce.

Under particular laser and gas jet conditions, we have shown these same atomic clusters to be a unique medium for high order harmonic generation. Thus we are able to use these clusters to produce coherent, soft x-rays of a short pulse duration (on order of 100 fs). Atomic clusters compare favorably to single atoms targets in that they show a more non-linear response to the laser field, they result in harmonics with higher maximum photon energy, their response to the laser field saturates at a higher laser intensity and they suggest higher energy efficiency in harmonic production.

Strong Incoherent Radiation

In our first experiments [3], we focus a short-pulse, high-intensity laser [4] (140 fs, 825 nm) onto the flow of a pulsed gas jet. X-ray radiation is collected using two spectrographs each of which incorporate a variable-line-spaced grazing-incidence grating with nominally 1200 lines/mm. One spectrograph (on-axis) images radiation onto a single-stage, multi-channel plate (MCP) electron multiplier that is fiber-optic coupled to a charge-coupled-device (CCD) camera to allow digital recording of the data; the other spectrograph (90° off axis) images radiation onto a streak camera which is also coupled to a CCD.

The interaction of the laser pulse with a non-clustering gas is well described and understood [5]: the interaction of the laser pulse with the single atom is dominated by optical field ionization (OFI). The interaction of the laser pulse with an atomic cluster is more complex. The leading edge of the laser pulse optically ionizes some small fraction of the cluster atoms, the resulting free electrons are then strongly driven in the laser field. These electrons will undergo collisions with the cluster atoms/ions resulting in collisional ionization of the cluster and very efficient heating of the free electrons [3]. Ion charge states that are reached by this process are significantly higher than those which are reached by OFI in the single atom case.

Fig. 1 compares helium (He) Ly-α (304 Å) emission from two types of He plasma. The first type of plasma is created by irradiating a He gas jet flow which is doped (10%) with a clustering gas (argon [Ar], krypton [Kr], or nitrogen [N_2]). The second type of plasma is either pure He, or He doped (10%) with neon (Ne); neither of these gases will form clusters under our experimental conditions. The He plasma that contains a clustering gas emits 10^2 more Ly-α photons than a pure He plasma, or a Ne doped He plasma. The explanation for this has to do with the hot electrons that are produced in the laser-cluster interaction. Cluster electrons are heated to high temperature (~ 1 keV) during the laser pulse and subsequently interact with the He ions. After the laser pulse passes, He^{1+} is formed through recombination and the Ly-α transition is then collisionally excited by the hot electrons. Thus, each He^{1+} will emit more than one Ly-α photon. This is in contrast to the case when only He is present in the plasma (or He and Ne) and no hot electrons are produced; here comparativly little collisional excitation occurs.

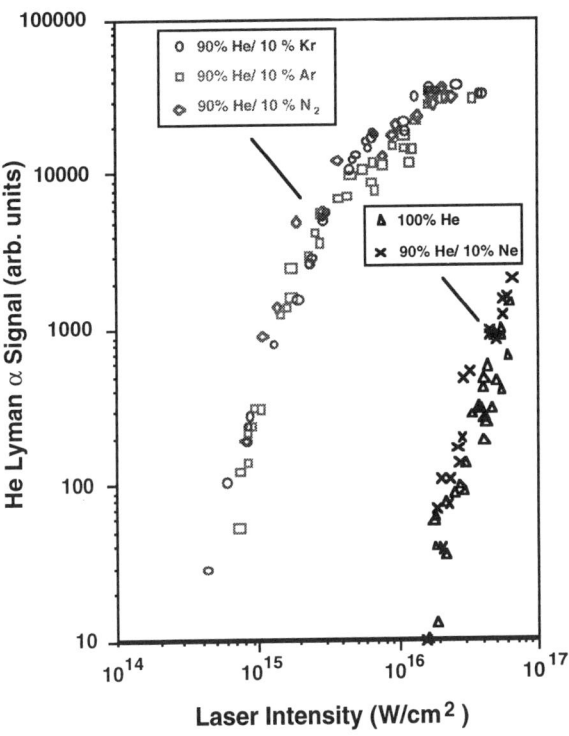

FIGURE 1. Ly-α emission as a function of laser intensity from a helium plasma. The helium plasma exhibits enhanced x-ray emission when a clustering gas is added to the helium gas jet flow. When there are clusters present in the gas jet flow Ly-α emission is present even for laser intensities below the OFI threshold for producing He^{2+}.

We have also time resolved the Ly-α emission. We have show that when clustering gases dope the He gas jet flow, the Ly-α emission begins a few ns after the laser pulse has passed, and lasts for over 100 ns. This is consistent with our picture of hot electrons leaving the vicinity of the heated cluster and then collisionally exciting the Ly-α transition. In the absence of any clustering gas in the He gas jet flow, the Ly-α emission begins promptly after the laser pulse and lasts for only a few ns; this scenario is consistent with OFI during the laser pulse followed by recombination.

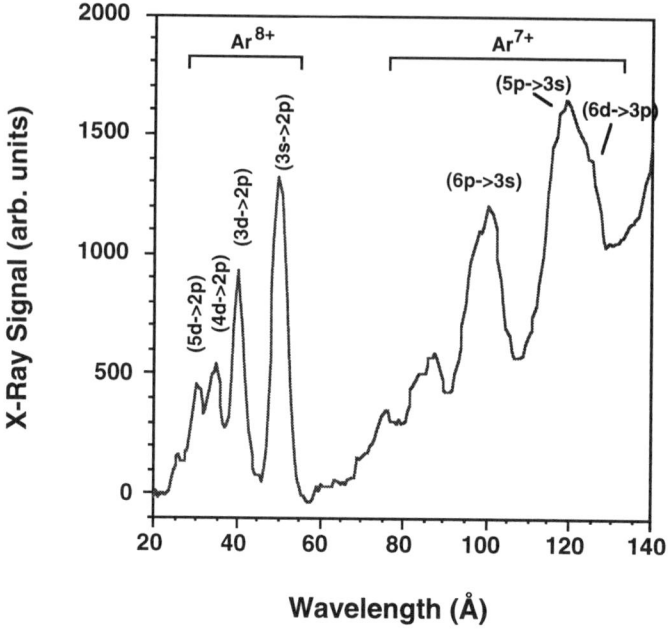

FIGURE 2. Strong x-ray emission from neon-like argon. This data was measured for a laser intensity of 8x10^{15} W/cm^2 and a gas jet backing pressure of 500 psi of pure Ar.

As mentioned above, the collisional ionization mechanism which is dominant in the cluster interaction (for large enough cluster size) results in ion charge states which are higher than can be accessed at a similar laser intensity by OFI. For example, Ly-α emission in Fig. 1 is present at a laser intensity below that which is required for OFI of He^{2+}. Fig. 2 shows a time integrated spectrum that resulted from the irradiation of atomic clusters of Ar by our high intensity laser pulse. We observe lines that are produced by Ne-like Ar; to reach Ne-like Ar by OFI requires a laser intensity of over 10^{18} W/cm^2 (more than two orders of magnitude greater than the laser intensity in this shot). We estimate that ~1% of the laser energy was converted into photons in the 20-60 Å wavelength band.

Coherent Radiation

High order harmonics have been produced in monomer gases [6], molecules [7], and at solid surfaces [8]. We report [9] the generation of high-order harmonics of intense laser radiation from atom clusters containing approximately 10^3 atoms, and show that clusters are a unique nonlinear medium with potentially useful properties.

Compared with the situation of a high-field interacting with the valence electron of a single atom, electrons in a cluster will be subject to (1) partial shielding of the laser field by both bound and free charges, (2) an initial binding potential that is modified from the single atom case by nearest atomic neighbors in the cluster, (3) the Coulomb field of all ions in the cluster (space-charge) as electrons are driven away from the cluster, and (4) scattering in the cluster during their oscillatory motion. For the work described here, the free electron oscillation amplitude (determined by the peak laser intensity) is calculated to be on order of the cluster radius (typically 2 nm).

In these experiments, we focus the laser [4] onto Ar targets including both a pulsed-gas jet (target gas is near 100% clustered) and a static-gas cell (target gas is in monomer form). Harmonic radiation is collected as described above using the MCP. The laser beam is clipped using an aperture at the focusing lens which produces a flat-top spatial beam-profile and an f/50 focusing geometry. We are in the weak focusing limit,[10] with a laser spot size of 50 μm at focus. Although the uncertainty in the determination of the absolute laser intensity is a factor of 2, relative intensities in experimental runs comparing monomer and cluster targets are accurate to several percent.

Figure 3 shows the yield of the 23rd harmonic as a function of laser intensity, I (W/cm^2), from (a) monomers and (b) clusters. Harmonic yield from the Ar monomers (at 40 Torr) is fit initially to the indicated power law, I^9, with saturation near an intensity of 1.5×10^{14} W/cm^2. Harmonic yield from the Ar clusters appears at higher intensities, and is fit (at a backing pressure of 115 psi) initially to the indicated power law, I^{17}, and then to a reduced power law, I^4, at intensities above 2×10^{14} W/cm^2, without an observation of saturation similar to that seen using the monomers.

Figure 4 shows the harmonic spectrum for various laser intensities and targets; spectrum were chosen from data generated in the high-laser-intensity regime for each target. We note that the static-gas cell and the gas jets produce similar average atom densities of about 10^{18} cm^{-3}.[11] Figure 4a shows harmonics for a super-sonic gas jet, with a backing pressure of 255 psi, and a laser intensity of 9×10^{14} W/cm^2; Fig. 4b shows harmonics for a sonic gas jet, with a backing pressure of 115 psi, and a laser intensity of 3×10^{14} W/cm^2; and Fig. 4c shows harmonics for a static-gas cell, with a pressure of 50 Torr, and a laser intensity of 6×10^{14} W/cm^2. The backing pressure of the super-sonic jet yields a cluster size of ≥ 3000 atoms,[2] corresponding to a cluster radius ≥ 3 nm. We indicate specific harmonic orders to draw attention to the increasing value of the harmonic order at which the cut-off appears in the three sets of data (monomer = 29th, sonic jet cluster = 31st, super-sonic jet cluster = 33rd). We define the cut-off to be the harmonic at which the signal drops to less than 1/4 of the signal of the previous harmonic. The monomer data (4c) is obtained at a peak laser intensity which is more than twice the monomer ionization intensity, demonstrating that the 29th harmonic is the cut-off harmonic. Additionally, the cut-off is observed to be fairly independent of static-gas pressure; Ar clusters have a consistently higher cut-off than monomers and the cluster cut-off is seen to increase with increasing cluster size.

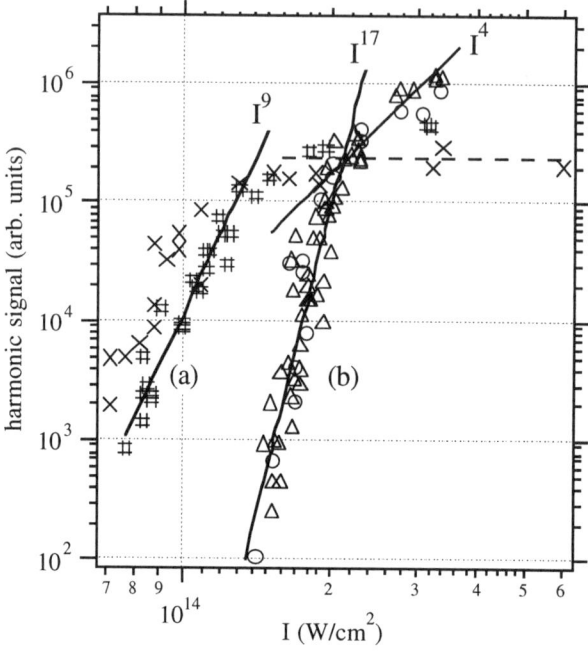

Figure 3. Intensity of the 23rd harmonic as a function of peak laser intensity from: (a) Ar monomers and (b) Ar clusters. Data are shown from: (Δ) sonic jet clusters (115 psi backing pressure), (O) sonic jet clusters (135 psi), (X) static-gas cell (70 Torr), and (#) static-gas cell (40 Torr). The dotted line (- - -) indicates saturation of the monomer harmonics. Best power law fits to the data are given.

First, we consider the increased appearance intensity in the case of the clusters. We verified that the increase is not due to defocusing of the laser beam by checking the appearance intensity of particular harmonics as a function of jet backing pressure. Increased appearance intensity can be explained by shielding of the laser field within the cluster [9]. The index of refraction of solid-density Ar is 1.2; this leads to a reduction in the laser intensity by a factor of 0.7 before the cluster is ionized. Once ionization of the cluster begins, and harmonic production maximizes, the contribution to the dielectric constant from the plasma electrons (determined from the Drude model) must be included. The time-dependent free-electron density and the resulting effect of shielding on the laser intensity within the cluster can be calculated [9]. This shielding results in a time-averaged decrease in the intensity of the laser within the cluster to approximately 60% of the laser's vacuum intensity, and may explain the increase in the appearance intensity of the cluster harmonics relative to the monomer harmonics.

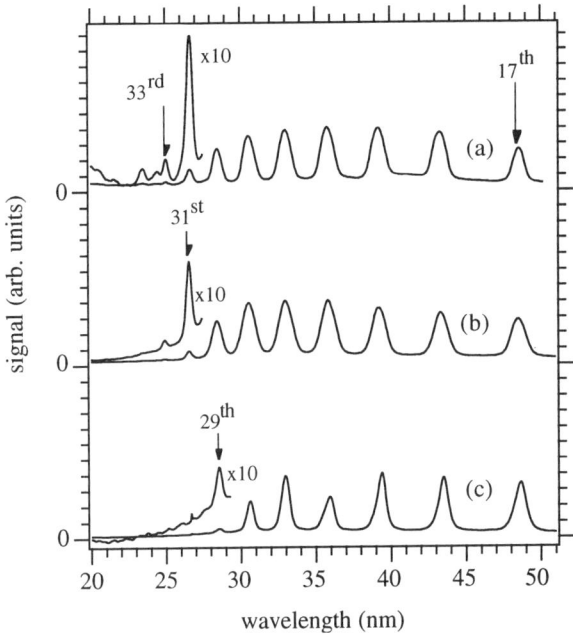

Figure 4. Intensities of the harmonic spectra from Ar monomers and Ar clusters. The harmonic cut-off wavelength is indicated. All spectra are scaled to unit height: Data are shown from: (a) super-sonic jet, (b) sonic jet, and (c) static-gas cell.

Next, we consider the power law dependence of cluster harmonic yield on laser intensity. We attribute the increased slope in the case of the clusters to the difference between the atomic potential which an electron experiences when bound by a single atom versus when it is bound in a solid. Using a semiclassical model to describe electron motion when driven by a laser pulse [9], we can determine the spectral density of the electron's emission.[12] From a series of such spectrum, generated for various laser intensities, we can compare the non-linearity of the harmonic yield versus laser intensity for the cluster and single atom potentials. Our result show that the cluster potential gives a more non-linear dependence (average power law of I^{20}) compared to the single atom potential (average power law of I^{15}) for the 19th through 23rd harmonics modeled. [9] We note that experimental power law fits of harmonics from various gases in the cut-off region are typically I^{12}.[10, 13]

Finally, we consider both the higher harmonic cut-off and reduced saturation in harmonic yield at high laser intensity for the clusters. We expect that Coulomb (space-charge) forces will result in an increase in the effective binding energy of electrons in the cluster compared with isolated atoms, resulting in a higher photon energy (shorter wavelength) cut-off in clusters than monomer harmonics. Due to space charge restraints on particle motion, the laser intensity may be allowed to rise to arbitrarily high limits without resulting in electron removal from the cluster, yielding harmonics at even shorter wavelengths. We also note that electrons may scatter from other atoms in the cluster.

In conclusion, we have shown that atomic clusters are a very strong and efficient source of x-rays that reach into the water window. We have also shown that these clusters can be used to generate coherent, short-wavelength radiation in the form of high-order harmonics.

Acknowledgements

We would like to acknowledge helpful conversations with, Keir Neuman, Ken Kulander and Sasha Rubenchik. This work was supported by the U. S. Air Force Office of Scientific Research and through a collaboration with Lawrence Livermore National Laboratory under Contract No. W-7405-ENG-48.

References

[1] S.P. Gordon, *et al.*, Opt. Lett **19**, 484 (1994); M. Murnane, *et al.*, Appl. Phys. Lett. **62**, 1068 (1993).
[2] O.F. Hagena and W. Obert, J. Chem Phys. **56**, 1793 (1972); O.F. Hagena, Surf. Sci. **106**, 101 (1981); J. Wormer, V. Guzielski, J. Stapelfeldt and T. Moller, Chem. Phys. Letters **159**, 321 (1989); C.L. Briant and J.J. Burton, J. Chem. Phys. **63**, 2045 (1975).
[3] T. Ditmire, T.D. Donnelly, R.W. Falcone and M.D. Perry, Phys. Rev. Lett. **75**, 3122 (1995); to be published in Phys. Rev. A, 1996.
[4] T. Ditmire and M.D. Perry, Opt. Lett. **18**, 426 (1993).
[5] T.E. Glover, *et al.*, Phys. Rev. Lett **73**, 78 (1994).
[6] A. L'Huillier and P. Balcou, Phys. Rev. Lett. **70**, 774 (1993); J.J. Macklin, J.D. Kmetec and C.L. Gordon, Phys. Rev. Lett. **70**, 766 (1993); T. Ditmire, *et al.*, Phys. Rev. A **51**, R902 (1995).
[7] Y. Liang, *et al.*, J. Phys. B. **27**, 5119 (1994).
[8] C.L. Carman, C.K. Rhodes and R.F. Benjamin, Phys. Rev. A **24**, 2649 (1981); B. Bezzerides, R.D. Jones and D.W. Forslund, Phys. Rev. Lett. **49**, 202 (1982).
[9] T.D. Donnelly, *et. al*, to be published in Phys. Rev. Lett., 1996.
[10] A. L'Huillier, *et al.*, in *Atoms in Intense Laser Fields*, edited by M. Gavrila (Harcourt Brace Jovanovich, Boston, 1992), p. 139.
[11] M.D. Perry, *et al.*, Opt. Lett. **17**, 523 (1992); personal communication by J.K. Crane.
[12] K. Kulander and B. Shore, Phys. Rev. Lett. **62**, 524 (1989).
[13] C.G. Wahlstrom, *et al.*, Phys. Rev. A **48**, 4709 (1993).

ATOMIC PHYSICS
IN COOL PLASMAS

Electron-Impact Ionization of Molecules of Interest to Processing Plasmas

Kurt H. Becker and Vladimir Tarnovsky

Physics Department, City College of C.U.N.Y.
Convent Avenue and 138th Street, New York, NY 10031 U.S.A.

Abstract. This paper reviews the recent developments in the measurement of absolute partial cross sections for the ionization and dissociative ionization of molecules and free radicals of interest to low-temperature processing plasmas. In addition to a discussion of selected recent results, this paper will highlight experimental challenges such as the need to develop techniques for the production of unstable target species such as free radicals and the influence of discrimination effects arising from the formation of energetic fragment ions on the measurement of dissociative ionization cross sections. Lastly, an attempt is made to relate measured molecular ionization cross sections to realistic plasma applications. As an example, we will discuss a collision-induced decomposition scheme of the Si-organic molecule tetramethylsilane (TMS), which is used in plasma deposition applications, based on recently measured partial ionization cross sections.

I. INTRODUCTION AND BACKGROUND

Low-temperature processing plasmas are important tools to advance many key technologies such as the etching of microstructures, the deposition of thin films and various surface modification applications. Up to now, low-temperature plasma technology has been developed largely empirically through a trial-and-error approach, while the understanding of the plasma processes in terms of the basic interactions in the plasma have been lagging behind. As more complex processes and plasma reactors are being used and higher demands are placed on the processed materials in terms of shrinking feature size, increased selectivity and uniformity, and as the need for faster throughput and lower rejection rates increases, it is highly desirable to improve our understanding of the basic plasma processes and our plasma diagnostics capabilities. These goals can only be achieved, if we drastically improve the existing data base on electronic and atomic collision processes relevant to the physics and chemistry that govern the properties of these plasmas.

© 1996 American Institute of Physics

Glow discharge processing plasmas are characterized by mean electron energies from 0.5 - 5 eV and charge densities of 10^8 - 10^{12} cm^{-3}. Fig. 1 shows the charge carrier density and mean electron energy of glow discharge processing plasmas and other plasmas ranging from extremely thin interstellar plasmas to the fusion goal and the extreme dense interior of stars. Collisions of interest in processing plasmas include a multitude of processes involving electrons, ions, and photons as projectiles and atoms, molecules, free radicals and excited species (metastables, electronically and vibrationally excited species) as targets. The list of specific targets of interest ranges from atomic hydrogen, oxygen and nitrogen to complex polyatomic molecules such as Si-organic and metal-organic compounds. Other species relevant in plasma-assisted processes include atoms such as Al, Si and the halogens, simple diatomics such as H_2, Cl_2, HBr, polyatomics, most notably partially and fully halogenated methane compounds, BCl_3, SF_6 and SiH_4, and free radicals such as CF_x and SiH_x (x=1-3).

Electron-impact ionization is the dominant process by which charge carriers in the plasma are formed and it thus determines the ionization balance in the plasma. Moreover, since technological plasmas commonly feature ployatomic molecules as feedgas constituents, the single-step dissociative ionization and the two-step dissociation followed by ionization of the dissociation fragment are two important mechanisms for the production of reactive ionic and neutral plasma radicals. The rest of this paper will focus on a discussion of selected aspects of ionization cross section measurements for molecules and radicals of interest to processing plasmas. We will limit the discussion to the measurement of absolute partial angle-integrated ionization cross sections which are more relevant to most plasma applications than angle-resolved doubly or triply differential ionization cross sections [1].

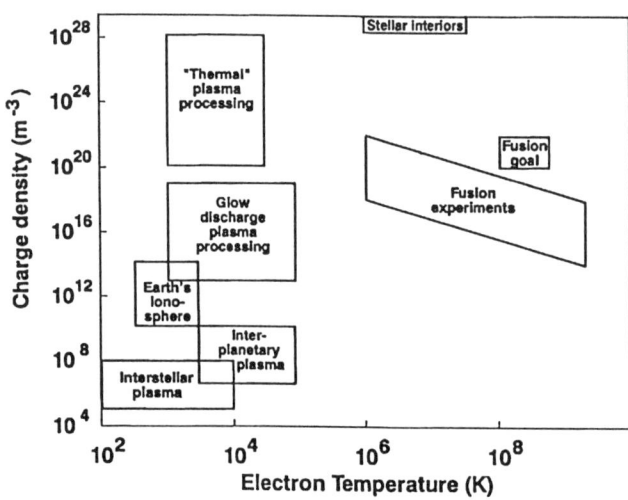

Fig. 1: A schematic diagram showing various plasmas in terms of their charge carrier density and mean electron energy (shown on a log-log scale).

II. THE STATUS OF MOLECULAR ELECTRON IMPACT IONIZATION CROSS SECTIONS

II.1 Survey and Experimental Challenges

In 1985, Märk [2] listed about 40 molecules for which absolute cross sections for ionization and dissociative ionization were available. Many cross sections had only been measured once using one experimental technique. Two recent reviews [1,3] put the number of molecular targets for which data are available now at about 100. Both reviews also state that a major driving force behind this development has been the need for reliable ionization cross section data for molecules and free radicals in numerous areas of application. The diverse data needs in various plasma applications were mentioned in particular with a special emphasis on low-temperature, non-equilibrium plasmas for materials processing. The reviews also stressed that not only did the number of targets for which data are available increase dramatically, but that the quality of data and the diversity of targets had also improved considerably. Two areas of significant progress deserve special consideration, (i) ionization studies involving reactive free radicals and (ii) the improved reliablility of cross section data for dissociative ionization processes.

Many technologically relevant plasmas use feedgas mixtures containing complex inert molecules (e.g. CF_4, CH_4, SF_6). The plasma chemical reactions in these plasmas involve reactive neutral and ionic radicals produced by the collisionally induced break-up of these feedgas molecules. Electron-impact ionization and dissociative ionization of the plasma species are important processes for the understanding and modelling of the key plasma chemical processes. Plasma diagnostics tools such as threshold ionization mass spectrometry (TIMS) also rely on the availability of reliable ionization cross sections for plasma radicals [4]. Based upon the pioneering work of Freund and collaborators [5,6] our group at CCNY has been utilizing the unique capabilities of the fast-neutral-beam technique in a comprehensive series of ionization studies of free radicals.

The fast-beam apparatus at CCNY and its performnce characteristics have been described in great detail in previous publications [5-8]. The apparatus utilizes a fast (3 kV) ion beam from a Colutron dc discharge source. The primary ion beam is passed through a Wien filter to isolate a single mass. Neutrals are subsequently formed by resonant or near-resonant charge transfer when the ion beam passes through a low-pressure charge exchange cell. The residual ions are then removed from the beam by electrostatic deflection. The charge exchange process also produces species in Rydberg states. Most Rydberg species are ionized and removed from the beam in a region of high electric field of at least 5 kV/cm. The remaining neutral beam is then crossed with a well-characterized electron beam of variable energy (3 - 200 eV). The product ions are focused into the entrance plane of an electrostatic hemispherical analyzer which separates the ions according to their mass-to-charge ratio. After leaving the analyzer the ions are detected by a channel electron multiplier (CEM) operated in the pulse counting mode. The neutral beam is monitored by a secondary emission detector and its absolute flux can be determined with a calibrated pyroelectric detector.

The two main advantages of our fast-beam apparatus are (i) the capability to generate target beams of unstable species such as free radicals which cannot be produced by conventional methods and (ii) the capability to collect and detect energetic fragment ions formed by dissociative ionization processes with 100% efficiency as long as the excess kinetic energy is not too large (typically less than 2-5 eV per fragment in most cases [8]). The main disdavantages of our apparatus are (i) the fact that the fast-beam technique results in weak ion signals and compartatively low signal-to-noise ratios which allow only the major ionization channels to be studied with reasonable accuracy and (ii) the fact that the charge neutralization can produce target beams contaminated with species in excited states (which can only be determined by time-consuming near-threshold studies).

The reliablility of dissociative ionization cross sections is an area where significant progress has been made in the past 10 years. Fragment ions produced by dissociative processes can carry translational kinetic energies of up to several electronvolts [8]. The collection of these energetic fragment ions and their subsequent detection with 100% efficiency or with a smaller, but known efficiency has presented a problem in many earlier experimental studies. For instance, cross sections for the formation of the light F^+ ion produced by dissociative ionization of CF_4 reported by two groups differed by a factor of 8, whereas their respective cross section values for the formation of the CF_3^+ ion showed excellent agreement [9,10]. Both group then carefully investigated discrimination effects in their respective apparati which influenced the extraction of energetic fragment ions from the ionization region and caused ion losses during the transport of the ions from the ionization region to the detector. Both groups determined appropriate correction factors depending on the excess kinetic energy of a particular process and fragment ion. Once these factors were applied, the F^+/CF_4 cross sections obtained by both groups agreed to within 20% which was well within the combined error margin of the two experiments. It has now become standard for any group in this field to use a combination of experimental studies and ion trajectory modelling calculations in an effort to quantify discrimination effects in their apparatus. Our fast-beam apparatus is especially suited to the detection of energetic fragment ions with 100% efficiency as long as the excess kinetic energy is not too large (less than 2-5 eV per fragment depending on the particular process [8]).

II.1 Some Selected Results

Methane (CH_4) is a commonly used constituent of deposition plasmas. Dissociation of the parent molecules results in the formation of the free radicals CH_x (x=1-3). Cross sections for the production of the various parent and fragment ions by electron collisional ionization and dissociative ionization of CH_x (x=1-4) are important for the understanding, modelling and diagnostics of CH4 containing processing plasmas. Threshold ionization mass spectroscopy (TIMS) has emerged as a particularly powerful plasma diagnostics tool for the determination of absolute radical concentrations in plasmas. The quantitative interpretation of TIMS measurements requires a knowledge of absolute radical ionization cross sections [4]. Most experimental studies to date focused on ionization cross section measurements for the CH4 molecule. We carried out a comprehensive series of measurements for CD4 and the CD_x radicals under a variety of experimental conditions. We used the deuterated species to ensure a better mass separation of the product ions. Ionization cross sections are insensitive to isotope effects [11].

Fig. 2 summarizes all measured partial ionization cross section for the CD_x (x=1-3) radicals. The figure also includes cross sections for the CD4 molecule, which are in good agreement with earlier measurements carried out for the CH4 molecule by other authors [12]. The CD_x results can be summarized as follows:
- for all four targets, parent ionization is the dominant channel (which is very different from earlier findings for fluorine-containing species [8])
- for all four targets, the parent ionization cross section has approximately the same value, 1.65×10^{-20} m^2 at 70 eV and all cross section shapes are very similar

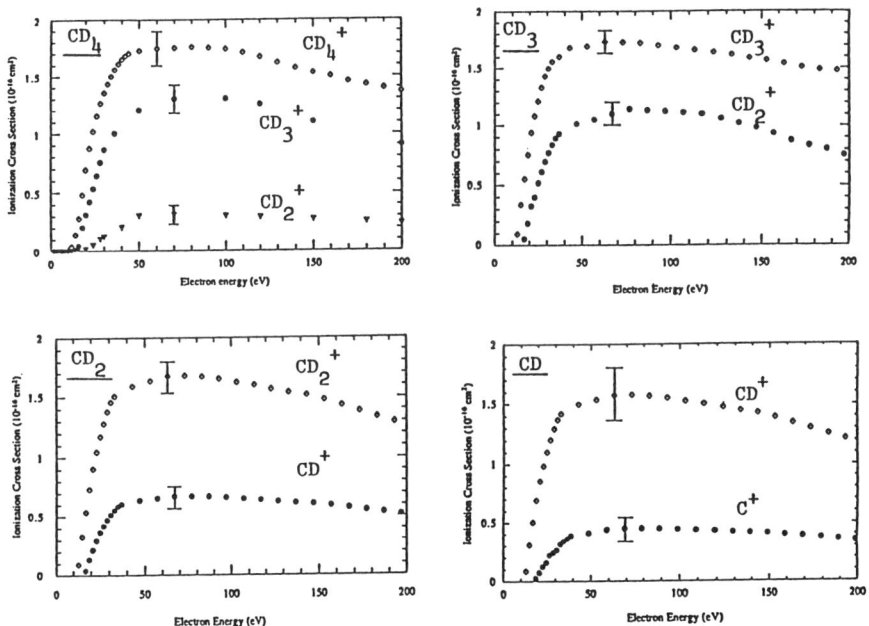

Fig. 2: Summary of the ionization and dissociative ionization cross sections for CD_x (x=1-4).

- our results for CD$_3$ and CD$_2$ agree well with the earlier results of Biaocchi et al. [13] for these species which were obtained under different experimental conditions
- dissociative ionization is generally weak with only one dissociative ionization channel of importance, the removal of a single D atom from the target
- dissociative ionization gets progressively weaker from CD$_4$ to CD; at 70 eV:

σ(CD$_3^+$/CD$_4$) = 1.4 x 10^{-20} m^2 (85% of the CD$_4^+$ parent ionization cross section)

σ(CD$_2^+$/CD$_3$) = 1.0 x 10^{-20} m^2 (60% of the CD$_3^+$ parent ionization cross section)

σ(CD$^+$/CD$_2$) = 0.5 x 10^{-20} m^2 (30% of the CD$_2^+$ parent ionization cross section)

σ(C$^+$/CD) = 0.35 x 10^{-20} m^2 (20% of the CD$^+$ parent ionization cross section)

- in CD$_4$, we found that vibrational excitation of up to 2 eV in the incident fast neutral CD$_4$ beam affected the measured ionization cross section only in the near-threshold region below approximately 25 eV

More details of the CD$_x$ results can be found in a recent publication [14].

Tetramethylsilane (TMS), Si(CH$_3$)$_4$, is the simplest Si-organic molecule. It is widely used in plasma polymerization applications both as a feedgas constituent and as a by-product in plasmas featuring more complex Si-organic molecules. Mass spectometric techniques were used to study the ionization and dissociative ionization of TMS [15]. Cross sections for fragment ions correponding to m/z values of 15 (CH$_3^+$), 28 (Si$^+$), 29 (SiH$^+$), 31 (SiH$_3^+$), 42 (SiCH$_2^+$), 43 (SiCH$_3^+$), 44 (HSiCH$_3^+$), 45 (H$_2$SiCH$_3^+$), and 74 (HSi(CH$_3$)$_3^+$) as well as the parent ionization cross section (m/z 88) were measured from threshold to about 90 eV. The largest ionization cross section of almost 11 x 10^{-16} cm^2 was measured for m/z 73 which corresponds to the Si(CH$_3$)$_3^+$ fragment ion. Maximum cross sections of close to 1.5 x 10^{-16} cm^2 were found for the ion signals at m/z = 43 and 45. The maximum cross section for the methyl ion reaches a value of 1 x 10^{-16} cm^2 at about 80 eV. All other maximum cross sections were below 1 x 10^{-16} cm^2. The parent ionization cross section is very small with a peak value of 0.16 x 10^{-16} cm^2.

A collision-induced decomposition scheme for the TMS molecule in low-temperature plasmas is shown in fig. 3 which summarizes the detected ions together with their measured appearance energies. The formation of the Si-containing fragment ions proceeds via three different mechanisms:
(i) the removal of a complete methyl group
(ii) the removal of a complete methyl group and an additional hydrogen atom (perhaps leading to the formation of the stable CH$_4$ methane molecule)
(iii) the removal of a CH$_2$ or CH group with one or two H atoms remaining with the ion (this process is most likely followed by a re-arrangement of the remaining H atoms within the residual ion).

The appearance energy of an ion formed via process (ii) is generally higher than that of an ion formed via process (i). Ions formed via process (iii) require less energy than ions formed via process (i) wih the exception of the HSi(CH$_3$)$_3^+$ ion. A possible explanation for this might be the remarkably small energy difference between the ionization energy of the parent ion (9.9 eV) and the 10.1 eV appearance energy of the most abundant fragment ion Si(CH$_3$)$_3^+$. Since the average Si - C

Fig. 3: Proposed collision-induced decomposition scheme of TMS in a low-temperature plasma. Shown are the various fragment ions, their m/z values and their measured appearance energies.

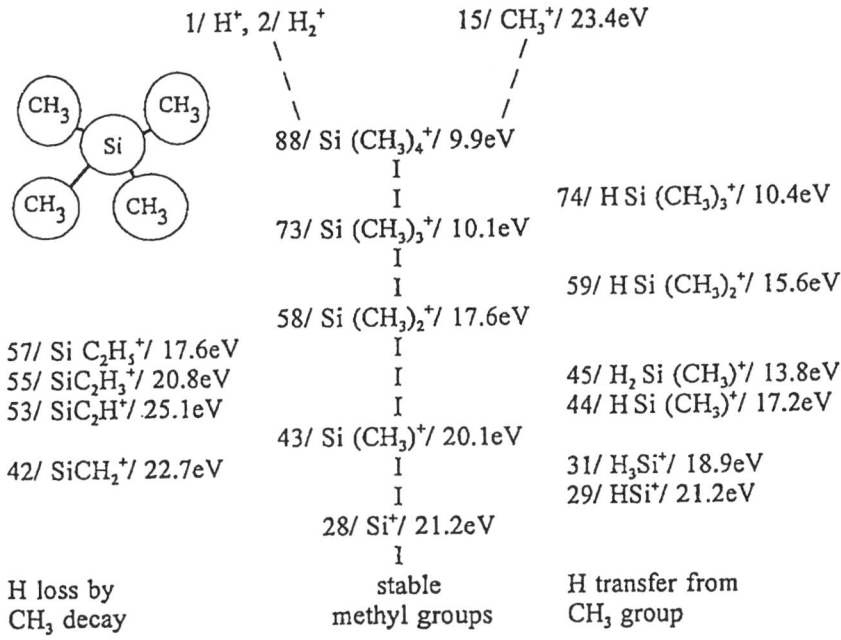

bond energy is about 3.2 eV, the 0.2 eV difference in the appearance energies for the two ions is most likely due to a lower energy of the planar Si(CH$_3$)$_3^+$ ion in comparison to the tetragonal Si(CH$_3$)$_4^+$ ion. Lastly, all fragment ions except for the CH$_3^+$ methyl ion, are formed with little or no excess kinetic energy. CH$_3^+$ fragment ions, on the other hand, are formed with a distribution of excess kinetic energies peaking at about 2 eV per fragment ion [15].

III. CONCLUDING REMARKS

Significant progress has been made in the past 10 years in the measurement of electron-impact ionization cross sections for molecules. Much of that progress can be attributed to application-driven data needs. Data needs in plasma-related applications were particularly important with special emphasis on low-temperature processing plasmas. Areas which deserve special mention include (i) measurements of ionization cross section for reactive free radicals, (ii) much more reliable cross sections for dissociative ionization processes, and (iii) ionization cross section studies of complex Si- and metal-organic compounds which contribute to a better understanding of the relevant plasma chemical processes in deposition plasmas featuring these molecules.

ACKNOWLEDGMENTS

I am grateful to Dr. R. Basner, Dr. R.S. Freund, Dr. M.Schmidt, Dr. S.K. Srivastava, Prof. H. Deutsch, and Prof. T.D. Märk for many helpful and stimulating discussions. The partial financial support of this work by the U.S. National Foundation (NSF) through grant PHY-9401874, the U.S. Department of Energy (DOE) through grant DE-FG02-94ER14476, and the U.S. National Aeronautics and Space Administration (NASA) through grant NAGW-4118 is gratefully acknowledged.

REFERENCES

1. K. Becker and V. Tarnovsky, Plasma Sources Sci. Technol. **4**, 307 (1995)
2. T.D. Märk, in "Electron Impact Ionization", edited by T.D. Märk and G.H. Dunn, Springer Verlag, Vienna (1985)
3. V. Tarnovsky and K. Becker, Invited Papers, XVIII. ICPEAC, Aarhus, Denmark (1993), American Institute of Physics Press, New York (1994), p. 234
4. R. Robertson, D. Hils, H. Chatham, and A. Gallagher, Appl. Phys. Lett. **43**, 544 (1983)
5. T.R. Hayes, R.C. Wetzel, and R.S. Freund, Phys. Rev. A **35**, 578 (1987)
6. R. Freund, R. Wetzel, R. Shul, and T. Hayes, Phys. Rev. A **41**, 3575 (1990)
7. V. Tarnovsky and K. Becker, Z. Phys. D **22**, 603 (1992)
8. V. Tarnovsky, P. Kurunczi, D. Rogozhnikov, and K. Becker, Int. J. Mass Spectrom. Ion Proc. **128**, 181 (1993)
9. K. Stephan, H. Deutsch, and T.D. Märk, J. Chem. Phys. **83**, 5722 (1985); H.U. Poll, C. Winkler, D. Margreiter, V. Grill, and T.D. Märk, Int. J. Mass Spectrom. Ion Proc. **112**, 1 (1992)
10. C. Ma, M.R. Bruce, and R.A. Bonham, Phys. Rev. A **44**, 2921 (1991); M.R. Bruce and R.A. Bonham, Int. J. Mass Spectrom. Ion Proc. **123**, 97 (1993)
11. T.D. Märk and F. Egger, J. Chem. Phys. **67**, 2629 (1977); T.D. Märk, F. Egger, and M. Cheret, J. Chem. Phys. **67**, 3795 (1977)
12. H. Chatham, D. Hils, R. Robertson, and A. Gallagher, J. Chem. Phys. **81**, 1770 (1984)
13. F. Biaocchi, R.C. Wetzel,, and R.S. Freund, Phys. Rev. Lett. **53**, 771 (1984)
14. V. Tarnovsky, A. Levin, H. Deutsch, and K. Becker, J. Phys. B **29** (1996), in press
15. R. Basner, R. Foest, M. Schmidt, F. Sigeneger, P. Kurunczi, K. Becker, and H. Deutsch, Int. J. Mass Specrtom. Ion Proc. (1996), in press

New Model for Electron-Impact Ionization Cross Sections of Atoms and Molecules

Y.-K. Kim,[*] W. Hwang,[*†] and M. E. Rudd[‡]

[*]*National Institute of Standards and Technology, Gaithersburg, Maryland 20899*
[†]*Present address: Ultraprecision Technology Team, Samsung Electronics Co., Suwon, Kyonggi-do, Korea*
[‡]*Department of Physics and Astronomy, University of Nebraska-Lincoln, Lincoln, Nebraska 68588-0111*

Abstract. A new theoretical model for electron-impact ionization cross sections for atoms and molecules is presented. The new model combines the binary-encounter theory and the Bethe theory for electron-impact ionization, and uses minimal theoretical data for the ground state of the target atom or molecule. Two versions of the model are presented. The first one, the Binary-Encounter-Dipole (BED) model, requires the knowledge of continuum oscillator strengths and produces the differential ionization cross section, i.e., energy distribution of ejected electrons. The differential cross section is then integrated over the ejected electron energy to obtain the total ionization cross section. The second version, the Binary-Encounter-Bethe (BEB) model, assumes a simple form of the continuum oscillator strength to obtain a compact and analytic form of the total ionization cross section. We found that both the BED and BEB models provide total ionization cross sections from threshold to several keV in incident energy within 5% to 15% of known experimental data for many neutral targets. The total ionization cross sections are expressed in compact analytic expressions suitable for use in modeling, e.g., of plasmas and radiation effects. We found that the BEB model is particularly effective in estimating total ionization cross sections of complex molecules.

1. INTRODUCTION

Most theoretical models for electron-impact ionization cross sections of atoms and molecules have difficulties in two aspects. The first difficulty is that the same model is rarely applicable to both atoms and molecules, because often the theoretical quantity needed, such as the continuum wave functions, can be calculated for atoms but not easily for molecules. The second difficulty is that a theoretical model may not be valid for all incident energies, T, particularly between the threshold and the ionization peak, which is located between 60 eV

© 1996 American Institute of Physics

and 150 eV for most neutral atoms and molecules. The collision of a slow incident electron with a target requires an approach that treats the incident electron and the bound electrons in the target on equal footing as a compound system, for instance by introducing strong coupling and exchange interaction between them. This is one of the major reasons for the failure of most theories, particularly those based on the perturbation approach, at low incident electron energies. At high T the plane-wave Born approximation provides accurate ionization cross sections when used with reliable initial- and final-state wave functions.

Strong coupling theories, such as the close-coupling or the R-matrix method, require a large basis set to describe the system, making it very difficult to treat ionizing collisions. Except for two recent theoretical methods [1,2], these strong coupling theories are mostly limited to discrete excitations and are difficult to extend to ionization of atoms and molecules with complex shell structures.

In this article, we describe a new theoretical method that provides reliable electron-impact ionization cross sections for neutral atoms and molecules using very simple input data, all of which can be obtained from standard atomic and molecular wave function codes for the ground state of the target. There are no adjustable or fitted parameters in our theory. The theory does not provide details of resonances in the continuum, vibrational and/or rotational excitations concomitant with ionization, multiple ionization, dissociative ionization, etc. It simply predicts the differential and total ionization cross sections as the sum of cross sections for ejecting one electron from each of the atomic or molecular orbitals. We will present examples to show that the theory is valid over the entire energy range of T, from the first ionization threshold to several keV.

As is outlined in Sec. 2, our theoretical model combines the Mott cross section [3] modified by the binary-encounter theory [4] for low T with the Bethe theory [5] for high T. Many models have been proposed to use this combination [6-9] but they all require either some empirical parameters or explicit knowledge of the continuum dipole oscillator strengths of the target. Some of these models require empirical parameters which are difficult to obtain, or use a large number of such parameters. The present model uses a new way to determine the ratio between the low-T and high-T cross sections without resorting to any empirical parameters, and ionization cross sections in analytic forms are derived for the entire range of T with three atomic/molecular constants per atomic/molecular orbital, which are available from atomic/molecular structure codes. Deep inner shells do not contribute appreciably to total ionization cross sections; hence they can be omitted for total ionization cross sections.

Because of its simplicity, the present theory can predict cross sections for complex molecules such as CH_4 and SF_6 as easily as for simple atoms such as H and He. Moreover, the theory can be applied not only to stable molecules but also to transient radicals. The applicability of the theory is limited only by the availability of an atomic/molecular structure code that can provide basic information on atomic/molecular orbitals. Judging from the examples presented in this article and our experience with other atoms and molecules, the present theory can provide total ionization cross sections accurate enough to be used in modeling of plasma chemistry, magnetic fusion plasmas, and radiation effects.

Two other theoretical methods to estimate ionization cross sections of molecules with comparable flexibility are the "DM approach" based on an additivity rule developed by Deutsch et al. [7] and the Weizsäcker-Williams method (WW method) as modified by Seltzer [9]. The DM approach constructs a molecular ionization cross section by adding ionization cross sections for the constituent atoms. The basic shape of these atomic cross sections is given by the classical theory of Gryzinski [10], while their absolute values are given in terms of (a) atomic orbital radii, which can be obtained from atomic wave function codes, (b) atomic orbital occupation numbers, which are derived from the Mulliken population analysis of the target molecule, and (c) atomic weighting factors which have been fitted to known atomic ionization cross sections. The present model uses far fewer, *ab initio* atomic/molecular parameters which are also standard output of atomic/molecular structure codes. The WW method [9] is very similar to the BED model described in the next section in that it requires explicit data on continuum dipole oscillator strengths of the target. However, the WW method is primarily designed for high-energy incident electrons, and may lead to unrealistic results for slow incident electrons of hundreds of eV or lower.

The underlying theory of the present model is outlined in Sec. 2, application examples are described in Sec. 3, and conclusions are presented in Sec. 4.

2. OUTLINE OF THEORY

Recently, we proposed the BED (Binary-Encounter-Dipole) model for electron-impact ionization cross sections of atoms and molecules [11]. The BED model combines the binary-encounter theory [4] and the Bethe theory [5]. The ratio between the binary-encounter theory and the Bethe theory is set by requiring the asymptotic form at high incident energy T of the former to match that of the latter both in the ionization cross section and in the stopping cross section. The stopping cross section, which is the integral of the product of the energy-loss cross section and the energy loss of the incident electron, is used to evaluate the stopping power of the target medium.

The BED model provides a formula to calculate the singly differential cross section, or the energy distribution of ejected electrons $d\sigma/dW$ with the ejected electron energy W, for each atomic/molecular orbital. With the binding energy

B, the average kinetic energy $U = \langle \mathbf{p}^2/2m \rangle$, $\mathbf{p}$ being the bound electron momentum and m its mass, and the orbital occupation number N, $d\sigma/dW$ is given by

$$\frac{d\sigma(W,T)}{dW} = \frac{S}{B(t+u+1)} \left\{ \frac{(N_i/N-2)}{t+1} \left[\frac{1}{w+1} + \frac{1}{t-w} \right] + [2-(N_i/N)] \left[\frac{1}{(w+1)^2} + \frac{1}{(t-w)^2} \right] + \frac{\ln t}{N(w+1)} \frac{df(w)}{dw} \right\}, \quad (1)$$

where $t = T/B$, $u = U/B$, $w = W/B$, $S = 4\pi a_0^2 NR^2/B^2$, $a_0 = 0.5292$ Å, $R = 13.61$ eV, and

$$N_i \equiv \int_0^\infty \frac{df(w)}{dw} dw, \quad (2)$$

df(w)/dw being the continuum dipole oscillator strength. Hence, to apply the BED model one needs for each orbital B, U, N, and df/dw.

The value of U for each orbital in the initial state (usually the ground state) of the target is a theoretical quantity evaluated in any atomic/molecular wave function code that calculates the total energy. However, both the initial- and continuum-state wave functions are needed to calculate df/dw and this is the only nontrivial data needed to apply the BED model. Alternatively, df/dw can be deduced from experimental photoionization cross sections, though partial cross sections are needed to deduce df/dw for each orbital. The total ionization cross section σ_i is then obtained by integrating $d\sigma/dW$ over the allowed range of W, i.e., from 0 to (T-B)/2:

$$\sigma_i(BED) = \frac{S}{t+u+1} \left[D(t) \ln t + \left(2 - \frac{N_i}{N}\right) \left[1 - \frac{1}{t} - \frac{\ln t}{t+1} \right] \right], \quad (3)$$

where

$$D(t) \equiv N^{-1} \int_0^{(t-1)/2} \frac{1}{w+1} \frac{df(w)}{dw} dw. \quad (4)$$

The BED model was found to be very effective in reproducing known values of $d\sigma/dW$ and σ_i for small atoms and molecules, demonstrating an agreement of better than 10% for the entire range of incident electron energies [11].

Although one can in principle calculate df/dw for each orbital, it is available only for a limited number of atoms and very few molecules. Hence, we also proposed a simplified version of the BED theory when no information on df/dw is available. In this case, which we refer to as the BEB (Binary-Encounter-Bethe) model [11], we assume a simple form for df/dw,

$$df/dw = N/(w+1)^2 \qquad (5)$$

which yields [11]

$$\sigma_i(BEB) = \frac{S}{t+u+1}\left[\frac{\ln t}{2}\left(1-\frac{1}{t^2}\right)+1-\frac{1}{t}-\frac{\ln t}{t+1}\right]. \qquad (6)$$

Equation (5) is accurate enough to give good results when integrated over w, but only generates reliable $d\sigma/dW$ for targets with simple shell structures [11] such as H, He, and H_2.

In Eqs. (3) and (6), the term associated with the first logarithmic function on the right-hand side (RHS) represents distant collisions (large impact parameters) dominated by the dipole interaction, and the rest of the terms on the RHS represent close collisions (small impact parameters) as described by the Mott cross section. The second logarithmic function originates from the interference of the direct and exchange scattering also described by the Mott cross section.

We present the values of B, U and N for H, He and H_2 in Table 1 and those for CH_3, CH_4, and SF_6 in Table 2. The data for He are from Kim and Rudd [11], those for H_2 from the correlated wave function of Kołos and Roothaan [12], and the rest of the data in Table 2 are from the molecular structure code GAMESS [13]. Since deep inner shells, such as the K shell of SF_6 contributes little to total ionization cross sections, we have omitted them from the table.

One can use either theoretical or experimental values of B, while U is a theoretical quantity that cannot be directly measured, though the sum of all U's is equal to the total energy of the target molecule according to the virial theorem. Since experimental values of B are often smaller than theoretical ones, the BEB cross sections obtained using experimental B values are usually higher (by 10--15% at the cross section peak) than those obtained using theoretical B values. Using the experimental values for the lowest electron binding energy (=first ionization potential) will not only assure that the cross section starts at the right threshold but also we found that the shape and magnitude of the BEB cross section near the threshold agree better using known experimental cross sections. We have used the experimental values for the lowest B, which are available for many molecules and radicals [14], and theoretical values for the remaining orbitals.

Table 1. Orbitals, Electron Binding Energy B in eV, Kinetic Energy U in eV, and Electron Occupation Number N for H, He and H_2. All B and U values are theoretical, except for those marked by an asterisk, which are experimental.

Target	Orbital	B	U	N
H	1s	13.61	13.61	1
He	1s	24.59*	39.51	2
H_2	$1\sigma_g$	15.43*	15.98	2

Table 2. Molecular Orbitals, Electron Binding Energy B in eV, Kinetic Energy U in eV, and Electron Occupation Number N for CH_3, CH_4, and SF_6. All B and U values are theoretical, except for those marked by an asterisk, which are experimental.

Molecule	MO	B	U	N
CH_3	$2a_1$	24.57	34.18	2
	$1e$	15.64	26.46	4
	$3a_1$	9.84*	30.40	1
CH_4	$2a_1$	25.73	33.05	2
	$1t_2$	12.51*	25.96	6
SF_6	$3a_{1g}$	256.18	510.35	2
	$2t_{1u}$	193.27	479.25	6
	$4a_{1g}$	50.93	85.62	2
	$3t_{1u}$	47.09	101.6	6
	$2e_g$	45.50	110.3	4
	$5a_{1g}$	30.32	98.55	2
	$4t_{1u}$	25.61	83.20	6
	$1t_{2g}$	23.03	75.55	6
	$3e_g$	20.35	86.84	4
	$1t_2$	20.06	90.74	6
	$5t_{1u}$	19.69	91.23	6
	$1t_{1g}$	15.33*	98.29	6

3. APPLICATIONS TO ATOMS AND MOLECULES

In this section, we compare our theoretical cross sections, mostly BEB and some BED cross sections, to atoms and molecules. Most older experiments measured the "gross" ionization cross section, which is determined by measuring the total ion current rather than the number of ions. On the other hand, most theoretical values are the "counting" ionization cross section, which accounts for the number of ions produced. When many multiply charged ions are produced, the gross ionization cross section will be significantly larger than the counting ionization cross section. The cross sections based on the BEB and BED models are counting ionization cross sections, and therefore should be considered as the lower limits to experimental gross ionization cross sections. When an atom is ionized, the resulting ion may be multiply charged if an inner shell is ionized. In molecules, however, molecular ions as well as their fragments are often produced. Since the BEB and BED cross sections are simple sums of cross sections for ejecting one electron from an atomic/molecular orbital, the theory cannot give a detailed account of dissociative ionization or fragments produced. Hence, comparisons of the theory with experiments on large molecules with diverse channels for dissociative ionization and fragmentation are not straightforward. For simplicity, we compared our theoretical cross sections for molecules to the simple sum of all experimental partial cross sections that produced an ion. Nevertheless, the comparisons presented here will clearly demonstrate the utility of our theory, which is applicable to a wide range of molecules [15].

3.1 Single Orbital Targets: H, He and H_2

In Fig. 1, we compare the BED and BEB cross sections for H with the experimental σ_i by Shah et al. [16], the distorted-wave Born cross section by Younger [17] and the classical cross section by Gryzinski [10]. Although $\sigma_i(BEB)$ agrees better with the experiment near the peak than $\sigma_i(BED)$, the BED model will provide better $d\sigma/dW$, particularly for more complex targets where electron correlation strongly affects the df/dw of valence shells. The classical cross section by Gryzinski [10] tends to overestimate the peak value not only for H but also for other targets, such as He and H_2 (see below).

In Fig. 2, we compare the BED and BEB cross sections for He with the experimental σ_i by Shah et al. [18] and by Montague et al., [19] as well as with the distorted-wave Born cross section by Younger [17]. In this case, the BED cross section is in better agreement with the experiments than the BEB cross section.

In Fig. 3, we compare our BED and BEB cross sections [11] for H_2 to the experimental data by Rapp and Englander-Golden [20], the data by Krishnakumar and Srivastava [21], those by Schram et al. [22,23], classical cross section by

Gryzinski [10], and the classical trajectory Monte Carlo (CTMC) cross section by Schultz et al. [24].

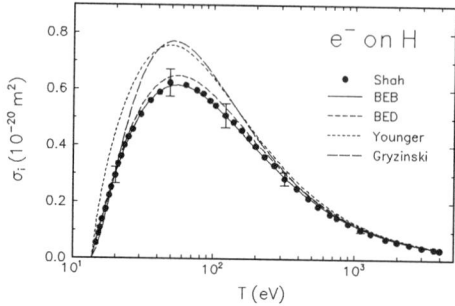

Fig. 1. Comparison of the BED and BEB cross sections to other theory and experiment for H. Solid curve, the BEB cross section [11]; medium-dashed curve, the BED cross section [11]; short-dashed curve, distorted-wave Born cross section by Younger [17]; long-dashed curve, classical cross section by Gryzinski [10]; circles, experimental data by Shah et al. [16].

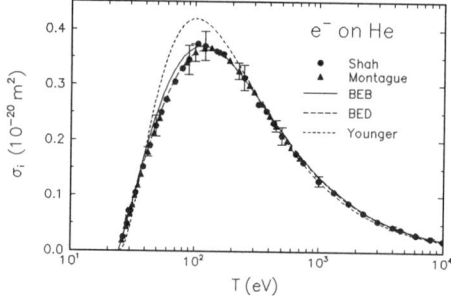

Fig. 2. Comparison of the BED and BEB cross sections to other theory and experiment for He. Solid curve, the BEB cross section [11]; medium-dashed curve, the BED cross section [11]; short-dashed curve, distorted-wave Born cross section by Younger [17]; circles, experimental data by Shah et al. [18]; triangles, data by Montague et al. [19].

The low uncertainty ($\pm 4.5\%$) claimed by Rapp and Englander-Golden implies that the excellent agreement between the BED cross section and the data by Krishnakumar and Srivastava near the cross section peak may be accidental. The CTMC cross section at high T falls short of experimental values because the CTMC theory lacks the dipole contribution, which is significant at high T.

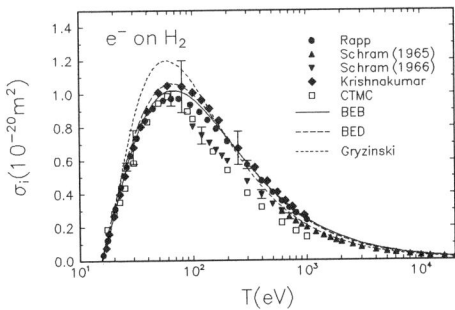

Fig. 3. Comparison of the BED and BEB cross sections to other theory and experiment for H_2. Solid curve, the BEB cross section [11]; medium-dashed curve, the BED cross section [11]; short-dashed curve, classical cross section by Gryzinski [10]; squares, CTMC theory [24]; circles, experimental data by Rapp and Englander-Golden [20]; triangles, data by Schram et al. [22]; inverted triangles, data by Schram et al. [23]; diamonds, data by Krishnakumar and Srivastava [21].

3.2 Molecules: CH_3, CH_4, and SF_6

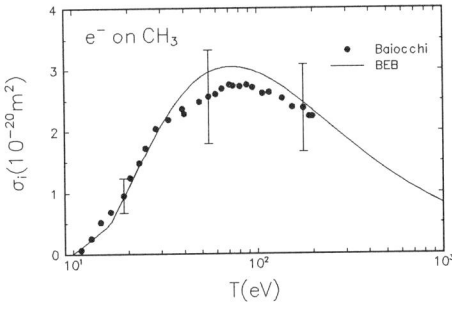

Fig. 4. Comparison of the BEB cross sections to experiment for CH_3 radical. Solid curve, the BEB cross section [15]; circles, experimental data by Baiocchi [25].

The BEB model can also be applied to radicals, such as CH_3 and SiF [15]. We found that the BEB model is particularly successful in reproducing known cross sections of hydrocarbons. The BEB cross section for CH_3 [15] is compared in Fig. 4 to the experimental data by Baiocchi et al. [25].

In Fig. 5, the BEB cross section for CH_4 [15] is compared to two versions of additivity rule cross sections by Margreiter et al. [26], experimental data by Rapp and Englander-Golden [20], the data by Orient and Srivastava [27], the data by Durić et al. [28], and to the data by Schram et al. [29]. The poor agreement between the BEB cross section and experiments at T < 60 eV is partly due to the high probability of neutral dissociation (=low ionization yield) when the molecule absorbs energy transfers greater than the ionization potential [30]. The discrepancy may also be an indication that a more accurate U value is needed for the valence orbital because of its large occupation number, N=6 (see Table 2).

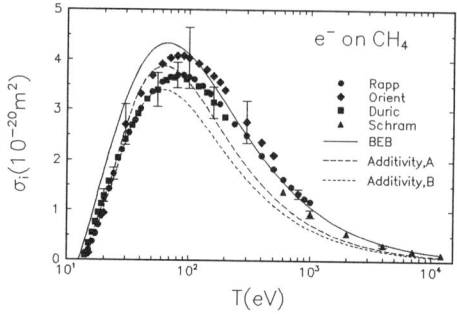

Fig. 5. Comparison of the BEB cross section to experiment for CH_4. Solid curve, the BEB cross section [15]; long-dashed curve, semiempirical additivity rule with 12 constants [26]; short-dashed curve, semiempirical additivity rule with 24 constants [26]; circles, experimental data by Rapp and Englander-Golden [20]; diamonds data by Orient and Srivastava [27]; squares, data by Durić et al. [28]; triangles, data by Schram et al. [29].

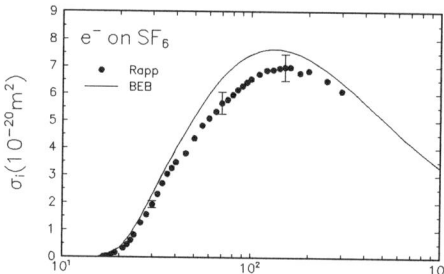

Fig. 6. Comparison of the BEB cross section to experiment for SF_6. Solid curve, the BEB cross section [15]; circles, experimental data by Rapp and Englander-Golden [20].

The BEB cross section for SF_6 [15] is compared to the experimental data by Rapp and Englander-Golden [20] in Fig. 6.

4. CONCLUSIONS

We have shown that the BEB and BED cross sections provide reliable electron-impact total ionization cross sections for atoms and molecules, from ionization threshold to high incident energies, T ~ 10 keV. The BEB cross section uses only a minimal set of atomic/molecular constants for the initial state of the target, which are readily available from public-domain structure codes.

Moreover, the BEB cross section consists of simple analytic expressions as functions of the incident energy for each atomic/molecular orbital that contributes to the ionization cross section, making the cross sections ideally suited for applications in modeling low-energy plasmas in plasma processing and fusion devices. When appropriate df/dw is available, the BED model provides reliable energy distributions of ejected electrons (singly differential cross sections) as well as total ionization cross sections. The BEB model uses far fewer constants than the additivity rules known as the DM approach. The latter also requires empirically fitted parameters, while the present model has *no adjustable parameters*. Examples of the atomic/molecular orbital constants needed to construct BEB cross sections for some atoms and molecules have been presented in Tables 1 and 2. Constants for many more molecules are listed in Ref. [15].

The success of the BEB model on a wide range of molecules is somewhat surprising, because our experience on atomic ionization cross sections clearly indicated that the BED model with appropriate df/dw was needed for good agreement with experiment [11]. We speculate that the break-up of atomic orbitals to many molecular orbitals in a molecule must act as a sort of "averaging" of atomic character and makes the BEB model adequate for most molecules.

Undoubtedly, a simple theory such as the BED and BEB models will require further refinements to expand its application to a wider class of atoms and molecules. Our preliminary study indicates that (a) we must modify our theory for fluorine and chlorine compounds [15]; (b) the BED model can be applied to rare gases with minimal modifications, while (c) more substantial modifications are necessary for targets with B values much lower than 10 eV, e.g., H^-, Li and metastable H(2s), or much higher than 30 eV, e.g., highly charged atomic ions. Nevertheless, we are confident that the BEB model will reliably predict the ionization cross sections of hydrocarbons and other molecules made of light atoms, particularly closed-shell molecules.

BEB cross sections near the ionization threshold are sensitive to the lowest values of B and U used. To insure the proper behavior near the ionization threshold, experimental values of the lowest ionization potential, which are well known for many atoms as well as molecules and radicals [14], should be used as the lowest B. In some cases, cross sections near the threshold are also likely to be influenced by unusually low ionization yields, resonances and autoionization peaks, making it difficult for a simple theory such as the BED or BEB model to be universally effective.

For those who are interested in representing a known cross section by a simple analytic formula, Eq. (6) can be used—for the total ionization cross section, not orbital cross sections—by treating the lowest value of U as a fitting parameter. For instance, using higher values of U for the outermost orbitals in CH_4 and SF_6 (while keeping other constants to the values in Table 2) will reproduce the experimental cross sections by Rapp and Englander-Golden [20] in Figs. 5 and 6 to a very high accuracy between the threshold and the peak without significantly altering the high-T part of the BEB cross section.

ACKNOWLEDGEMENTS

We are grateful to the creators of the GAMESS code (see Ref. 12), without which we could not have carried out this work, to Professor M.A. Ali for valuable advice on molecular wave functions, and to Dr. S.K. Srivastava and Professor T.M. Märk for providing tables of their experimental data. This work at NIST was supported in part by the Office of Fusion Energy of the U.S. Department of Energy, and at the University of Nebraska-Lincoln by the National Science Foundation Grant No. PHY-9119818.

REFERENCES

1. I. Bray, J. Phys. B **28**, L247 (1995).
2. D. Kato and S. Watanabe, Phys. Rev. Lett. **74**, 2443 (1995).
3. N.F. Mott, Proc. Roy. Soc. (London) **A126**, 259 (1930).
4. L. Vriens, in *Case Studies in Atomic Physics*, Vol. 1, ed. E.W. McDaniel and M.R.C. McDowell (North Holland, Amsterdam, 1969), p. 335.
5. H. Bethe, Ann. Physik **5**, 325 (1930).
6. S.P. Khare and W.J. Meath, J. Phys. B **20**, 2101 (1987), and references therein.
7. H. Deutsch, C. Cornelissen, L. Cespiva, V. Bonacic-Koutecky, D. Margreiter, and T.D. Märk, Int. J. Mass Spectrom. Ion Processes **129**, 43 (1993), and references therein.
8. H. Deutsch, T.D. Märk, V. Tarnovsky, K. Becker, C. Cornelissen, L. Cespiva, and V. Bonacic-Koutecky, Int. J. Mass Spectrom. Ion Processes **137**, 77 (1994), and references therein.
9. S.M. Seltzer, in *Monte Carlo Transport of Electrons and Photons*, ed. T.M. Jenkins, W.R. Nelson, and A. Rindi (Plenum, New York, 1988), p. 81, and references therein.
10. M. Gryzinski, Phys. Rev. **138**, A305 (1965); **138**, A322 (1965); **138**, A336 (1965).
11. Y.-K. Kim and M.E. Rudd, Phys. Rev. A **50**, 3954 (1994).

12. W. Kołos and C.C.J. Roothaan, Rev. Mod. Phy. **32**, 205 (1960).
13. We have used the version of GAMESS described by M.W. Schmidt, K.K. Baldridge, J.A. Boatz, S.T. Elbert, M.S. Gordon, J.H. Jensen, S. Koseki, N. Matsunaga, K.A. Nguyen, S.J. Su, T.L. Windus, M. Dupuis, and J.A. Montgomery, J. Comput. Chem. **14**, 1347 (1993).
14. S.G. Lias, J.F. Liebman, R.D. Levin, and S.A. Kafafi, *NIST Positive Ion Energetics Database, Version 2.0*, Standard Reference Database 19A, National Institute of Standards and Technology, Oct. 1993.
15. W. Hwang, Y.-K. Kim and M.E. Rudd, J. Chem. Phys. **104**, 2956 (1996).
16. M.B. Shah, D.S. Elliott, and H.B. Gilbody, J. Phys. B **20**, 3501 (1987).
17. S.M. Younger, J. Quant. Spectrosc. Radiat. Transfer **26**, 329 (1981).
18. M.B. Shah, D.S. Elliott, P. McCallion, and H.B. Gilbody, J. Phys. B **21**, 2751 (1988).
19. R.G. Montague, M.F.A. Harrison, and A.C.H. Smith, J. Phys. B **17**, 3295 (1984).
20. D. Rapp and P. Englander-Golden, J. Chem. Phys. **43**, 1464 (1965).
21. E. Krishnakumar and S.K. Srivastava, J. Phys. B **27**, L251 (1994). See also, E. Krishnakumar and S.K. Srivastava, Abstracts, *Sixteenth International Conference on the Physics of Electronic and Atomic Collisions* (New York, 1989), p. 326.
22. B.L. Schram, F.J. de Heer, M.J. van der Wiel, and J. Kistemaker, Physica **31**, 94 (1965).
23. B.L. Schram, H.R. Moustafa, J. Schutten, and F.J. de Heer, Physica **32**, 734 (1966).
24. D.R. Schultz, L. Meng, and R.E. Olsen, J. Phys. B **25**, 4601 (1992).
25. F.A. Baiocchi, R.C. Wetzel, and R.S. Freund, Phys. Rev. Lett. **53**, 771 (1984).
26. D. Margreiter, H. Deutsch, M. Schmidt, and T.D. Märk, Int. J. Mass Spectrom. Ion Processes **100**, 157 (1990).
27. O.J. Orient and S.K. Srivastava, J. Phys. B **20**, 3923 (1987).
28. N. Durić, I. Čadež, and M. Kurepa, Int. J. Mass Spectrom. Ion Processes **108**, R1 (1991).
29. B.L. Schram, M.J. van der Wiel, F.J. de Heer, and H.R. Moustafa, J. Chem. Phys. **44**, 49 (1966).
30. C. Backx, G.R. Wright, R.R. Tol, and M.J. Van der Wiel, J. Phys. B **8**, 3007 (1975).

SPECTROSCOPY AND DIAGNOSTICS

Critical Tests of Line Broadening Theories by Precision Measurements

Siegfried H. Glenzer

Institut für Experimentalphysik V, Ruhr-Universität, 44780 Bochum, Germany
present address: Lawrence Livermore National Laboratory L-399, University of California, P.O. Box 808, Livermore, CA 94551, USA

Abstract. We describe recent measurements of spectral line profiles of a z-pinch experiment employing precision plasma diagnostic techniques. In particular, the electron-collisional-broadened 2s – 2p transitions in B III have been investigated because their line profiles provide an excellent test for electron-impact line shape theories and electron collision strength calculations. Although we find good agreement with semiclassical calculations, a factor of two discrepancy with the most elaborate quantum-mechanical five-state close coupling calculations is observed. We discuss the experimental error estimates of the various measured quantities and show that the observed discrepancy can not be explained by experimental shortcomings. We further discuss measurements of non-isolated spectral lines of some $\Delta n = 1$ transitions in C IV – O VI. For these transitions ion broadening dominates. Excellent agreement for the whole line profile with line broadening calculations is obtained for all cases only when including ion dynamic effects. The latter are calculated using the frequency-fluctuation model and account for about 10 – 25 % of the line width of the considered ions.

INTRODUCTION

The spectral line profiles of ionized emitters in plasmas play an important role in the calculation of opacity (1,2), for short-wavelength laser studies (3), and for the diagnostics of inertial confinement fusion plasmas (4-6). Sophisticated theoretical methods and modeling have been advanced and applied in recent years (7-9) to calculate spectral line profiles in the limits where broadening by electron collisions or by ion microfields dominates.

Electron collisional broadening dominates the line broadening of isolated spectral lines of nonhydrogenic emitters. In most cases the impact approximation is valid over the frequency range of the spectral line profiles. The criterion (10) for the validity of the impact approximation is that the duration of the collision of the perturber with the emitter τ is much smaller than the inverse of the half-width at half maximum ϖ or the inverse of the angular frequency separation $|\Delta\omega|$ from the line center

$$\tau = \frac{\rho}{v} \ll min\left(\frac{1}{|\Delta\omega|}, \frac{1}{\varpi}\right); \qquad (1)$$

© 1996 American Institute of Physics

ρ is the impact parameter and v is the velocity of the electron. The general solution is a Lorentzian with a full-width at half maximum w given by the rates of effective (electron) collisions (11). Baranger used the optical theorem to derive the following expression

$$w = \left\{ n_e v \left(\sum_{u' \neq u} \sigma_{uu'} + \sum_{l' \neq l} \sigma_{ll'} + \int |\Phi_u - \Phi_l|^2 \, d\Omega \right) \right\}_{av}. \quad (2)$$

This equation shows explicitly the contributions to the linewidth from inelastic electron collisions by summing over the cross sections of electron collisions between levels involving the upper ($\sigma_{uu'}$) and lower ($\sigma_{ll'}$) states of the transition of interest, and from a term taking into account elastic scattering by subtracting the scattering amplitudes Φ of the initial and final level and integrating over the scattering angle $d\Omega$. In most practical cases the average is over a Maxwell-Boltzmann velocity distribution function.

Equation (2) shows the correlation between spectral line shape calculations and atomic collision theory. Some recent progress on line broadening calculations came from this field where collision strength or cross sections are now routinely calculated with a variety of advanced methods. In particular, within the Opacity Project close-coupling calculations have been performed to calculate collision strengths and line widths for a large number of transitions. These calculations are expected to be the most accurate theoretical data because semiempirical (12) or semiclassical (10,13,14) approximations use *ad hoc* estimations for the effective Gaunt factor to account for elastic electron collisions. Further, similar crude assumptions are also necessary within a semiclassical theory to account for strong collisions (8,10,13). On the other hand, there are a lack of reliable experiments testing close-coupling calculations. It is obvious from Equation (2) that measurements of spectral line widths from isolated nonhydrogenic ions in well-diagnosed plasmas can provide critical tests of calculations of effective (elastic and inelastic) cross sections. Some experiments (15,16) have been performed in discharge tubes which were diagnosed with interferometry. But, as pointed out by Seaton (7) the experimental data were not accurate or convincing enough to provide a critical test for close-coupling calculations.

For that reason we have performed new experiments (17,18) with the well-diagnosed gas-liner pinch where electron densities and temperatures are determined independently with Thomson scattering. Furthermore, no assumptions about homogeneity of the plasma or radiative transport effects are necessary, because we have been able to directly measure electron and emitter density distributions in the plasma. Our measurements include the $2s - 2p$ resonance transitions in B III since their line profiles are excellent test objects of close-couplng calculations (18). It is the simplest system to calculate since there is only one electron outside the first shell. Moreover, for these transitions between lowly excited states the inclusion of only a small set of perturbing levels should lead to high accuracy.

For especially broad spectral lines the duration of a given perturbation assumes the same order as the decay time of the autocorrelation function of the light amplitude, and no general solution of the spectral line profile similar to Equation (2) can

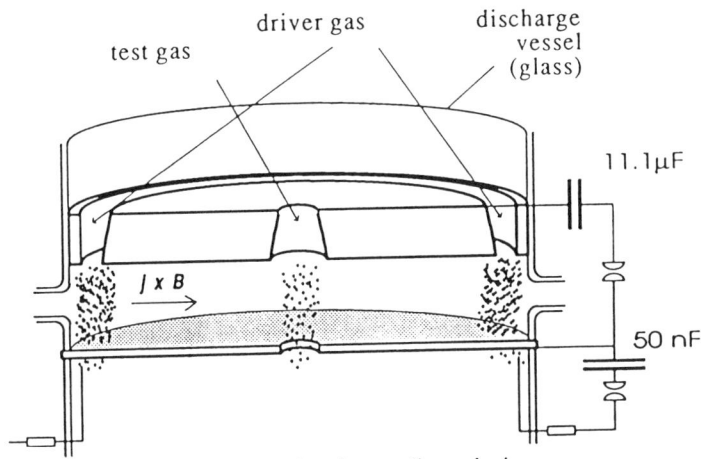

Figure 1. Schematic of a gas-liner pinch.

be given. In particular, broadening by ion microfields becomes more important and in the limit where Equation (1) is reversed, the quasi-static (ion) approximation can be applied (10,19). More often, however, Equation (1) is only reversed for the wings of the line profile and for the central part of the profile ion dynamic corrections have to be taken into account. For example, this is the case for some non-isolated $\Delta n = 1$ transitions of the lithium-like ions C IV, N V, and O VI for transitions of more highly excited states which have close-lying perturbing levels (20,21). We tested complete spectral line profile calculations which were performed for the independently measured plasma parameters of our experiment.

EXPERIMENT AND DIAGNOSTIC TECHNIQUES

GAS-LINER PINCH. A gas-liner pinch resembles a large aspect ratio z-pinch characterized by two independent fast gas inlet systems (22). Figure 1 shows the experimental setup. For the present investigations we used hydrogen or helium as driver gas. It is injected through an annular nozzle into the vacuum chamber by a fast electromagnetic valve. The diameter of the vacuum chamber is 18 cm and the electrode separation is 5 cm. The gas forms a hollow gas cylinder near the wall before we preionize it by discharging a 50 nF-capacitor (charged to 20 kV) between 50 annually mounted needles and the lower cathode. Finally, the discharge of the main capacitor (capacitance 11.1 μF, voltage 25-35 kV) compresses the gas on axis to a plasma column of 1-2 cm diameter and 5 cm length. Typical electron densities and temperatures reached on the axis are between $0.5 < n_e < 4 \times 10^{18} \mathrm{cm}^{-3}$ and $7.5 < k_B T_e < 50$ eV which is sufficiently hot and dense to produce multiply ionized atoms with Stark broadening as the dominant broadening mechanism of their emission lines. The compression time and the life time of the plasma depend on the discharge conditions and for the present studies they are about 2.5 μs and 0.5 μs, respectively.

The atomic species of interest for spectroscopic measurements are introduced as (test) gas into the discharge tube by a second independently operating fast electromagnetic valve. The gas is injected through a nozzle in the center of the upper electrode and is dissociated and ionized by the imploding driver gas and by ohmic heating. For the present studies we used BF_3, CH_4, N_2, CO_2, a mixture of 10 % SF_6 in hydrogen or Ne as test gas in order to produce the lithium-like ions B III, C IV, N V, O VI, F VII, and Ne VIII. Their radial emission is observed with several spectrometers for the visible and vacuum ultraviolet spectral range. Gated microchannel plates and optical multichannel analyzers or charge-coupled devices were used to perform all measurements with a time resolution of 20 - 30 ns.

THOMSON SCATTERING. An independent and accurate measurement of all relevant plasma parameters is a prerequisite for a critical test of line shape theories. For this purpose we focussed a pulse of a Q-switch driven Ruby laser (2 J, 30 ns) into the center of the plasma column and observed the scattered light at an angle of $\theta = 90°$. This arrangement gives typical values for the scattering parameter $\alpha = 1/(k\lambda_D) > 1$. In this regime light is predominantly scattered into a narrow ion feature which could be detected spectrally resolved with a 1-m spectrometer and a gated optical multichannel analyzer. Scattering occurs on electrons which are bunched in the Debye spheres of the ions, and from the width of the scattering spectrum the temperature of the ions is obtained. Furthermore, on the wings of the ion feature, heavily-damped ion acoustic waves determine the shape of the scattering spectrum. Since the phase velocity and the damping of ion acoustic waves depend on the ion and on the electron temperature, an accurate measurement of the shape of the scattering spectrum also yields the electron temperature.

By calibrating the detection system absolutely by Rayleigh scattering on propane the electron density is deduced from the intensity of the scattering spectrum. For the present investigations multiply ionized test gas atoms are added to the plasma column which is predominantly formed by the driver gas. Hence, we fit the theoretical form factor of Evans (23) to the measured scattering spectra. This form factor calculates the scattering spectra for a plasma composed of different ionic species. We gave a complete discussion applying this form factor to deduce the plasma parameters in Ref. (24).

In general, we obtain that the temperatures of all species are equal within the stated error of about 15 %. This finding is expected from estimates of the electron-ion collision time (25) resulting in typical values of about 3 ns. This is much smaller than typical time scales on which plasma parameters change after stagnation on the axis, i.e. after the maximum compression of the plasma. Figure 2 shows an example of a measured Thomson scattering spectrum along with the fitted form factor. This measurement was performed with hydrogen as driver gas and borontrifluoride as test gas. An impurity peak from test gas ions, which was easily observed in Refs. (26,27) when using larger amounts of test gas and more highly ionized species, can not be identified in Fig. 2. This is because only very small amounts of borontrifluoride have been used for the investigation of the line profiles of the $2s - 2p$ resonance transitions

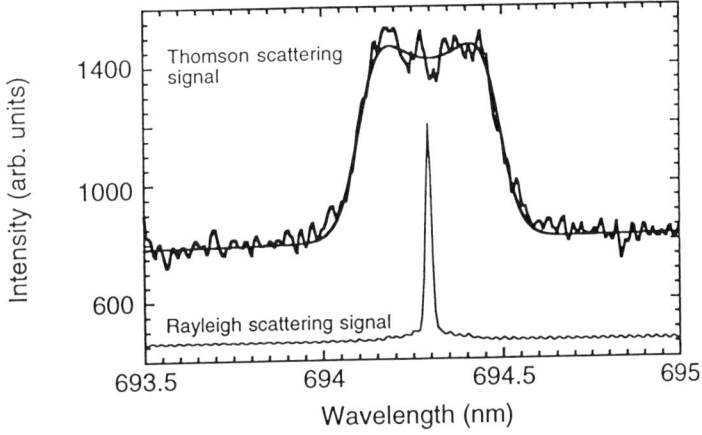

Figure 2. Example of a Thomson scattering spectrum detected 50 ns after maximum pinch compression along with the fit.

in order to avoid self-absorption. An upper limit for the test gas ion concentration n_t can be determined: $n_t < 0.3$ % of the electron density.

In Fig. 2 we also show a Rayleigh scattering spectrum obtained from light scattering on propane. We performed these type of calibration (about 20 measurements) after each shot day and monitor the small changes of the sensitivity of the detection system over several months resulting in a high reliability of the calibration. These measurements are carried out in the pulsed mode of the detector, or in other words in the same configuration as the Thomson scattering or the line broadening setup. Therefore, the Rayleigh scattering signal directly provides a measurement of the instrument function of the spectrometers for the visible spectral range: a Voigt function with 0.0071 nm Lorentzian FWHM and 0.0049 nm Gaussian FWHM. As usual we perform Rayleigh scattering for a variety of different propane gas densities, and besides checking the linearity of the scattering and of the detection system, the instrument function is obtained for the full dynamic range of the detector. We should mention that the instrument function obtained in this way is in excellent agreement with the measurement performed in the cw mode of the detector employing cold spectral lamps and Fe and Al hollow cathode lamps.

PLASMA HOMOGENEITY AND RADIATIVE TRANSPORT. Very favorable plasma conditions for line broadening studies are achieved when the injection of the gases into the discharge chamber is properly timed and only when very small amounts of test gas (about 1 % of the density of the driver gas) are used. The test gas ions become confined in the center of the plasma column where the plasma is homogeneous in radial and axial direction.

Experimentally we have verified the plasma homogeneity with various methods (17,27,28). Electron densities from Thomson scattering as a function of the radius

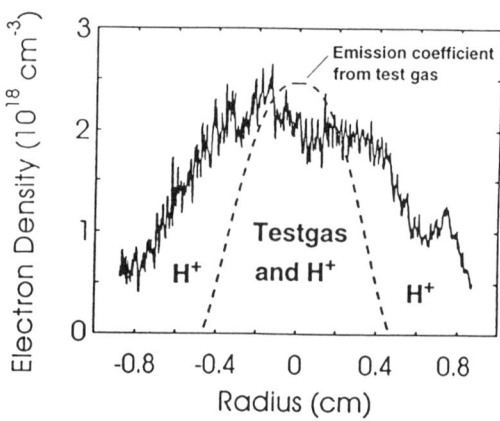

Figure 3. Electron density as a function of the radius of the discharge. The dashed line represents the emission coefficient for line radiation emitted from test gas ions.

show that the emission from test gas ions originates from a homogeneous center of the plasma column with about 1 cm diameter. Figure 3 shows a result obtained shortly after maximum pinch compression. Also shown is the emission coefficient of line radiation from test gas ions obtained after Abel inversion.

It is further of interest to verify the homogeneity of the plasma column in axial direction because magnetohydrodynamic instabilities could arise. We measured the homogeneity along the axis of the discharge in two different experiments (25). In Fig. 4 we show the Stark-broadened $n = 5$ to $n = 4$ transitions of F VII at $\lambda = 82.5$ nm. These linewidths are very sensitive to changes of the electron density but insensitive to temperature variations. Temporally and axially resolved spectra have been detected from single discharges at various times in the discharge with a MCP-CCD system. Figure 4 shows an example of a measurement at maximum pinch compression. About 1 cm of the 5 cm long plasma column is shown with a resolution of 0.17 mm. In order to derive the electron density from the line profiles we determined the full-widths at half maximum of the transitions as a function of the height of the plasma with a resolution of 0.5 mm. We converted the linewidth obtained in this way into electron densities using spectral line broadening calculations (see Ref. (9)) and which are tested below. The rms value is 14 % of the mean value of the electron density. The homogeneity of the discharge is evident.

For transitions with highly populated lower levels self-absorption is a serious problem which leads to line profile distortions. In particular, this is true for resonance transitions. Since at our experimental conditions the plasma is homogeneous and a cold boundary layer of the investigated ions is effectively absent, radiative transport effects of the emission lines are easily controlled by varying the amount of test gas. In fact, optically thin plasma conditions for all line broadening measurements were achieved. This is expected from the measurement of the emitter density with Thomson scattering which is, e.g., about 0.3 % of the electron density for the

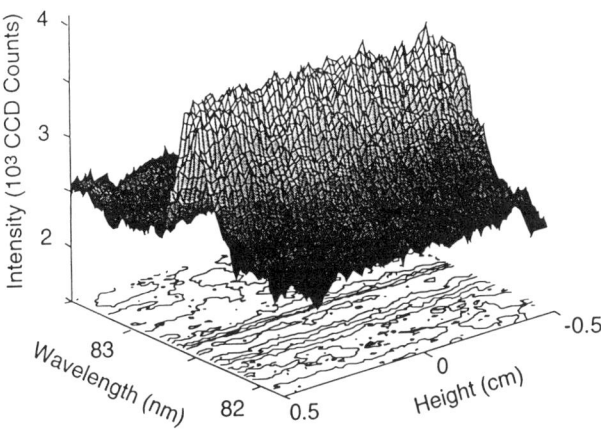

Figure 4. Example of a MCP-CCD measurement of the $n = 5$ - $n = 4$ transitions in F VII as a function of the height of the pinch.

boron measurements. We further verified that radiative transport is negligible by measuring relative line intensities within multiplets for different plasma conditions and test gas ion concentrations. Since the high particle densities of a gas-liner pinch plasma result in sufficiently high collision rates, level population densities within a multiplet are given by the Boltzmann statistics. Hence, we compare measured line intensities with the predictions of the LS-coupling approximation.

We find for the investigated Li-like ions B III – Ne VIII that the measured line intensity ratios of the $2s - 2p$ or $3s - 3p$ transitions are in good agreement with the predictions of the LS-coupling approximation (see also Figure 5). The predicted ratios are within the measured uncertainties of about 3 %. Also, increasing the test gas concentration by a factor of 2 did not affect this result.

EXPERIMENTAL RESULTS AND DISCUSSION

Figure 5 shows an example of a spectrum of the $2s - 2p$ resonance transitions of B III detected 50 ns after maximum pinch compression. Also shown is a fitted Voigt function which takes into account the instrument function and Doppler broadening in the following way: the measured instrument profile (see Fig. 2) is convoluted with a Doppler profile which was calculated for each plasma condition according to the measured temperature. The resulting profile is convoluted with a Lorentzian with a variable width giving the Stark broadening. Both multiplet lines are fitted independently giving for each spectrum the same width for both components within 5 %. For each spectrum the relative intensities of both components is 2 : 1 ± 4 %, in excellent agreement with the LS-coupling approximation predicting 2.00 : 1.00. Both facts show that there are no unwanted detector distortions for the high intensity line and that the measurement is not affected by self-absorption. For the condition in Fig. 5 we obtain for the Stark broadening $w_m = 0.022$ nm ± 11.5 %.

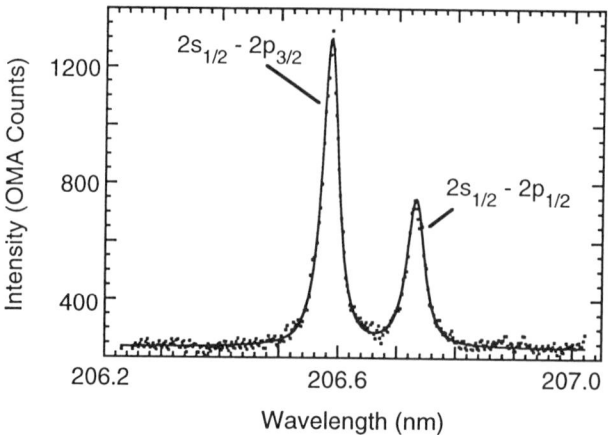

Figure 5. Experimental spectrum of the $2s - 2p$ multiplet of B III measured 50 ns after maximum pinch compression; ■, experimental data; —, Voigt function best fit. Electron densities and temperatures are obtained from Thomson scattering.

For the narrow resonance transitions of B III sufficiently high resolution of the detection system is obtained by going to the 6th spectral order and using an interference filter to suppress the radiation from all other orders. The filter has equal transmission over the wavelength range of interest as proven with hydrogen continuum measurements. For our experimental conditions the instrument and the Doppler profile together account for 33 % of the measured linewidth. Hence, uncertainties in the measurement of these quantities increases the error of the Stark broadening measurement. The instrument profile is more critical because it contributes directly to the Lorentzian contribution of the measured spectra. Fortunately, the instrument profile is measured very accurately and on a frequent basis, and therefore, only the experimental error in the determination of the temperature of the plasma results in an uncertainty of the Gaussian contribution of the measured line profiles. Since temperatures are also measured with high accuracy with Thomson scattering and since the Gaussian contribution is less important for the width of the resulting Voigt functions, the additional error from the Gaussian contribution is only 2 %. This error is added to the rms value of the fitted Stark profiles of 10 spectra measured at the same plasma condition. It results in a total error for the experimental Stark width of 13.5 %.

Stark width are usually calculated for a series of temperatures and for one value of the electron density. This is because it is a fundamental concept of the impact theory that the contribution to the linewidth owing to electronic or ionic collisions is proportional to their number density (10,11) (see Equation (2)). Indeed, a linear scaling of linewidths of nonhydrogenic ions has been proven in a number of experiments (10,15). In practical situations, deviations from a linear scaling may be possible due to Debye shielding effects (10,29) which are, however, unimportant for this study (30). For that reason we can scale the experimental or theoretical Stark

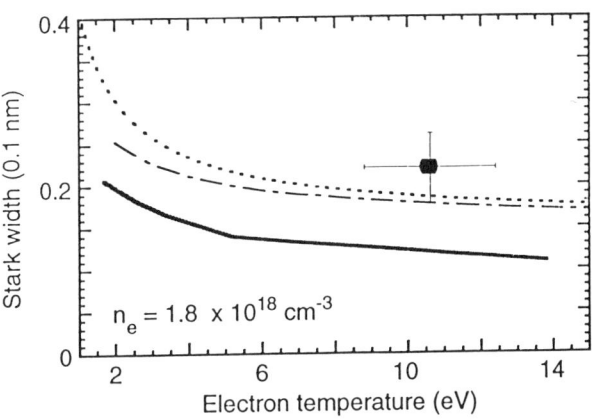

Figure 6. Comparison of the experimental Stark width (FWHM) of the $2s - 2p$ transitions in B III with the results of the theoretical approximations after Griem (10): ····, by Hey and Breger (13): - · - · -, and from Seaton (7): —— . The vertical error bar takes into account the errors of the measured line profiles, of the convoluted Doppler profile, and of the electron density determination.

width linearly to one value of the electron density. Hence, for a comparison with theory the experimental errors of the measured electron density have to be taken into account. Plasma parameters are measured independently with Thomson scattering, and for the B III study the rms values result in a relative error in the density and in the temperature of 13 % and 16 %, respectively. Since the measurements of the linewidths and of the electron densities are independent and statistical, combining both errors for a comparison with theory amount to 20 % of the Stark width.

In Figure 6 we compare our experimental result to various theoretical data which are plotted as a function of the electron temperature. The measured electron density is $n_e = 1.81 \times 10^{18} cm^{-3}$. The results of two semiclassical approximations according to Griem (10) and to Hey and Breger (13), and close-coupling calculations from Seaton (7) are shown. These theoretical approximations calculate only electron-impact widths. The semiclassical approximation after Griem (10) calculates inelastic cross sections with good accuracy for impact parameters larger than the extent of the wavefunction of the perturber, i.e. in a regime where the semiclassical approximation is valid. Collisions with smaller impact parameters are estimated with a Lorentz-Weisskopf term and give the so-called strong collision term which is 12 % of the total calculated width. A maximum impact parameter of the order of the Debye length, and elastic contributions are included according to an extrapolation procedure below threshold. Hey and Breger (13) chose different procedures from those of Griem (10) to calculate the minimum and maximum impact parameters of the electron collision process. Although their strong collision term is appreciably larger, even for the relatively low charge state of the studied ion, their resulting widths differ only slightly from that of Griem. Both semiclassical calculations agree with the experiments within the 20 % error bar.

On the other hand, the close-coupling calculations give results which are smaller than the experiment by a factor of about two. This discrepancy can not be explained by other broadening mechanisms besides electron-impact broadening because those broadening mechanisms can account for only a few percent of the measured width. Ion broadening is almost negligible for the resonance lines of B III. Griem gives a relatively large rough estimate of the ion quadrupole broadening (Eq. 218b, Ref. (10)). This term is about 10 % of the calculated electron-impact width but is not included in the comparison of Figure 6 because it is too roughly known. More recent detailed calculations of the ion quadrupole broadening of the $3s - 3p$ transitions of the Li-like ions C IV – Ne VIII from Ref. (31) show that this additional broadening contribution can be appreciably smaller than Griem's estimate and is smaller than 5 % of the electron-impact width. Dipole-allowed proton collisions are completely negligible for the B III experiment (8,10), as is Zeeman broadening.

Moreover, we consistently find that semiclassical Stark width calculations are in better agreement with our experimental results than close-coupling calculations. We performed a detailed study for the $3s - 3p$ transitions in the Li-like ions C IV – Ne VIII (17,27). Our experimental results scaled to one value for the electron density and temperature, i.e. $n_e = 1.8 \times 10^{18}$ cm^{-3} and $k_B T_e = 12.5$ eV, are shown in Fig. 7. The error bars include the error from the width measurement, the electron density and temperature. Also, an error estimate for the scaling of the data is taken into account (17). For Ne VIII two experiments have been performed because for this transition proton collision broadening plays a role. One data point is obtained from experiments with hydrogen and one with helium as driver gas. The difference between both values gives an estimate for the magnitude of the proton collision contribution to the linewidth. We find for all transitions that the five-state close-coupling calculations give results which are too small to explain the measured widths. This finding might be partially explained by the fact that only five states are included in the calculations ($n = 2$ and $n = 3$ levels). While the $n = 4$ states do not contribute to the Stark broadening of the $2s - 2p$ resonance transitions in B III, they are expected to play a role for the $3s - 3p$ transitions in Li-like ions. Indeed, the nine-state close-coupling calculations from Burke (33) give an improved value for the Stark width of C IV which is marginally inside of the error bar of the experiment. However, the overall comparison of quantum-mechanical calculations with our experiments is not satisfactory, while comparisons with semiclassical calculations show generally good agreement. Until recently there have been problems explaining the experimental Stark width of the higher ionized emitters. A simplified Z^{-2} scaling as predicted by some theoretical approximations could not be verified by our experiments (see also Ref. (28)). On the other hand, the new semiclassical method of Alexiou (32,34) gives good agreement with the experiment for all investigated transitions. As opposed to other theoretical approximations Alexiou calculates the collision operator exactly for impact parameters in the region where unitarity is violated but the semiclassical approximation is still valid. This is done by a fully numerical solution of the Schrödinger equation. In fact, this method gives excellent agreement with our experimental data.

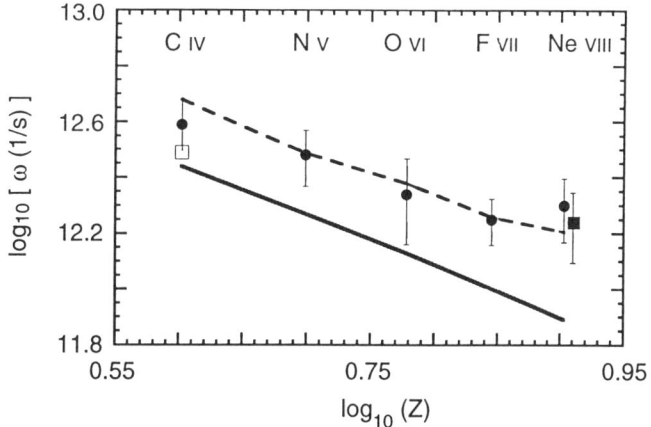

Figure 7. Experimental and theoretical Stark width in frequency units of the $3s\,^2S - 3p\,^2P^o$ transition in C IV – Ne VIII as function of Z: —, calculations by Seaton (7) (the values for N V and F VII are interpolated); □, calculated by Burke (33) ; - - -, calculations by Alexiou (8,31,32); experimental values (17,27): •, measured with hydrogen as driver gas; ■, measured with helium as driver gas.

Inaccuracies when deriving Stark widths from close-coupling calculations may arise because of large cancellations of direct and mixed terms (16) when calculating $(1 - S_a S_b)$. Further approximations are the use of Hartree-Fock target states and the so-called top-up procedure where contributions from orbital quantum numbers $L > 16$ are only roughly estimated. An indication that these features indeed lead to problems in the calculation is the fact that two close-coupling calculations of the Stark broadening of the $2s - 2p$ resonance transitions in Be II performed by two different authors differ by a factor of 1.4 for low temperatures. On the other hand, we have the success of the semiclassical calculations. For example, from a large number of experiments with a broad range of experimental errors Konjević and Wiese (15), and Griem (10) find that the semiclassical approximation of Griem describes the Stark broadening of isolated spectral lines of two and three times ionized nonhydrogenic emitters within 50 %. From 28 independent experiments (17,28,30) with the well-diagnosed gas-liner pinch we find an even better agreement with the semiclassical calculations after Griem (10)

$$\frac{w_m}{w_G} = 0.92 \pm 26\ \% \tag{3}$$

and after Hey and Breger (13)

$$\frac{w_m}{w_{HB}} = 1.15 \pm 29\ \%. \tag{4}$$

So far we were concerned with isolated spectral lines of nonhydrogenic ions. Re-

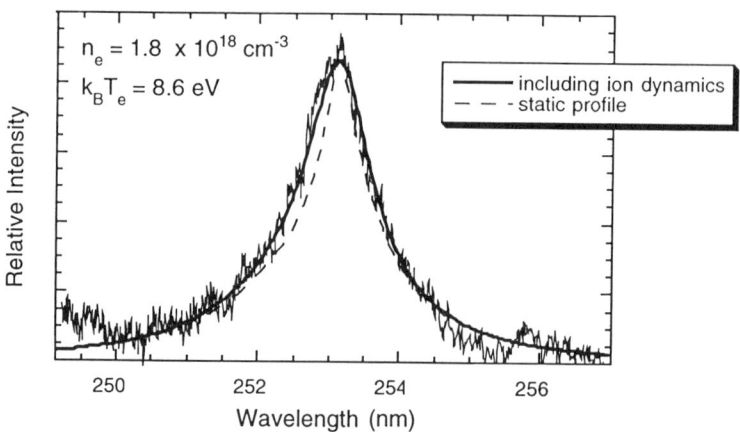

Figure 8. Experimental spectrum of the $n = 5$ to $n = 4$ transitions in C IV compared with independent calculations including static (- - -) and dynamic (—) ions. Electron densities and temperatures are obtained from Thomson scattering.

cently, we measured detailed line shapes of non-isolated spectral lines of lithium-like ions going to $\Delta n = 1$ transitions of more highly exited states with close-lying perturbing levels (21). In particular, we measured the $4f - 3d$ and $4d - 3p$ transitions of C IV and N V, and the $n = 5$ to $n = 4$ transitions in C IV, N V, and O VI. Due to the electric fields in the plasma the wave functions of the atomic states mix and become degenerate. Therefore, ion broadening dominates the line broadening. For these spectra forbidden transitions and overlapping of various $(n - 1, \ell') - (n, \ell)$ transitions occur. Furthermore, ion-collisional effects at the center of the transitions, where the quasi-static ion approximation breaks down, have to be taken into consideration. Independent line profile calculations for the electron densities and temperatures measured with Thomson scattering have been performed taking into account these effects. A comparison with our experimental line profiles is shown in Figure 8 where the $n = 5$ to $n = 4$ transitions of C IV at 252.7 nm are plotted together with a calculated static and dynamic line profile. It is obvious that the latter, where ion dynamic effects are calculated using the field fluctuation model, is in much better agreement than the static profile. In this case the ion dynamic effect accounts for 25 % of the linewidth. From nine experiments performed with the gas-liner pinch we find that the field fluctuation model agrees excellent with the experimental line widths

$$\frac{w_m}{w_{ffm}} = 0.99 \pm 5 \ \%. \tag{5}$$

In particular, in the central part of the transitions, agreement between the calculations and the experiments is obtained only when including ion dynamics, which accounts for 10 - 25 % of the linewidths. While ion dynamic effects have been found earlier to be important for hydrogen spectral lines this is the first decisive verification that ion dynamic effects can be also important for more highly ionized emitters.

CONCLUSIONS

For electron collisional broadened spectral lines we find that our experimental data of several studies clearly favor semiclassical calculations over five-state quantum-mechanical close-coupling calculations. In particular, the semiclassical calculations of Griem (10) and of Hey and Breger (13) agree well with the experiment for lowly ionized species. For more highly ionized emitters only the recent improvements of the semiclassical approximation (32,34) lead to satisfying agreement with our experimental data (17,34). Our experiments also include the best test case for close-coupling calculations: the $2s - 2p$ resonance transitions in the lithiumlike ion B III where a factor of two discrepancy between the close-coupling calculations and experiment is found (18). Since *a priori* the close-coupling calculations are considered to be most accurate it is imperative to understand this failure of close-coupling calculations before continuing to calculate more complicated systems.

For transitions where ion broadening dominates extensive studies show that the field fluctuation model developed in Refs. (4,9) agrees well with our experiments (20,21). In particular, we find for the first time that ion dynamic effects are important to describe the line profiles of more highly ionized emitters (21).

ACKNOWLEDGMENTS

I would like to thank H.-J. Kunze for many valuable discussions. I further thank C. A. Back, R. W. Lee, Th. Wrubel, and S. Büscher for helpful criticism. This research was supported by the Sonderforschungsbereich 191 of the DFG.

REFERENCES

1. Seaton M. J., J. Phys. B **20**, 6363–6378 (1987).
2. Rogers F. J. and Iglesias C. A., Astrophys. J. Suppl. Series, **79**, 507–568 (1992)
3. Koch J. A., MacGowan B. J, Da Silva L. B., Matthews D. L., Underwood J. H, Batson P. J., and Mrowka S., Phys. Rev. Lett. **68**, 3291–3294 (1992).
4. Keane C. J., Lee R. W., Hammel B. A., Osterheld A. L., Suter L. J, Calisti A., Khelfaoui F., Stamm R., and Talin B., Rev. Sci. Instrum. **61**, 2780–2783 (1990).
5. Hammel B. A., Keane C. J., Cable M. D., Kania D. R., Kilkenny J. D., Lee R. W., and Pasha R., Phys. Rev. Lett. **70**, 1263–1266 (1993). Hammel B. A., Keane C. J., Dittrich T. R., Kania D. R., Kilkenny J. D., Lee R. W., and Levedahl W. K., J. Quant. Spectrosc. Radiat. Transfer **51**, 113–124 (1994). N. Woolsey et al., *submitted to Phys. Rev. E*
6. Griem H. R., Phys. Fluids B **4**, 2346–2361 (1992).
7. Seaton M. J., J. Phys. B **21**, 3033–3053 (1988).
8. Alexiou S., Phys. Rev. A **49**, 106–119 (1994).
9. Talin B., Calisti A., Godbert L., Stamm R., and Lee R. W., Phys. Rev. A **51** 1918–1928 (1995).
10. Griem H. R., *Spectral Line Broadening by Plasmas* (Academic, New York, 1974).

11. Baranger M., in *Atomic and Molecular Processes*, edited by Bates D. R. (Academic, New York, 1962).
12. Griem H. R., Phys. Rev. **165**, 258–266 (1968).
13. Hey J. D. and Breger P., S. Afr. J. Phys. **5**, 111–121 (1982), J. Quant. Spectrosc. Radiat. Transfer **23**, 311–321 (1980), **24**, 349–364 (1980), **24**, 427–439 (1980), also in *Spectral Line Shapes*, edited by B. Wende (Walter de Gruyter, Berlin, 1981) pp. 191–200.
14. Dimitrijević M. S. and Sahal-Bréchot S., Astron. Astrophys. Suppl. Series, **93**, 359–371 (1992), **95**, 109–120 (1992), **107**, 349–351 (1994).
15. Konjević N. and Wiese W. L., J. Phys. Chem. Ref. Data **19**, 1307–1385 (1990).
16. Sanchez A., Blaha M., and Jones W. W., Phys. Rev. A **8**, 774–780 (1973).
17. Glenzer S., Uzelac N. I., and Kunze H.-J., Phys. Rev. E **45**, 8795–8802 (1992), also in *Spectral Line Shapes*, edited by Stamm R. and Talin B. (Nova Sci. Commack, New York, 1993), pp. 119–120
18. Glenzer S. and Kunze H.-J., Phys. Rev. A (1996) *in print*.
19. Lee R. W., in *Atomic Processes in Plasmas* eds. Marmor E. S. and Terry J. L. (AIP Conference Proceedings No. 257, New York, 1991) pp. 39–57.
20. Godbert L., Calisti A., Stamm R., Talin B., Glenzer S., Kunze H.-J., Nash J., Lee R. W., and Klein L., Phys. Rev. E **49**, 5889–5892 (1994).
21. Glenzer S., Wrubel Th., Büscher S., Kunze H.-J., Godbert L., Calisti A., Stamm R., Talin B., Nash J., Lee R. W., and Klein L., J. Phys. B **27**, 5507–5515 (1994).
22. Kunze H.-J., in *Spectral Line Shapes*, ed. Exton R. J. (Deepak, Hampton, Virginia, 1987) pp. 23–35.
23. Evans D. E., Plasma Phys. **12**, 573–584 (1970).
24. Wrubel Th., Glenzer S., Büscher S., and Kunze H.-J., J. Atmos. Terr. Phys. (1996) *in print*.
25. Spitzer L., Jr., *Physics of Fully Ionized Gases* (Interscience, New York, 1962).
26. DeSilva A. W., Baig T. J., Olivares I., and Kunze H.-J., Phys. Fluids B **4**, 458–464 (1992).
27. Glenzer S., in *Spectral Line Shapes*, eds. May D., Drummond R., and Oks E. (AIP Conference Proceedings No. 328, New York, 1995) pp. 134–150.
28. Glenzer S., Hey J. D., and Kunze H.-J., J. Phys. B **27**, 413–422 (1994).
29. Büscher S., Glenzer S., Wrubel Th., and Kunze H.-J., J. Quant. Spectrosc. Radiat. Transfer **54**, 73–80 (1995).
30. Uzelac N. I., Glenzer S., Konjević N., Hey J. D., and Kunze H.-J., Phys. Rev. E **47**, 3623–3630 (1993).
31. Alexiou S. and Ralchenko Yu., Phys. Rev. A **49**, 3086–3088 (1994), **50**, 3553 (1994).
32. Alexiou S., Phys. Rev. Lett. **79**, 3406–3409 (1995).
33. Burke V. M., J. Phys. B **25**, 4917–4928 (1992).
34. Wrubel Th., Glenzer S., Büscher S., Kunze H.-J., and Alexiou S., Astron. Astrophys. (1996) *in print*.

Spectroscopic Temperature Measurements of Non-Equilibrium Plasmas

C. A. Back, S. H. Glenzer, R. W. Lee, B. J. MacGowan,
J. C. Moreno, J. K. Nash, L. V. Powers, and T. D. Shepard

Lawrence Livermore National Laboratory, P.O. Box 808, Livermore, CA 94551

The characterization of laser-produced plasmas has required the application of spectroscopic techniques to non-standard conditions where kinetics models have not been extensively tested. The plasmas are produced by the Nova laser for the study of inertial confinement fusion, can be mm in size, and evolve on sub-nanosecond time scales. These targets typically achieve electron temperatures from 2 - 4 keV and electron densities of 10^{20} - 10^{22} cm^{-3}. We have measured the electron temperature of two types of targets: bags of gas and hohlraums, Au cylinders with laser entrance holes in the flat ends. By comparing data from different targets, we examine the time-dependence of spectroscopic plasma diagnostics.

Spectroscopy has evolved as an important diagnostic of laser-produced plasmas. Discrete transitions allow identification of different ionic species formed in the plasma by the wavelength of their emission. As the description of plasma constituents and their atomic details have become better understood, emission and absorption spectroscopy have been used to measure electron temperature, ionic temperature, electron density, and ionic populations (1).

In inertial confinement fusion research, lasers can produce extreme plasma conditions such as the high temperatures and densities found in stars. Since the hydrodynamic evolution of the plasma takes place in extremely short time scales, diagnostics must be applied to difficult, non-steady state conditions. Furthermore, radiation fluxes of laboratory plasmas produce a regime in which collisional-radiative plasma models have not been well-tested. In such plasmas, the spectroscopic techniques must be carefully applied.

In these experiments, spectroscopy is used to diagnose the electron temperature, T_e, of mm-size plasmas. Measurements of T_e enable a check of energy absorption, electron conduction, and overall energy balance in the hydrodynamic simulations. For example, parametric plasma processes are dependent on T_e and scattering of laser light can lead to a significant loss of energy over the mm-long laser path lengths (2). Hence, measurements of T_e give a direct measure of our success in modeling the hohlraum environment (3).

Time-dependent analysis of laser-produced plasmas can be important because the plasmas are created by ns-long pulses. These plasmas ionize rapidly and can easily achieve $T_e > 1$ keV. However, the heating can create plasmas having

© 1996 American Institute of Physics

densities too low to assure that a steady state model of the plasma kinetics is sufficient. The densities of the plasmas in this study can be varied by changes in the gas fill, thus diagnosing a variety of targets allows us to test the time-dependence of the models.

MEASUREMENT OF T_e

Two types of target geometries are used in this study: bags of gas, which may be treated with a virtually time-independent analysis, and hohlraums, which require time-dependent analysis. The gasbags are formed by inflating CH membranes which are glued to both sides of a 2.75 mm diameter washer. The gas-filled hohlraums are Au cylinders 2.5 mm long and 1.6 mm in diameter. Thin CH membranes are placed over the endcaps of the cylinder to prevent the gas from escaping. Figure 1 shows a schematic of the two target geometries.

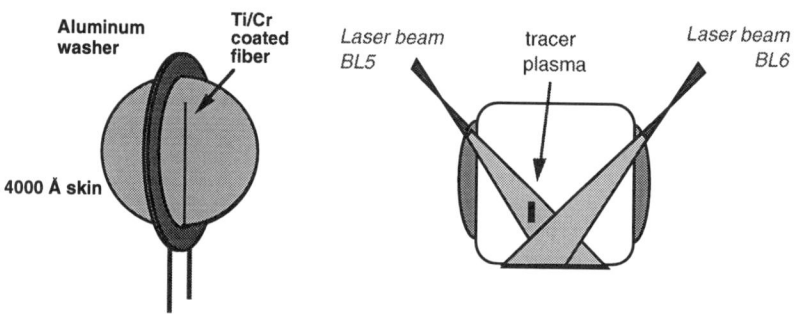

FIGURE 1. The gasbag is formed by confining gas between two thin CH membranes. The plasma is formed by overlapping five laser beams on each membrane. The gas-filled hohlraum is an Au cylinder in which the laser beams enter through the endcaps which are covered by a thin CH membrane to enclose the gas.

The plasmas are produced by laser-irradiating the targets with approximately 20 kJ of laser energy. The laser ionizes the gas-filled target to form a plasma with an electron density, n_e, determined by the gas fill. For instance, neopentane, C_5H_{12}, filled to a pressure of 1 atm provides a density of 10^{21} cm^{-3} when it is completely ionized. Time-resolved x-ray diagnostics are used to monitor the plasma's volume, position, and spectra. These diagnostics provide a check of the density in the target because the X-ray images enable a measure of the expansion of the spectral emitter.

X-ray spectra used to diagnose the target are emitted from mid-Z dopants. These are deposited on a thin CH substrate that is inserted into the target at a chosen position. The dopant coatings are typically only 2000 Å to insure that their emission for the wavelength of interest is optically thin. To diagnose T_e, we use the ratio of the He-like lines from two different elements or the He-like lines and their Li-like satellites. An example of the x-ray spectrum of a mixed Ti and Cr dopant inside these targets is shown in Fig 2. The spectral lines identified in the figure are the two He-like n =3-1 transitions that are used in forming an isoelectronic line intensity ratio.

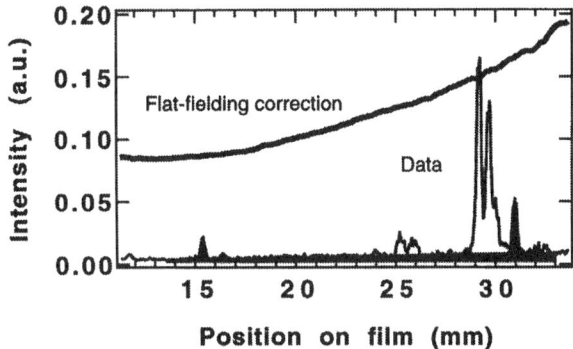

FIGURE 2. X-ray spectra emitted by the plasma of a typical dopant located inside a gas-filled target. The flat-fielding of the spectrometer is an important correction to the spectra and is measured for each instrument.

Because the temporal dependence is important, the spectra are recorded on gated microchannel plates which provide resolution in space and time (4). These instruments depend on fast electronics and this introduces two considerations for the measurements. First, the dynamic range of the instrument is approximately 40. This must be taken into consideration for choosing the optimal spectroscopic diagnostic. Second, the instrument detector response changes over the length of the data recording strip (5). A typical instrument can decrease by a factor of 2 in gain. For this reason, calibration of the instrument is performed for each microchannel plate and the response is unfolded from the measurement. An example of the detector response is shown in Figure 2.

Analysis of the spectrum relies on a collisional-radiative model of the plasma. In these examples we use the code, FLY, which sets up the rate matrix and solves for populations of K-shell ions (6). To look at the details more in-depth, we will consider a hypothetical test case to illustrate the behavior of the spectroscopic diagnostics, then we will show the results from three different targets. Depending on the target, differences of up to 30 % are observed between the steady-state and non steady-state analysis.

MODEL T_e SLOPE

For the test case, we consider a plasma that has a constant rise in T_e and a constant electron density which is a simplified representation of plasma conditions expected from the targets we will discuss later. Figure 3 shows a temperature slope that rises from 500 to 3500 eV in 600 ps. Shown on the left hand abssissa is the ratio predicted for three possible spectroscopic diagnostics. In the figure a commonly used ratio, the He-like $n=2-1$ resonance line to its Li-like satellites, He-like α /jkl , shoots up from 0 to 45 for Ar. Here, the plasma is hot enough to quickly strip through the Li-like ionization stage. This results in a very large ratio that requires a dynamic range that can surpass that of gated microchannel plate detectors for the highest temperatures. Another typical ratio, the Ly α to the He α, has a significant temporal lag during the time frame of interest because of the temporal ionization lag to the hydrogen-like stage. The third ratio shown is the isoelectronic line ratio of Cr He-like $n=3-1$ to the Ti He-like $n=3-1$ transitions. It follows the input T_e temporal history rise very well and is easily within the sensitivity of the diagnostic. This case illustrates some of the advantages of using an isoelectronic line ratio which is discussed in other papers (7).

The test case was also calculated for different densities. The shaded area in figure 3 show the limits for densities of 2×10^{21} - 5×10^{20} cm^{-3} for the isoelectronic ratio only. For the laser-produced plasmas in this study, the range of densities expected varies by a factor of 2. Over this range, the ratio changes by a value of 10 % which translates to a change of over 15 % for the T_e itself. This relatively small change enables us to investigate the temporal dependence of the line ratios without introducing significant errors due to possible density changes.

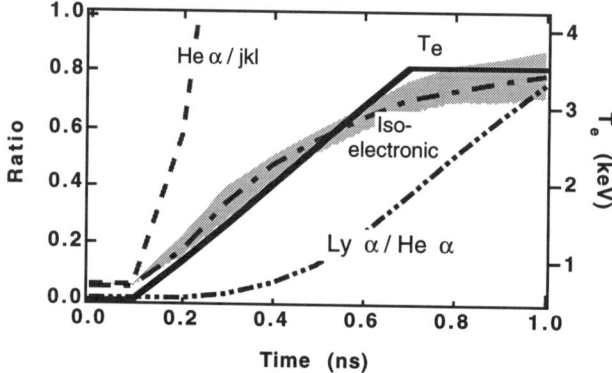

Figure 3. Three possible ratios that can be used to diagnose T_e. The isoelectronic ratio tracks the T_e slope quite well. The shaded area correpsonds to a factor of 2 variation of density for the isoelectronic ratio. The He α / jkl and the Ly α/ He α are not ideal because they do not follow the T_e slope very well.

GASBAGS

To demonstrate the increasing time-dependence of plasmas, we begin with the gasbags. The gasbag targets are similar to the model plasma because the T_e is similar and the density is virtually constant. After the initial ablation of the CH membrane, the gasbags form a quiescient plasma ~ 2.5 mm in diameter (8). According to computer simulations, the temperature rises very quickly during the 1 ns square heating laser pulse to achieve a peak temperature of 3-4 keV. X-ray images of the plasma volume have confirmed that the dopant in the neopentane-filled gasbags expands and equilibrates to a volume consistent with a density of 10^{21} cm^{-3} and remains stable during the second half of the 1 ns laser pulse. The disassembly of the gasbag plasma is due to its expansion, therefore in the central region where the temperature measurements are performed, it is reasonable to assume a constant density during the time of the measurements.

In gasbags, an iterative process is used to determine the T_e. First steady state values are determined from the measured ratios and are used to create the inital time history. Then a time-dependent run is performed to calculate ratios. These ratios are compared with the measured ratios and, based on these results, a corrected T_e time history is produced. This processes is continued until it converges to a time-dependent history that is consistent with the measured ratios. Since the plasma density remains constant in this case, the process converges in only 5 or 6 iterations.

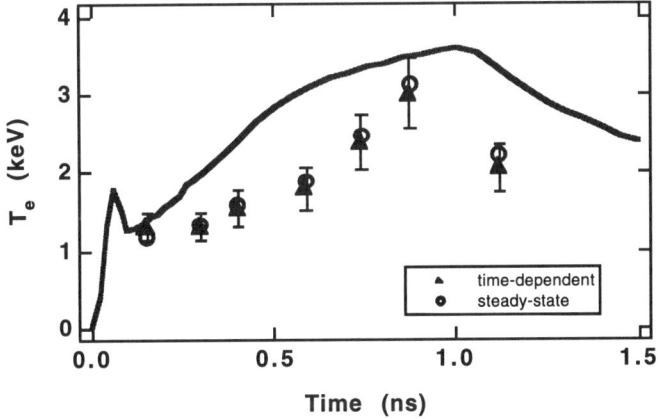

FIGURE 4. Graph of T_e vs time for gasbags targets. The plasmas are homogeneous after 300 ps and the volume at the center being measured is at constant density.

Figure 4 shows the Lasnex calculated temperature as well as the temperature deduced from steady state and time-dependent analysis. Temporal error bars for all measurements are $\pm$ 0.10 ns. As there were a large number of shots, each point represents an average over many shots. There is only a 4% different from the peak temperatures in this case. The small difference between the time-independent and time-dependent analysis is not surprising since the density does not significantly change, and the electron density remains fairly high, 10^{21} cm^{-3}. Although the peak temperature is consistent with the calculations, the rising temperature slope is inverted when compared to the calculations, a discrepancy which is still under investigation.

HOHLRAUMS

Gas-filled hohlraums are more complex than gas bags because the electron density can change over the time of interest. Unlike gasbags, the hohlraum wall is thick and confines the plasma inside. The laser energy that is not absorbed by the gas is incident on the wall of the hohlraum where it produces an expanding Au plasma. Large hohlraums are 2.5 mm long by 2.5 mm in diameter and have roughly the same gas volume as the gasbags. The temperature history is similar, however, the density history is no longer constant and it increases towards the end of the 1.6 ns pulse (9).

Lasnex simulations of these targets predict a change in the density history of a factor of 2. For this reason, a simple fit to the ratios is no longer sufficient to deduce the temperatures. In these cases, it is more appropriate to initialize the time-dependent analysis with the time-dependent hydrodynamic temperature and density histories. The data and plasma model results can be iterated to convergence by varying the T_e as in the gasbag case. As discussed in the test T_e slope case, the spectroscopic diagnostic is more sensitive to T_e than to n_e. Hence, in this process, we assume the n_e history does not vary, however, after the convergence on T_e, variations of the n_e are performed to verify this assumption.

For hohlraums we use isoelectronic line ratio measurements. This technique is well-suited to diagnose harsh plasma environments because it is better able to track the temporal history of ionizing plasmas. The ratio is temperature sensitive because of its exponential dependence on $1/T_e$. In these measurements, we use the $n=3$-1 transitions because they are not optically thick and do not have underlying satellites for our plasma conditions, therefore they are reliable diagnostics.

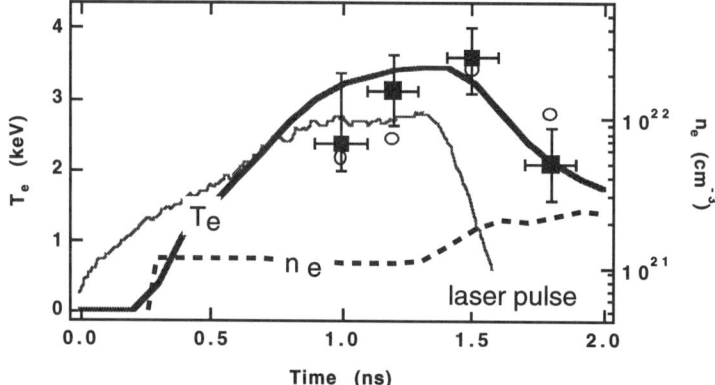

FIGURE 5. T_e versus time for a large hohlraum. The circles denote the steady state values and the squares denote the time-dependent values. The calculated temperature history is similar to the gasbag history, but the density increases as the plasma is confined.

Figure 5 shows the T_e vs time for the large hohlraums. The ratios of the spectral lines are only measured after 1 ns to allow for sufficient ablation and equilibration of the tracer with the plasma formed by the gas. As in Fig. 4, we show the time-dependent and time-independent T_e along with the Lasnex calculation. Since these targets were designed to be large scale-length plasmas, the T_e history is similar to that of the gasbags. However, the steady-state T_e is now in error by up to approximately 9 % because both the density and temperature change as a function of time. By this method, we have measured the T_e of large scale hohlraums to $\pm$ 15% of the calculated Lasnex value.

The smaller hohlraums are often referred to as Scale-1 hohlraums because they are the standard-sized targets used to implode a spherical capsule for laser fusion studies on Nova. These plasmas are very time-dependent and can be subjected to very strong radiation fields. The use of isoelectronic Ti/Cr ratios is advantageous because it is able to track changing ionization balances, and, it is insignificantly sensitive to high radiation fields (10).

The analysis of these targets is more time-dependent because the typical gas-fill in these targets is lower in n_e. Methane produces an initial density of 0.6×10^{21} cm^{-3} in the hohlraum. Furthermore, the heating of the hohlraum depends more on electron conduction and radiative heating than the other targets because less surface area of the hohlraum is directly irradiated by the beams.

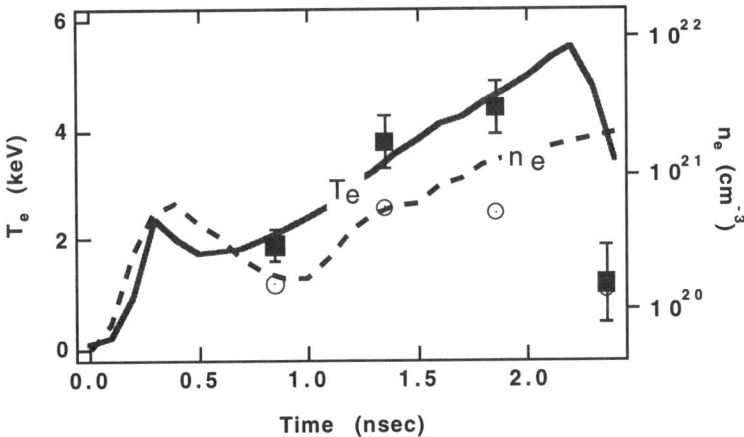

FIGURE 6. Measurements for scale-1 gas-filled hohlraums, T_e vs time. The calculated T_e history is monotonically increasing, however, the measured T_e falls after 1.8 ns. The circles denote the steady state values and the squares denote the time-dependent values.

For the analysis we show the temperature in a volume of the hohlraum directly irradiated by the lasers which is more analgous to the other two cases already presented. The data track the calculated history for ~ 1.8 ns, but then turn over and the temperature drops even though the hohlraum heating pulse is still on. Part of this discrepancy is due to scattering of laser light by the plasma (11). Measurements of scattering do increase during the second half of the pulse and the Lasnex temperature profile here has not yet been corrected for this. For these history, the n_e changes by a factor of 10. Images of the plasma dopant show that it does become more confined radially later in time which increases the density. The electron temperature due to a steady state calculation in this case would be in error by over 30 % if no time-dependent analysis was performed.

CONCLUSIONS

Time-dependent measurements and analyses are important to correctly diagnose large laser-produced plasmas. Experiments in which a steady-state analysis is used to determine T_e may introduce an error of up to 30 %. By a method of iteration starting from steady state values, we find that we can measure the temperature of gasbags in the 2 - 3 keV T_e range. Gas-filled hohlraums require a more involved iteration process that uses a temperature and density history from calculations as the initial conditions. Electron temperature measurements of large gas-filled hohlraums have verified the hydrodynamic calculations which peak at 3.7 keV. The smaller scale-1 hohlraums are more challenging because the temporal histories vary more and current work shows that these hohlraums attain peak temperatures of $T_e \sim 4.5$ eV, though the measurements do not agree with simulations at late times. These measurements allow us to assess the simulations and help us understand other related plasma processes occurring in the plasma.

ACKNOWLEDGEMENTS

We thank O. Landen and B. Hammel for useful discussions. This work was performed under the auspices of the U.S. Department of Energy by the Lawrence Livermore National Laboratory under Contract No. W-7405-ENG-48.

REFERENCES

1. Kauffman, R. L., "X-ray Radiation from Laser Plasma" in *Handbook of Plasma Physics*, vol. 3, eds. Rubenchik and Witkowski, pp. 111-162 (Elsevier Science, North-Holland, 1991) and references therin.
2. J. C. Fernandez, et al *Phys. Rev. E* **53**, 2747-2750 (1996).
3. Lindl, J. D., *Phys. of Plasmas* **2**, 3933-4024 (1995).
4. Back, C. A. Kauffman, R. L., Bell, P. M., Kilkenny, J. D., *Rev. Sci. Instrum.* **66**, 764-766 (1995).
5. Kilkenny, J. D., *Laser and Part. Beams* **9**, 49-69 (1991).
6. Lee, R. W., Whitten, B. L. and Strout, R. E., *J. Quant. Spectrosc. Radiat. Transfer* **32**, 91-101 (1984); also see the FLY users manual available from R. W. Lee.
7. Marjoribanks, R. S., Richardson, M. C., Jaanimagi, P. A. and Epstein, R., *Phys. Rev. A* **46**, R1747-R1750, (1992)
8. Kalantar, D. H., Klem, D. E., MacGowan, B. J., Moody, J. D., et al., *Physics of Plasmas* **2**, 3161-3168 (1995).
9. Powers, L. V., Berger, R. L., Kauffman, R. L., MacGowan, B. J., et al, *Physics of Plasmas* **2**, 2473-2479 (1995).
10. Shepard, T. D., Back, C. A., Kalantar, D. H., Kauffman, R. L., et al., *Rev. of Sci. Instruments* **66**, 749-751 (1995).
11. MacGowan, B. J., *Physics of Plasmas*, to be published May 1996.

Various Applications of Atomic Physics and Kinetics Codes to Plasma Modeling

J. Abdallah, Jr., R. E. H. Clark, D. P. Kilcrease, G. Csanak, and C. J. Fontes

Los Alamos National Laboratory
P. O. Box 1663, MS B212
Los Alamos, New Mexico 87545

Abstract

A collection of computer codes developed at Los Alamos have been applied to a variety of plasma modeling problems. The CATS, RATS, ACE, and GIPPER codes are used to calculate a consistent set of atomic physics data for a given problem. The calculated data include atomic energy levels, oscillator strengths, electron impact excitation and ionization cross sections, photoionization cross sections, and autoionization rates. The FINE and LINES codes access these data sets directly to perform plasma modeling calculations. Preliminary results of some of the current applications are presented, including, the calculation of holmium opacity, the modeling of plasma flat panel display devices, the analysis of some new results from the LANL TRIDENT laser and prediction of the radiative properties of the plasma wakefield light source for extreme ultraviolet lithography (EUVL). For the latter project, the simultaneous solution of atomic kinetics for the level populations and the Boltzmann equation for the electron energy distribution is currently being implemented.

1 Introduction

Theoretical calculations are important for predicting and diagnosing plasma properties. Atomic physics and kinetics codes[1] developed at Los Alamos have been applied to a number of plasma problems. These include aluminum opacity calculations[2-5], high resolution spectroscopy[6-8] of laser produced plasmas, neonlike X-ray laser plasmas[9-10], diagnostics for inertial confinement fusion plasmas[11], radiative power loss calculations for magnetic fusion devices[12-16], and for simulations of spectra observed in pulsed power devices such as the X-pinch[17].

© 1996 American Institute of Physics

The purpose of the present paper is to report preliminary results for work that is in progress for several applications. Section 2 briefly describes each of the computer codes, Sec. 3 presents holmium opacity calculations, Sec. 4 describes an effort to model xenon gas for the modeling of plasma display panels, Sec. 5 describes the effort to model recent experiments at the LANL TRIDENT laser, Sec. 6 describes the modeling of the LANL EUVL radiation source, and Sec. 7 provides conclusions and acknowledges our many collaborators and contributors.

2 The Codes

Figure 1 is a schematic diagram of the atomic physics and kinetic codes which have been developed in recent years at Los Alamos. The codes on the left produce atomic data while the codes on the right are atomic data users. The two sets of codes communicate via binary random access files that form a consistent atomic physics data base which can be used for plasma modeling.

A brief description of each code is provided below.

The CATS code is an adaptation of the atomic structure codes of R. D. Cowan[18]. The input to CATS basically consists of an the element, its stage of ionization, and a list of electron configurations. CATS solves the Hartree-Fock equations with relativistic corrections for each orbital of each input configuration. Fine structure levels are generated for each configuration and intermediate coupling and configuration interaction are included using perturbation theory. Energy levels and their designations, wavefunctions, Slater integrals, mixing coefficients, oscillator strengths for transitions between levels are stored on the atomic data files. Plane-wave Born (PWB) electron impact excitation cross sections may also be optionally mass produced.

The ACE code is used to provide higher quality electron impact cross sections than those calculated by CATS. The distorted wave method with CATS atomic structure is used to calculate the improved cross sections. The GIPPER code uses CATS atomic structure to compute cross sections for ionization processes. The processes considered include photoionization, electron impact collisional ionization, and autoionization. Various methods of calculation are available and distorted wave continuum functions are employed. CATS, ACE, and GIPPER can operate in both configuration average or detailed fine structure mode. All of these codes write files for the database.

The RATS and GRATS codes are relativistic structure codes. RATS is based on the Dirac-Fock-Slater code of Sampson et. al.[19] while GRATS is an adaptation of the multi-configuration Dirac-Fock code of Grant and coworkers[20]. The input and output from RATS and GRATS are similar to CATS, the list of relativistic configurations involving the κ quantum number are generated internally from the input list of non-relativistic configurations that would be input to CATS. The relativistic capabilities will not be discussed further in the present paper.

The TAPS code is used to access and display atomic physics quantities from the database. The LINES code is used to construct fine structure spectra based on a LTE population model. Populations are calculated using the energy levels and statistical information stored on the database. The spectra is assembled using these populations with stored transition energies and oscillator strengths. LINES has the capability to process an almost unlimited number of calculated spectral lines.

The FINE code is used for modeling non-equilibrium plasmas. The necessary data is read directly from the atomic physics files created by CATS, ACE, GIPPER, RATS, and GRATS. FINE can perform kinetics calculations based on fine structure, configuration average, or composite state models. Rate coeficients for collisional ionization, photoionization, autoionization, collision excitation, radiative excitation are calculated directly from calculated cross sections, while the inverse processes three-body recombination, radiative recombination, dielectronic cature, collisional deexcitation, and radiative decay are calculated using the principle of detailed balance. Rate coefficients can be calculated using arbitrary electron distributions and radiation fields. The population models include LTE, coronal equilibrium, collisional radiative, steady-state, quasi-steady state, time-dependent, and effective schemes. FINE can produce various types of spectra using different choices of line shapes and widths. Other output includes ionization balance, radiative power, effective ionization and recombination rates, line ratios, gain coefficients, and DCA models. Further capabilities include continuum lowering, escape factors, Stark line profiles, unresolved transition array theory, and solution of the Boltzmann equation.

3 Holmium Opacity Calculations

Recent holmium (Z=67) transmission experiments[21] present a new challenge for the models used to calculate high-Z opacities. In the range of density and temperature of the experiment holmium is approximately 5-8 times ionized. The electron configurations needed to describe these stages of ionization involve near half-filled f shells which make accurate calculations of the atomic structure very time comsuming. The main feature in the spectral range studied are 4d-4f transitions. In the present section, the method of Ref. 5 is used to simulate the experimental observations. The method involves LTE configuration average modeling for populations and the use of unresolved transition array (UTA) theory for the spectral simulation.

Fig. 2 is a comparison of the current calculations with experiment and previous STA calculations[21]. Note that the current calculations are in better agreement with experiment especially near the transmission minimum near 70 angstroms. Also note that the best fit temperature is different for both calculations resulting in signifcantly different ion stage distributions. Neither theory does well in the range from 40 to 60 angstroms. The sensitivity of the current calculations was studied by increasing the number of configurations per ion stage from approximately 200 to 4000. Calculations with the expanded basis produced negligible differences in the calculated transmission. The current calculations also indicate that the bound-free contribution is significant. Future plans include the study of relativistic, fine structure, and configuration interaction effects.

4 Plasma Display Panels

Atomic data for xenon have been generated for one and two-dimensional modeling of plasma display panels[22]. Here, an electrical discharge is used to excite a gas and create radiation which is used to drive a display panel. Since the computer simulations are very time consuming, only a few (6-10) atomic levels can be included in the calculations.

For this case, the detailed fine structure levels including the effects of configuration interaction were calculated by CATS for 13 configurations of xenon through $5p^58s^1$. These 13 configurations resulted in 87 fine structures levels. ACE and GIPPER were used to calculate the corresponding excitation and ionization cross sections. The cross sections were tested by comparison with experiment[23-24]. The calculated and measured cross sections are in general agreement depending on the transition, the impact energy, and the scattering angle. Since the number of levels calculated went beyond the capability of the modeling codes, the data set was reduced by forming composite states by combining levels and cross sections appropriately from the raw fine structure data. A 7-level and an 8-level model were constructed. The integrity of the reduced data sets was tested independently by calculating a time integrated spectra in FINE for an electron density and temperature near actual pixel conditions. The results for the 7-level, 8-level, and the full fine structure calculation are shown in Fig. 3. Note that the reduced models reproduce the full calculation only approximately, and that care must be exercised when constructing composite states.

5 TRIDENT Experiments

Several high resolution spectroscopy experiments were recently performed at the Los Alamos TRIDENT laser facility. TRIDENT has both long pulse ($100-1000ps$) and ultrashort pulse ($500fs$) capabilities. The laser intensity is about $5 \times 10^{15} W/cm^2$ for the long pulse and $1 \times 10^{19} W/cm^2$ for the short pulse. Both $1054nm$ and $527nm$ laser light are available in the short pulse mode. In the experiments performed, the $527nm$ radiation had high contrast causing the laser to interact mainly with the solid surface. Experiments with the $1054nm$ light had considerable prepulse, and the laser interacted mainly with a preformed plasma. The observed spectra were very different from the $527nm$ case having very broad features. Experiments were also performed using TRIDENT's $1ns$ capability. The discussion here is restricted to the $1054nm$ short pulse experiments with magnesium targets.

The observed spectra show considerable evidence for the influence of hot electrons. Fig. 4 shows the observed satellite lines which accompany the He-like resonance line of Mg. The three major peaks from 1330-1338 eV are the Li-like satellites jkl, qr, and $abcd$, respectively. The peak near 1343 eV is the spin-forbidden He-like intercombination line, and the lines above 1348 eV are additional Li-like satellites. Extensive steady-state calculations with FINE show anomalous behavior of the experimental results. The relative intensities of the jkl, qr, and abcd lines are indicative of electron densities near or greater than or equal to 10^{23} while the strong intercombination line is indicative of electron densities below 10^{21}.

Model calculations using an electron distribution consisting of a thermal Maxwellian part and a non-thermal high energy Gaussian part[17] were performed to determine if hot electrons could be responsible for the anomalous results. Fig. 5 (top) shows the effect of 10% by number of hot electrons on the calculated spectrum at a temperature and density near $200eV$ and $10^{22}cm^{-3}$ compared to the same calculation without hot electrons. Note the significant difference between the relative intensities predicted by the two different cases. Fig. 6 (bottom) shows a comparison of the hot electron spectrum compared to experiment. Not only are the relative intensities of the jkl, qr, $abcd$, and intercombination line in good agreement but so are the other Li-like satellite lines above $1348eV$.

6 EUVL Radiation Source

Recently the codes have been used to study the proposed Los Alamos radiation source[25] for extreme ultraviolet lithography (EUVL). The purpose of the work is to produce light in the 13-15 nm range where mirrors have optimal efficiency.

The method involves injecting relativistic electron micropulses into preionized (.1%) near neutral gas. Under the proper conditions the plasma electrons will absorb most of the kinetic energy of the beam, leaving the electrons in a hot and very non-Maxwellian state which can be predicted by Los Alamos paticle-in cell codes. The FINE code is used to study the conversion of this energy into radiation through electron collisions with the plasma ions.

For this purpose, the simultaneous solution of kinetics equations and the Boltzmann equation has been incorporated into the FINE code. The solution of the Boltzmann equation will allow propogation of the electron energy distibution with time, which in turn will be used to provide realistic estimates of the output radiation based on the calculated populations. Electron energy space is divided into user specified bins, and the number of electrons in each bin is computed at each time step. The Boltzmann package currently includes the effects of ionization, excitation, de-excitation and electron-electron collisions for the study of the early time development of the electron distribution.

In this paper, the solution of a model problem is presented. The initial conditions are a neon gas with a density of $2 \times 10^{19}/cm^3$, 50% neutral, and 50% singly ionized. The initial electron density is $1 \times 10^{19}/cm^3$ with a monoenergetic distribution near $900eV$. Electron enery space is divided into 100 equal bins from 0 to $1000eV$. The atomic model is confiuration average and includes over 100 congiurations in each of the first five ion stages of neon. Fig. 7 shows the evolution of the electron density distribution as a function of energy for various times between 0 and 100 ps. At early times the initial distribution spreads due to electron-electron collisions. At the same time low energy electrons are being created at a slow rate through ionization and excitation followed by ionization. As time increases, the high energy tail vanishes and the distribution attains a low energy form. The dominant ion stages at the later times is Ne IV and Ne V. Fig. 8 shows the corresponding time integrated spectrum using the UTA theory as described above. Note that the spectrum shows strong line emission in the wavelength range of interest.

7 Summary and Acknowledgments

The recent applications of the Los Alamos atomic physics and plamsa kinetics codes have been discussed. These codes have been applied to a wide variety of problems. Future work includes enhacement of the Boltzmann package, and integration of the kinetics package with hydrodynamics and radiation transport codes.

The authors extend a special thanks to Anatoly Faenov, Tania Pikuz, and Mark Wilke for providing the TRIDENT data discussed above. We would also like to thank our collaborators and acknowlege individuals who have provided data for us. These include, Bob Cowan, Joe Mann, Jim Peek, Mark Wilke, George Kyrala, Doug Fulton, Gordon Olson, Greg Pollak, John Goldstein, Mike Jones, Dan Prono, Bruce Carlsten, Anatoly Faenov, Sergei Pikuz, Tania Pikuz, Igor Skobelev, B. Bryunetkin, Dave Hammer, Peter Nickles, Mike Kalashnikov, Steve Davidson, Colin Smith, John Foster, G. Winhart, K. Eidmann, Chris Keane, Tom Sheppard, Larry Suter, Tina Back, Ted Perry, Frank Serduke, Doug Post, Russel Hulse, Bob McGrath, Ramana Veerasingam, Bob Campbell, Merle Riley, Sandor Trajmar, Morty Khakoo, and Roberto Mancini.

This work has been supported in part by the U.S. Department of Energy contract no. W-7405-ENG-36.

References

[1] J. Abdallah,Jr. and R.E.H. Clark, Los Alamos National Laboratory reports, LA-11436-M, I-V(1988)

[2] J. Abdallah, Jr. and R. E. H. Clark, J. Appl. Phys. 69, 23 (1991)

[3] J. Abdallah, Jr., R. E. H. Clark, and J. M. Peek, Phys. Rev. A 44 4072(1991)

[4] T. S. Perry, S. J. Davidson, F. J. D. Serduke, D. R. Bach, C. C. Smith, J. M. Foster, R. J. Doyas, R. A. Ward, C. A. Iglesias, F. J. Rogers, J. Abdallah, Jr., R. E. Stewart, J. D. KIlkenny, and R. W. Lee, Phys. Rev. Letters. 67, 3784(1991)

[5] D. P. KIlcrease, J. Abdallah, Jr., J. J. Keady, and R. E. H. Clark, J. Phys. B 26 1717 (1993)

[6] J. Abdallah,Jr., B. A. Bryunetkin,M. P. Kalashnikov, R. E. H. Clark, P. Nickles, S, A, Pikuz, I. Yu. Skobelev, A. Ya. Faenov, and M. Schnuerer, Quantum Electrodynamics (Russian) 20, 1159 (1993)

[7] B. A. Bryunetkin, A. Ya. Faenov, M. Kalashinov, P. Nickles, M. Schnuerer, I. Yu. Skobelev, J. Abdallah, Jr., and R. E. H. Clark, JQSRT 53, 45 (1995)

[8] A. Ya. Faneov, I. Yu. Skobelev, S. A. Pikuz, G. A. Kyrala, R. D. Fulton, J. Abdallah, Jr., and D. P. Kilcrease , Physical Review A 51 , 3529 (1995)

[9] J. Abdallah, Jr., R. E. H. Clark, and J. M. Peek, Phys. Rev. A 45, 3980(1992)

[10] J. Abdallah, Jr., R. E. H. Clark, J. M. Peek, and C. J. Fontes, JQSRT 51, 1(1994)

[11] J. Abdallah, Jr., R. E. H. Clark, C. J. Keane, T. D. Shepard, and L. J. Suter, JQSRT 50, 91(1993)

[12] R. E. H. Clark and J. Abdallah, Jr. ,Los Alamos National Laboratory report LA-UR-92-995 (1992)

[13] R. E. H. Clark and J. Abdallah, Jr. ,Los Alamos National Laboratory report LA-UR-95-2906 (1995)

[14] J. Abdallah, Jr. and R. E. H. Clark, J. Phys. B 27, 3589(1994)

[15] R. E. H. Clark, J. Abdallah,Jr. and D. Post, Journal of Nuclear Materials 220-222, 1028(1995)

[16] D. Post, J. Abdallah, Jr., R. E. H. Clark,and N. Putvinskaya, Phys. Plasmas 2, 2328(1995)

[17] J. Abdallah, Jr., A. Ya. Faenov, D. Hammer, S. A. Pikuz, G. Csanak, and R. E. H. Clark, Physica Scripta xx, xxxx(1996)

[18] R. D. Cowan, Theory of Atomic Spectra, University of California Press, Berkeley, CA (1981)

[19] D. H. Sampson, H. L. Zhang, A. K. Mohanty, and R. E. H. Clark, Phys. Rev. A 40, 604(1989)

[20] I. P. Grant, B. J. McKenzie, P. H. Novington, D. F. Mayers, and N. C. Pyper, Comp. Phys. Commun. 21, 207 (1980)

[21] G. Winhart, K. Eidmann, C. A. Iglesias, A. Bar-Shalom, E. Minguez, A. Rickert, and S. J. Rose, JQSRT 54, 437(1995)

[22] R. Veerasingam, R. B. Campbell, R. T. McGrath, M. Riley, J. Alford, J. W. Shon, J. Abdallah, Jr., R. E. H. Clark, and C.J. Fontes,"Ionization and excitation rates for a multi-species He/Xe gas mixture determined from Boltzmann calculations," Gaseous Electronics Conference, Berkeley CA (October,1995)

[24] M. A. Khakoo, S. Trajmar, L. R. LeClair, I. Kanik, G. Csanak, and C. J. Fontes,to be published.

[25] M. A. Khakoo, S. Trajmar, S. Wang, I. Kanik, A. Aguirre, C. J. Fontes, R. E. H. Clark, and J. Abdallah,Jr. ,to be published.

[25] D. S. Prono, M. E. Jones, B. Carlsten, and J. Abdallah, Jr.,"rf Electron-beam plasma XUV source for projection lihtography systems,"OSA Proceedings on Soft X-ray Projection Lithography, Volume 18, Proceedings of the topical conference, Monterey CA (May,1993)

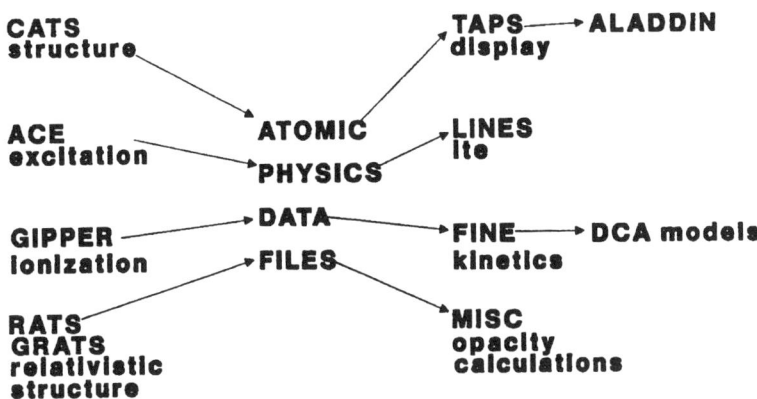

Figure 1: A schematic diagram of the Los Alamos suite of atomic physics and kinetics codes.

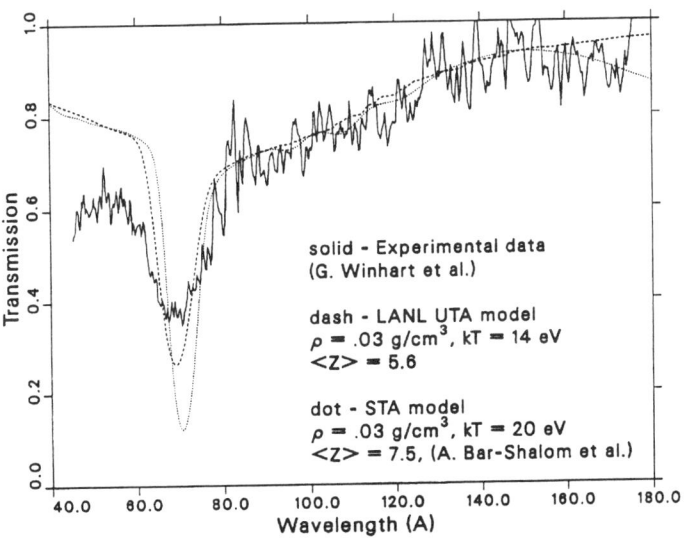

Figure 2: A plot of transmission versus wavelength for holmium. The current calculations (dashed line) are compared to experiment (solid line) and previous calculations (dotted line).

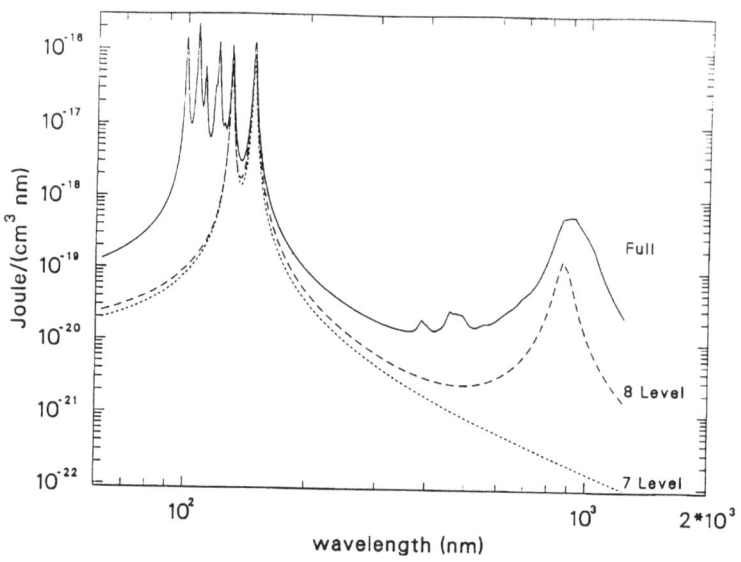

Figure 3: The calculated time integrated spectrum for xenon as a function of wavelength for the 7-level, 8-level, and full detailed models.

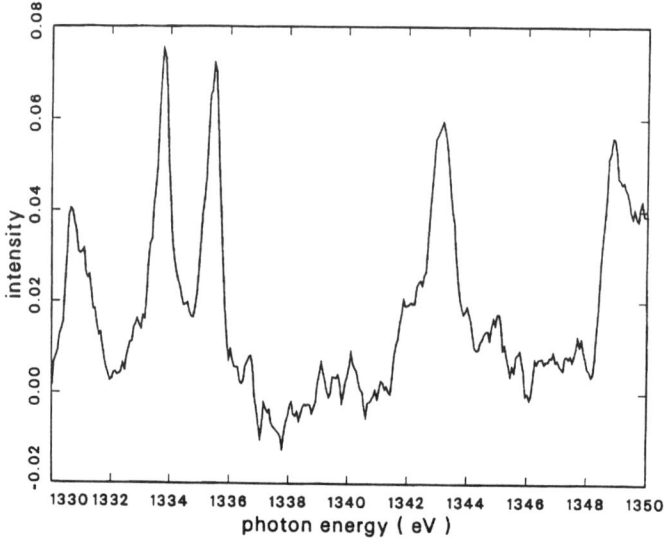

Figure 4: The experimental spectrum corresponding to a TRIDENT shot using a wavelength of $1054 nm$ with a $500 fs$ pulse duration on a magnesium target plotted as a function of photon energy (eV).

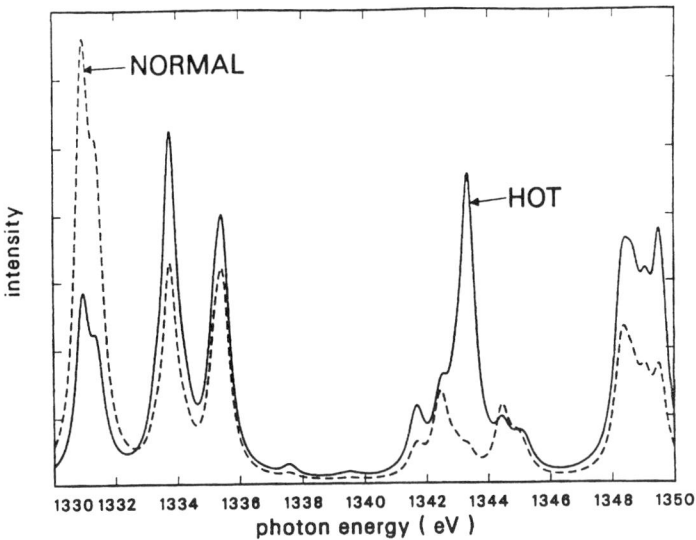

Figure 5: The calculated spectrum as a function of photon energy (eV), the curve labeled normal has no hot electron component while the curve labeled hot has a 10% hot electron component.

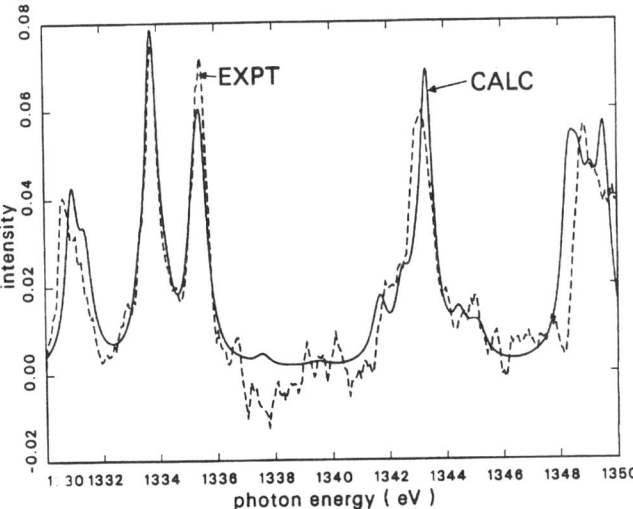

Figure 6: A comparison of the observed and theoretical spectrum as a function of photon energy (eV).

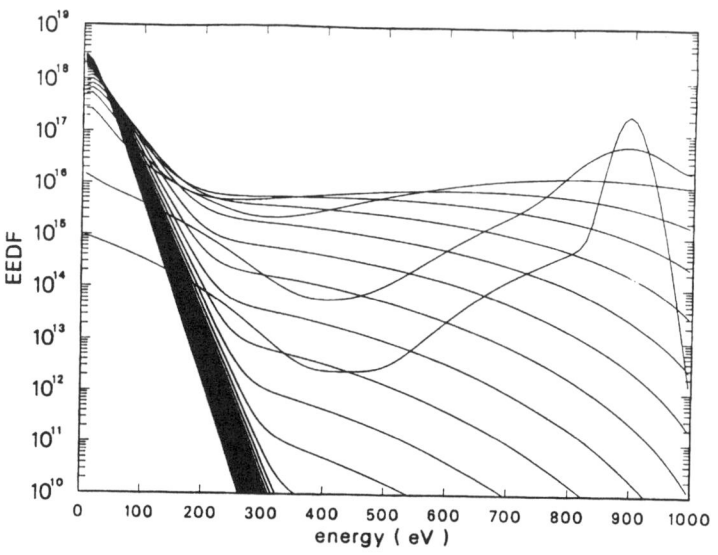

Figure 7: A plot of the electron energy distribution function as a function of electron energy for various times between 0 and 100 *ps* corresponding to the model problem described in the text.

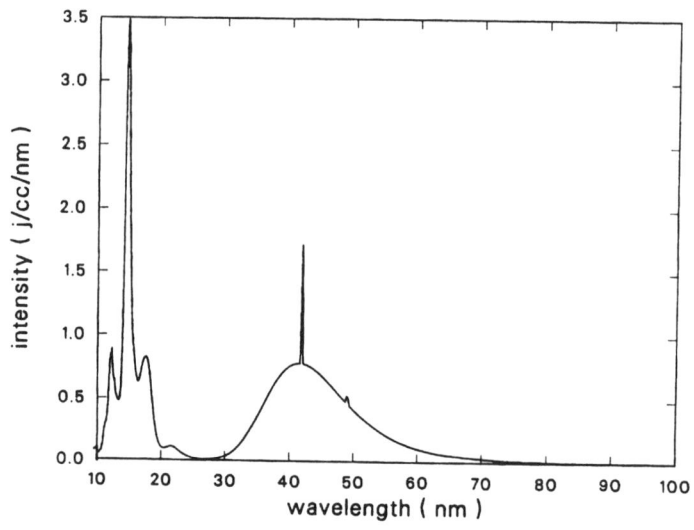

Figure 8: A plot of the predicted time integrated spectrum as a function wavelength.

The Motional Stark Effect Diagnostic on TFTR

F. M. LEVINTON

Fusion Physics & Technology, Inc.
3547 Voyager St., Torrance, CA 90503.

Abstract. The propagation of a neutral beam across a magnetic field induces an electric field, $\mathbf{E} = \mathbf{V} \times \mathbf{B}$, in the rest frame of the atomic beam. The induced Lorentz electric field results in the motional Stark effect (MSE) that causes a wavelength splitting and polarization of the emitted radiation. Collisions of the neutral beam with the background plasma will cause excitation of the beam atoms leading to emission of radiation. Depending on the collision and mixing processes the Stark spectrum can be significantly affected. Comparison of the Stark spectrum from neutral beam injection into plasma agrees well with a statistical weight model, but the spectrum from injection into a gas filled torus does not agree with either the statistical or dynamical intensity models. The Stark spectrum, which determines the polarization pattern, can affect the calibration of the diagnostic in measuring the direction of the local internal magnetic field.

INTRODUCTION

The q-profile, which plays a key role in determining the plasma equilibrium, stability, and transport in tokamaks, was quite difficult to measure until the development of the motional Stark effect (MSE) diagnostic[1]. The principle of the measurement relies on the Stark effect[2] from the Lorentz electric field, $\mathbf{E} = \mathbf{V} \times \mathbf{B}$, induced on an atom as an energetic neutral beam propagates across a magnetic field. The Stark effect causes a wavelength splitting and linear polarization of the emitted radiation. The diagnostic measures the direction of polarization of the emitted radiation to determine the magnetic field pitch angle. When viewing a hydrogen or deuterium beam the Stark shift is linear with the electric field. This occurs because the field-free atom has degenerate energy levels for different l-values and fixed n. The perturbation matrix has non-zero elements between states of odd and even l removing the degeneracy. The eigenfunctions with fields are then a superposition of the field-free functions with different l-values and produces an interaction which is linear with the electric field. Complex atoms, which are not degenerate in l, have a vanishing matrix element and the interaction is given by second order perturbation theory that is quadratic in the electric field. For beam energies of ~ 50 keV/amu the spectral shift from the Stark effect is much larger than from the Zeeman effect. When viewed transverse to the electric field the $\Delta m = 0$ transition, or π component, is linearly polarized parallel to the electric field, which is perpendicular to the magnetic field. Similarly, the $\Delta m = \pm 1$ transitions, or σ components are linearly polarized perpendicular to the electric field or parallel to the magnetic field. Polarimeter measurements[3, 4] of the direction of the linearly polarized π or σ component emission give the magnetic field pitch angle, $\gamma_p = \tan^{-1}(B_p/B_t)$, where γ_p is the magnetic field pitch angle, B_p is the poloidal magnetic field, and B_t is the toroidal magnetic field. The pitch angle is used to determine the safety factor on axis, $q(0)$, and the safety factor profile. Also in conjunction with a free-boundary equilibrium reconstruction code, such as VMEC[5], the Grad-Shafranov equation can be solved numerically to determine the current density profile and other equilibrium quantities. The reconstruction code uses the data from MSE as well as the external magnetics data, such as coil currents and flux measurements, the pressure profile and total plasma current to solve the non-linear force balance equation using a least-squares minimization to reduced the χ^2 error from all the input data.

THE MSE DIAGNOSTIC ON TFTR

The diagnostic on TFTR views one of the deuterium neutral beams used for heating the plasma. The beam energy can be varied from 85 to 120 keV, with a typical energy of ~ 95 keV when taking MSE data. The deuterium Balmer-alpha transition, at 656.1 nm, is used for the measurement. With the viewing geometry on TFTR, as shown in Fig.1, the emission is Doppler shifted about 4.0 nm towards the red. As the neutral beam propagates through the plasma, spectral line emission will occur due to collisional excitation of the beam with the plasma ions and electrons. Light emission from the neutral beam is collected by 10 centimeter diameter optics via an in-vessel re-entrant mirror and imaged through a polarimeter onto a fiber optic array at $f/2$. The fibers are separated into individual sightlines with a spatial resolution of ~ 3 cm and a separation between sightlines of 3-5 cm. The optics coverage of the plasma is from inboard of the magnetic axis to the outboard edge of the plasma. The number of sightlines has recently been increased from 12 to 21 to improve the spatial resolution of the measured q-profile. The fiber optics bundle transmits the collected light to a location outside the test cell that is accessible during TFTR operation. The light from the fibers for each sightline is collimated through a narrow band interference filter and focused

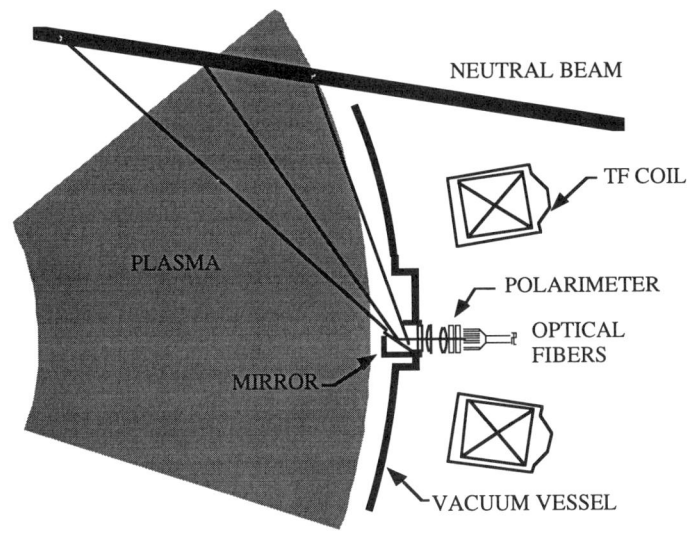

Figure 1: Layout of the MSE diagnostic on TFTR

onto a photomultiplier detector. The filter is sufficiently narrow, ~ 0.6 nm, to select the main group of π or σ lines. The output signal from each detector is amplified before going into lock-in amplifiers with a time constant of 2 milliseconds and then is digitized at 2 kilo-samples per second. Further details of the diagnostic can be found in references 3 and 4.

RESULTS

Under typical plasma conditions in a tokamak, with magnetic fields of 1-10 Tesla and densities of $0.1 - 1.0 \times 10^{20}\,\mathrm{m}^{-3}$, collisional Stark broadening and fine structure can be neglected. The deuterum spectral emission will be dominated by the Zeeman or motional Stark effects. In particular with beam velocities larger than 10^6 m/s the motional Stark effect will dominate the Zeeman effect. The spectrum of the Doppler shifted D_α light from a neutral beam is complicated by the three energy components from the beam. There is a full energy component in addition to components at the half and third energies. For each of the three Doppler shifted energies nine Stark components are emitted. The measured MSE spectrum is shown in Fig. 2, with the various π and σ components identified. The data is taken with a 0.5 meter Czerny-Turner spectrometer and a back illuminated low noise CCD detector[6].

There are two simple techniques to calculate the relative intensities of the Stark spectrum. The first assumes that the occupation of each Stark effect

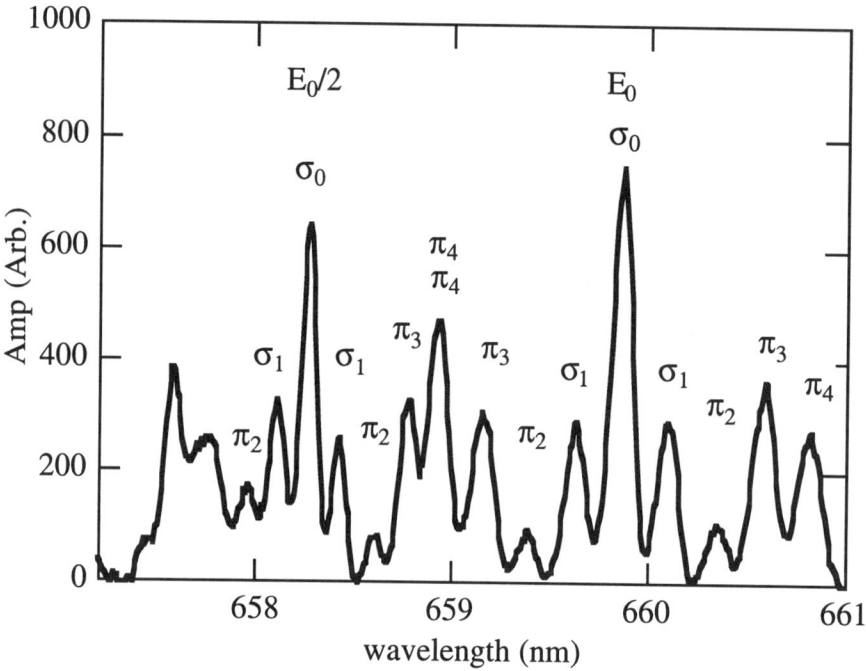

Figure 2: Spectrum of the σ and π components for the full and half beam energies.

level is proportional to its statistical weight. This calculation, known as the statistical intensity, is determined from the product of the transition probability and the statistical weight of the initial state[7]. Another method to calculate the intensities, know as the dynamical intensity, is to assume the excitation of each level is proportional to its statistical weight. The latter calculation assumes no mixing in the excited state, whereas the former case assumes complete statistical mixing. For the case of neutral beam injection into plasma, the background ions and electrons, through collisional excitation, will mix the excited state levels, so the statistical intensity estimate should be a good approximation. The issue of l-mixing has previously been addressed for charge exchange spectroscopy. Estimates of l-mixing, $n, l \longrightarrow n, l'(l' = l \pm 1)$, for ion-ion collisions in nearly degenerate levels for hydrogenic ions have found this to be an important effect[8] at densities and temperatures found in TFTR. Indeed, a comparison of the measured spectrum to the calculated statistical intensities shows good agreement as seen in Fig. 3. The calculated statistical intensities are normalized such that the σ_0 intensity matches the experimen-

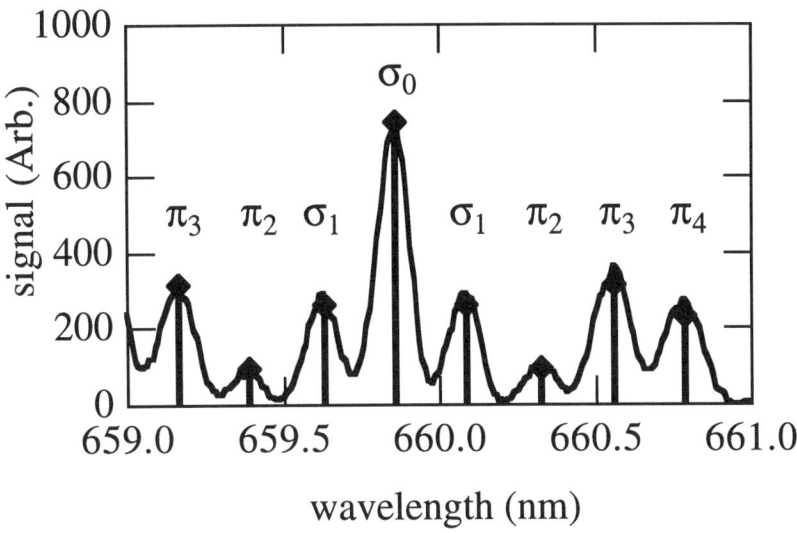

Figure 3: Motional Stark effect spectrum viewing the emission of a neutral beam injected into a plasma. The vertical bars represent the calculated statistical intensities normalized to σ_0.

tally measured amplitude. In the case of neutral beam injection into a neutral gas the measured spectra is different. Shown in Fig. 4 is a comparison of the calculated statistical and dynamical intensities to the measured MSE spectra. The spectrum shown is from a deuterium neutral beam of $\sim 95\,\text{keV}$ injected into the torus with a fill pressure of 1×10^{-4} torr of deuterium and the same toroidal field, 4.8 Tesla, as used with plasma data. In this case neither the statistical nor dynamical intensity models adequately describe the data. This may due to the collisional coupling effects of the magnetic and electric fields or beam directionality which are not taken into account[9, 10]. A summary of the results for beam injection into plasma and neutral gas is given in Table 1.

The difference in the Stark spectral patterns between neutral beam injection into neutral gas and plasma manifests itself in the MSE diagnostic as a difference in the measured polarization fraction, which can be expressed as $P_f = I_\pi - I_\pi / I_\sigma + I_\pi$, where P_f is the polarization fraction, and I_σ and I_π are the measured intensities of the σ and π components respectively. The difference is due to the finite spectral bandwidth of the interference filters used and the relative amounts of signal from the π and σ lines transmitted through the filter. The neutral beam injection into cold gas has been used for calibration purposes. With known toroidal and vertical vacuum fields the instrumental calibration can be determined[11]. However with the difference in polarization

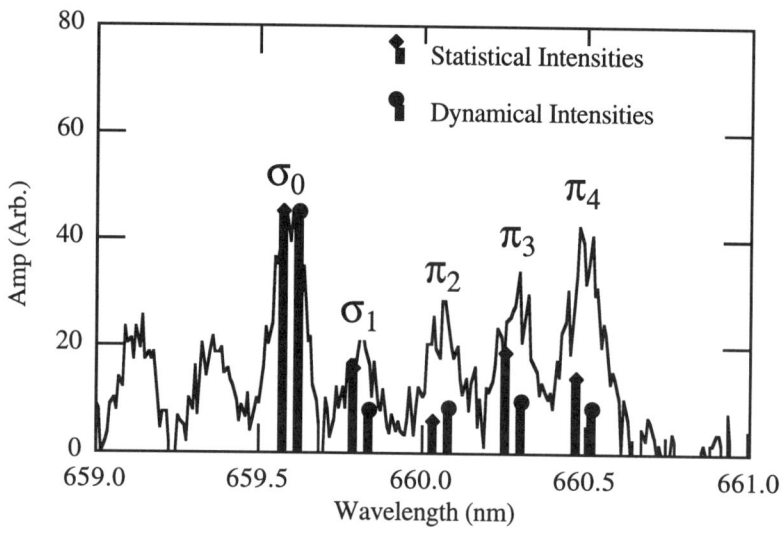

Figure 4: Motional Stark effect spectrum viewing the emission of a neutral beam injected into neutral gas. The vertical bars represent the calculated statistical (diamond) and dynamical (circle) intensities normalized to σ_0.

fraction between neutral gas and plasma data, systematic offsets are introduced into the data due to the polarimeter and optics response to unpolarized light. This is shown clearly by comparison to a second calibration technique which utilizes the known poloidal magnetic field at the plasma edge. By positioning the plasma edge to coincide with a given sightline at several different values of plasma current, the instrumental calibration can be determined. This is repeated for each sightline to map out the complete diagnostic calibration. The results of the two calibrations are shown in Fig. 5. They both result in similar slopes, but are slightly offset due to differences in the Stark spectra.

Table 1: Measured and calculated intensities normalized to one for the σ_0 component.

Transition	Data-plasma	Data-neutral gas	Statistical	Dynamical
σ_0	1.00	1.00	1.00	1.00
σ_1	0.39	0.47	0.35	0.18
π_2	0.14	0.54	0.13	0.19
π_3	0.45	0.70	0.42	0.22
π_4	0.37	0.88	0.31	0.19

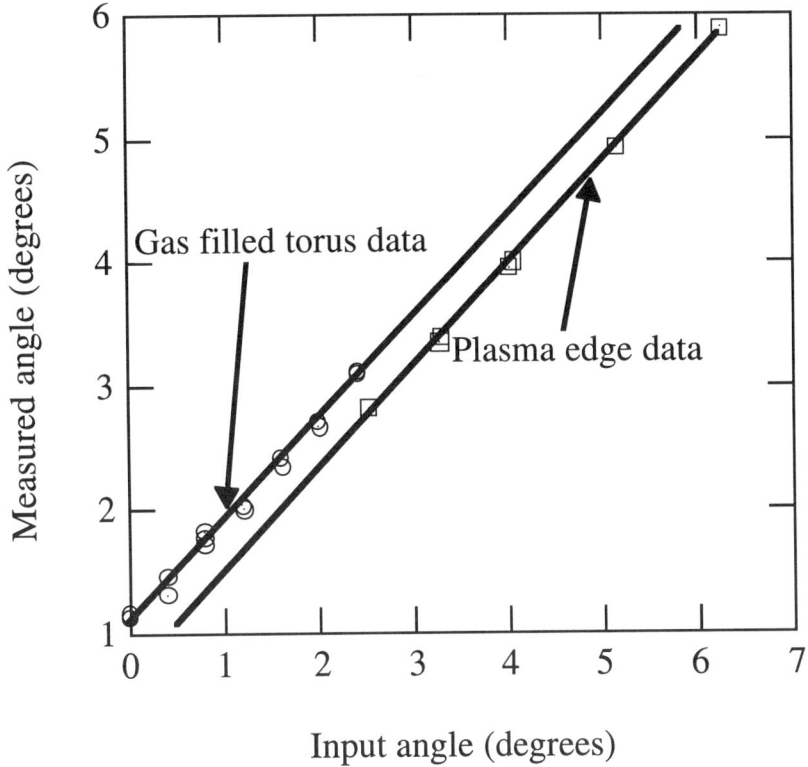

Figure 5: Results of gas filled torus and plasma edge data calibrations.

ACKNOWLEDGMENTS

I would like to thank S. Batha for analysis of the calibration data. Data from the high throughput survey spectrometer by G. R. McKee, B. C. Stratton, and R. J. Fonck is greatly appreciated.

REFERENCES

[1] F. M. Levinton et al., Phys. Rev. Lett. **63**, 2060 (1989).

[2] E. U. Condon and G. H. Shortley, *The Theory of Atomic Spectra*, Cambridge University Press, Cambridge, 1963.

[3] F. M. Levinton, G. M. Gammel, R. Kaita, H. W. Kugel, and D. W. Roberts, Rev. Sci. Instrum. **61**, 2914 (1990).

[4] F. M. Levinton, Rev. Sci. Instrum. **63**, 5157 (1992).

[5] S. P. Hirshman et al., Phys. Plasmas **1**, 2277 (1994).

[6] G. R. McKee, R. J. Fonck, T. A. Thorson, and B. C. Stratton, Rev. Sci. Instrum. **66**, 643 (1995).

[7] H. A. Bethe and E. E. Salpeter, *Quantum Mechanics of One- and Two-Electron Atoms*, Academic Press, New York, NY, 1957.

[8] R. J. Fonck, D. S. Darrow, and K. P. Jaehnig, Phys. Rev. A **29**, 3288 (1984).

[9] R. C. Isler, Phys. Rev. A **14**, 1015 (1976).

[10] C. Breton, C. D. Michelis, M. Finkenthal, and M. Mattioli, J. Phys. B **13**, 1703 (1980).

[11] F. M. Levinton, S. H. Batha, M. Yamada, and M. C. Zarnstorff, Phys. Fluids B **5**, 2554 (1993).

Polarization Spectroscopy of Ionized Gases

Sergei A. Kazantsev

Institute of Physics, St.Petersburg State University
7/9 Universitetskaya nab. 199034 St.Petersburg, Russia

Abstract. The principles and the present day achievements of the polarization spectroscopy as a new remote sensing technique for ionized gases of various physical nature have been reviewed.

POLARIZATION OF SPECTRAL LINES EMITED BY IONIZED GASES

Spectral distribution of the emission intensity of ionised gases is the basis of elaborate and widely used spectroscopic methods of the remote sensing of different plasmas and ionized entities (1). Until recently, polarization of a spectral line polarization as a source of quantitative information on the physical properties of ionized gases of various physical nature was not widely exploited. The reason was the complexity of physical mechanisms of formation of a spectral line polarization, even for a simple gas discharge, difficulties of experimental observations and detections, the lack of basic theoretical relations of the polarization of radiation with the kinetic characteristics of a plasma and relevant atomic parameters. These difficulties have been continuously overcome during last two decades, and to the beginning of 90th the principles of the polarization spectroscopy have been put forward and analysed in a very pragmatic way, basing on original spectropolarimetric studies of low temperature gas discharge and solar flare plasmas. The development of the polarization spectroscopy of ionized gases were decribed in the books (2,3).

The polarization of a spectral line is effectively a basic spectroscopic characteristic, which, within the most simple approach, is connected to the nonequilibrium distribution of populations of the magnetic substates of excited atomic particles (atoms, ions, molecules) and reflects the ordering of the total angular momenta of electron shells in an ensemble, or, in other words, the polarization of an ensemble. The polarization of an atomic ensemble in an ionized gas could be induced by different anisotropies of excitation processes inherent to a real entity, such as resonance optical excitation by the resonance field inside a plasma, collisional processes, distribution of internal fields, energy trans-

© 1996 American Institute of Physics

port by precipitating particles or by heat conduction, gradients of parameters, boundary effects. The polarization of the plasma emission spectrum induced by collisional processes is directly connected to anisotropies in the space of relative velocities of atoms, ions and light particles, such as electrons or protons, responsible for excitation and optical emission.

FORMALISM OF POLARIZATION MOMENTS

Theoretical fundamentals of the polarization spectroscopy are very closely connected to the quantum mechanical theory of atoms and to the collision theory, making use of the general formalism of the quantum theory of the angular momentum and of the group theory. Physical principles of the quantum mechanical theory of the ordering of atomic states and of its spectroscopic manifestations are based on the fact that excited atomic particles acquire the memory of anisotropies of the excitation process. For inelastic collisions it is the symmetry of velocity distribution of the exciting particles, for radiation processes it is the symmetry of the tensor of resonance radiation field, accounting for local intensity anisotropy and polarization. As a result of excitation, the angular momenta of the electron shells possess a particular ordering, that corresponds to the kinetic aspects of the collisional excitation or the anisotropy of the resonance optical field. On the mathematical side, this ordering is characterized by spherical tensors of different ranks or by the polarization moments $\rho_q^{(k)}(J)$, being the coefficients of expansion of the density matrix of an excited atomic ensemble in terms of the irreducible representation of the three dimensional rotation group:

$$\rho_q^{(k)}(J) = \sum_{m,m'} (-1)^{J'-m'} (2k+1)^{1/2} \begin{pmatrix} J & J & k \\ m' & -m & q \end{pmatrix} \rho_{m'm}(J)$$

where $\rho_{m'm}(J)$ is the density matrix elements of the state J of atomic particles in the representation of magnetic quantum numbers m, $\begin{pmatrix} J & J & k \\ m' & -m & q \end{pmatrix}$ is Clebsh-Gordan coefficient.

Zero order moment $\rho_0^{(0)}$ is a scalar, proportional to the population of the state. The three elements $\rho_{0,\pm1}^{(1)}$ may be regarded as the cyclic components of a certain vector known as the orientation vector. This vector defines the dipole ordering of the excited state angular momenta. Five components, $\rho_{0,\pm1,\pm2}^{(2)}$, form the alignment tensor, determining the quadrupole ordering of angular momenta of atomic particles. Component $\rho_0^{(2)}$ is called the longitudial alignment. It is expressed in terms of populations of the Zeeman substates $\rho_{mm}(J)$:

$$\rho_0^{(2)}(J) = \sqrt{\frac{5}{(2J+3)(2J+1)(J+1)J(2J-1)}} \sum_m \left[3m^2 - J(J+1)\right] \rho_{mm}(J)$$

Types of ordering of angular momenta represented by the polarization moments of excited atomic ensembles are directly reflected in the spectropolarimetric characteristics of the emission light or the full set of Stokes parameters I, P, Q, V.

IMPACT SELFALIGNMENT

The first qualitative studies in the simple laboratory gas discharges, performed in the middle of 70th (4) indicated, that the spectropolarimetric effects under the direct electron impact excitation are very informative, directly reflecting the structural features of a plasma. In such a way the spectropolarimetric studies allowed to detect a very fine kinetic effect of the formation of cones of losses in the velocity distribution of fast electrons, resulting from the interaction of the plasma with the wall of the discharge tube, to sense orientation of these cones of losses and the radial profile of the internal plasma electric field. The analysis of the spectropolarimetric effects in the most simple case of the low temperature gas discharge plasma allowed to infer, that, in absense of the external electric, magnetic fields, beams of precipitating particles, gradients of environmental parameters, the observed polarization is deteremined by the selfalignment of particles or the quadrupole orientation of momenta of ensembles of excited atomic particles, induced by internal excitation and relaxation processes in an ionized gas, $\rho_q^{(2)}(\vec{r})$. Selfalignment of atomic ensembles reflects one of the fundamental properties of a real ionized gas entity such as deviations from the thermal equilibrium and spatial confinement, and in such a way, it could be regarded as one of general features of ionized gas entities.

The most promising in respect of diagnostics is the selfalignment caused by impact processes in a plasma and specifically due to anisotropies in the relative velocity space of the colliding particles. Historically, the most clear way to understand and to demonstrate practical potentialities of the impact selfalignment was to analyse this effect under the direct excitation of atomic particles by fast electrons in a simple laboratory gas discharge, because electron atom collisions is a typical mechanism of excitation and ionization of atoms in low temperature diluted plasmas. Understanding of the basic propertiers of the impact selfalignment of atomic ensembles and formulation of the basic integral equations, connecting the observable spectropolarimetric characetristics with the kinetic parameters of fast exciting electrons, allowed to regard the polarization spectroscopy as a principally new method of the quantitative remotre sensing. The most important result in view of the new contribution of spectropolarimetry to remote sensing of ionized gases is that the Stokes parameters of the emited spectral lines by a plasma are expressed in terms of the quadrupole moment of the velocity distribution function of fast electrons $f^{(2)}(v)$ (5), which is the third coefficient of the multipole representation of the velocity distribution function in the velocity space:

$$f(\vec{v}) = \sum_{k=0}^{\infty} \sum_{q=-k}^{k} Y_q^{(k)}(\vec{n}) f_q^{(k)}(v)$$

where $Y_q^{(k)}(\vec{n})$ ($\vec{n} = \vec{v}/v$, $v = |\vec{v}|$) are spherical harmonics determined in the velocity space of electrons.

By its nature, the quadrupole momenmt of the velocity distribution function, being the anisotropic pressure tensor of the electron component, characterizes the local momentum flow density of electrons and, as a result, the quantitative characteristics of the energy exchange. Indeed, the principle processes indespensable for understanding of nature of an ionized gas entity and connected to the energy exchange with the environment are directly reflected in the kinetics of the most mobile electron component. Analysing the energy aspect of the maintenance of an ionized gas entity, it is possible to conclude, that the energy transport into an ionized medium is implemented, as a rule, in a nonthertmal form and it is accompanied by strong gradients of parameters and the spatially located anisotropy of the electron motion, especially in the fast part of the velocity distribution. Interaction of the fast electrons with other plasma components, transforms the transmittered energy, associated with the ordered motion into the energy of the chaotic thermal motion. Energy dissipation at the boundaries of an ionized medium is also reflected in the anisotropy of the electron motion and provides the increase of the local momentum flow tensor. All the anisotropies in the velocity space of electrons directly connected to the structural features of an ionized entity are by these means reflected in the polarization of emission (degree and polarization plane orientation).

POLARIZATION SPECTROSCOPY EQUATIONS

In terms of the multipole moments of the velocity distribution function, Stokes parameters of the spactral lines (I, P, Q, V) are expressed (3):

$$I = \frac{2}{3}(2\kappa + 1)^{-1/2} B(\omega) N_a \bar{\tau} \mid \langle J \parallel \hat{d} \parallel J_1 \rangle \mid^2 I_0$$

$$P = \frac{4\pi w(JJ_1)}{5I_0\Gamma_2}\int_{v_t}^{\infty}dvv^2\check{F}_2(\nu)f^{(2)}_{2+}(v) \quad Q = \frac{-i4\pi w(JJ_1)}{5I_0\Gamma_2}\int_{v_t}^{\infty}dvv^2\check{F}_2(\nu)f^{(2)}_{2-}(v)$$

$$I_0 = \frac{4\pi}{\Gamma_0}\int_{v_t}^{\infty}dvv^2\check{F}_0(\nu)f^{(0)}_0(v) + \frac{8\pi w(JJ_1)}{5\sqrt{6}\Gamma_2}\int_{v_t}^{\infty}dvv^2\check{F}_2(\nu)f^{(2)}_0(v)$$

where $f^{(2)}_{2\pm}(\nu) = f^{(2)}_2(v) \pm f^{(2)}_{-2}(v)$ is calculated in the detector frame of reference, $B(\omega)$, $w(JJ_1)$, $\tilde{\tau}$, N_a-parameters determined in (3), Γ_2-alignment relaxation constant.

The excitation rate may be expressed in terms of the cross section of impact excitation of atomic polarization moments $\check{F}_\kappa(v) = v\sigma^{(\kappa)}(v)$, where

$$\sigma^{(k)}(v) = (2k+1)^{1/2}\sum_m(-1)^{J-m}\begin{pmatrix} J & J & k \\ m & -m & 0 \end{pmatrix}\sigma_m(v)$$

and $\sigma_m(v)$ is the cross-section of excitation of magnetic sublevels.
The cross-section of the electron impact alignment takes the form:

$$\sigma^{(2)}(v) = \sqrt{\frac{5}{(2J+3)(2J+1)(J+1)J(2J-1)}}\sum_m\left[3m^2 - J(J+1)\right]\sigma_m(v)$$

Under the Born approximation the impact alignment cross section could be calculated and expressed following (3) in terms of the inelastic differential collision cross section $\sigma(\theta,v)$:

$$\sigma^{(2)}(v) = 3\pi\alpha(S)\int_0^\pi d\theta\,\sin\theta\sigma(\theta,v)\frac{1+\delta\cos^2\theta - 2\sqrt{\delta}\cos\theta}{1+\delta - 2\sqrt{\delta}\cos\theta}$$

Regarding one of the most typical case of an axially symmetric plasma, it is possible to write the following system of integral equation for the relevant Stokes parameters (ε_t - threshold excitation energy):

$$P = \frac{8\pi w(JJ_1)}{5m^2\Gamma_2 I_0}\int_{\varepsilon_t}^{\infty}d\varepsilon\varepsilon\sigma^{(2)}(\varepsilon)f^{(2)}_{2+}(\varepsilon), \quad Q = \frac{-i8\pi w(JJ_1)}{5m^2 I_0\Gamma_2}\int_{\varepsilon_t}^{\infty}d\varepsilon\varepsilon\sigma^{(2)}(\varepsilon)f^{(2)}_{2-}(\varepsilon),$$

$$R(JJ_1) = \frac{\frac{1}{5}\sqrt{\frac{3}{2}}w(JJ_1)\frac{\Gamma_0}{\Gamma_2}\int_{\varepsilon_t}^{\infty}d\varepsilon\varepsilon\sigma^{(2)}(\varepsilon)\,\tilde{\tau}f^{(2)}_0(\varepsilon)}{\int_{\varepsilon_t}^{\infty}d\varepsilon\varepsilon\sigma^{(0)}(\varepsilon)f^{(0)}_0(\varepsilon)},$$

$$I_0 = \frac{8\pi}{m^2\Gamma_0}\int_{\varepsilon_t}^{\infty}d\varepsilon\varepsilon\sigma^{(0)}(\varepsilon)f^{(0)}_0(\varepsilon) + \frac{16\pi w(JJ_1)}{5\sqrt{6}m^2\Gamma_2}\int_{\varepsilon_t}^{\infty}d\varepsilon\varepsilon\sigma^{(2)}(\varepsilon)f^{(2)}_0(\varepsilon)$$

Analysing the spatial dependencies of the degree of polarization on the transitions from the levels with different excitation energy and knowing the electron impact alignment cross sections and the radiative and collisional constants of the upper levels it is possible to determine the components of the anisotropic pressure tensor of fast electrons in different parts of plasma, to infer and understand the role of these parts in the energy balance and as a result to get information on the structural features of an ionized medium of different densities as a whole (6).

DRIFT SELFALIGNMENT

Polarization effects could also arise due to the orderings in the relative velocity space of atomic particles, which result from the drift or diffusion of neutral and charged particles in an ionized medium. Collisions between atomic particles having isotropic distribution of relative velocities play purely destructive role, destroying the ordering of the angular momenta and providing the isotropic collisional relaxation of atomic polarization moments. Polarization momets with different ranks undergo independent decay, resulting to independent collisional depolarization of unpolarized, linearly and circularly polarized radiation, described by the depolarization constants $\Gamma_{0,1,2}$. The kinetics of collisional relaxation is changed crucialy when the anisotropy of relative motion of atomic particles or the preferential direction of collisions exists. Collision symmetry transformation from the spherical to the axial one leads to the common collisional relaxation of atomic polarization momets with the same projection on the collision axis and the transformation of polarization moments with the different ranks. The most important place during these effects is taken by the formation of collisional selfalignment from the unequal populations of atomic and ionic levels.

The generation of collisional selfalignment under anisotropic relaxation process of atomic particles in a plasma is possible only if there exist also the quadrupole moment of the relative velocity distribution function. Kinetics of polarization moments of an atomic ensemble under anisotropic collisional relaxation is expressed by the equation:

$$\frac{d}{dt}\rho_0^{(\kappa)}(J) = \sum_{\kappa_1} R_0^{\kappa\kappa_1}(JJ_1)\rho_0^{(\kappa_1)}(J_1)$$

where $R_0^{\kappa\kappa_1}(JJ_1)$ is the matrix of the partial anisotropic relaxation of the longitudinal polarization moments $\rho_0^{(\kappa_1)}(J)$.

This type of selfalignment reflects directly the internal processes in the studied object and could be used for spectropolarimetric remote sensing of different ionized gases in the extraterrestrial, outer space media as well as laboratory and technical plasmas, for example for the plasma etching and deposition techniques. It was observed and analysed for ions with the narrow fine structure, drifting in the atmosphere of neutral component of the gas discherge plasma. The degree of anisotropy of relaxation of excited ions, affected by well expressed anisotropic collision with atoms, the effectivity of the drift selfalignment and the resulting degree of polarization of ionic lines are dependent on the drift velocity of ions in the plasma. Drift selfalignment of ions was exploited to sense the local drift velocity of ions and the distribution of electric field inside the low temperature plasma of the hollow cathode (7).

NEW DEVELOPMENTS

Polarization spectroscopy as a new field of remote sensing of ionized media is developing continuously. New techniques are being put forward, expanding considerably frontiers of possible applications. A typical case of formation of an ionized gas media in technical installations, in the close extraterestrial space and the outer space is when beams or jets of charged particles are transporting through a gas or an ionized media. The emission spectra in these cases are mainly governed by the charge exchange process. Spectropolarimetric effects under charge exchange have been analysed recently (8) and these results formed the basis for spectropolarimetric sensing of the boundary regions of the hot fusion plasma as well as non thermal processes in the solar atmosphere. The most effective results on the spectropolarimetric effects of the recharged ions were obtained with the help of the pseudolevel technique. The wave function in the field of two centers was represented as an expansion in terms of the full system of functions of the recharged ion and, introducing the pseudolevel, the system of equations of the impact parameter method was reduced to the second order system:

$$i\hbar \frac{\partial}{\partial t} b_m(t) = W_{00} b_m(t) + W_{0m} c_m(t) exp(-i\omega_m t)$$

$$i\hbar \frac{\partial}{\partial t} c_m(t) = W_{0m} b_m(t) exp(i\omega_m t) + W_{mm} c_m(t)$$

The matrix elements from this system could be calculated with the help of the hydrogenlike wave functions, using the technique, similar to the method of the quantum defect. Using this technique the spectropolarimetric effects on ionic lines have been computed for the one electron charge exchange process ArIX, KrIX + Li for energies of (60-120) KeV (9).

Another example is the spectropolarimetric remote sensing of the strong electric field in a plasma (10). A set of allowed and forbidden transitions from an excited level, subjected to noticeable Stark effect were proposed to be used.

For axially symmetric excitation process the polarization of emission is determined by the polarization moments, connected to the excitation anisotropies and by the electric field. The electric field contribution to the polarization degree for an optically allowed transition was expressed in terms of polarization moments as well as the degree of polarization for optically forbidden transitions from the same state:

$$P_{forb} = \sqrt{6} \sum \rho_0^{(k)}(J,J) \sum_{J_1 J_2} \begin{Bmatrix} J_1 & J_2 & 2 \\ 1 & 1 & J \end{Bmatrix} <J_0 \| D \| J_1><J_0 \| D \| J_2>^* \cdot$$

$$\cdot \sum_M C_{J_1}(M) C_{J_2}^*(M) \begin{pmatrix} J & J & k \\ M & -M & 0 \end{pmatrix} \begin{pmatrix} J_1 & J_2 & 2 \\ M & -M & 0 \end{pmatrix} (I_\|^{forb} + I_\perp^{forb})^{-1}$$

Measurements of the degree of polarization on the allowed and forbidden transitions provide a system of equations for polarization moments and the electric field strength.

REFERENCES

1. Griem, H.R., *Plasma Spectroscopy*, New York: McGraw-Hill, 1964.
2. Kazantsev, S.A., and Subbotenko, A.V., *Spectropolarimetric Diagnostics of Gas Discharges*, St.Petersburg: State University Publ. House, 1993.
3. Kazantsev, S.A., and Henoux, J.-C., *Polarization Spectroscopy of Ionized Gases*, Dordrecht: KLUWER Academic Publishers, 1995.
4. Kazantsev, S.A., *Sov. Phys. Usp. (USA)* **26**(4), 328-351 (1983).
5. Kazantsev, S.A., *JETP Letters (USA)* **37**, 158-160 (1983).
6. Kazantsev, S.A., Polynovskaya, N.Ya., Piatnitsky, L.N., Edelman, S.A., *Sov. Phys. Usp. (USA)* **31**(9), 785-809 (1988).
7. Kazantsev, S.A., Petrashen, A.G., Rebane, V.N., Rebane, T.K., Funtov, V.N., Neureiter, C., Windholz, L., *Physica Scripta* **52**, 572-587 (1995).
8. Kazantsev, S.A., Petrashen, A.G., *Optics and Spectroscopy (USA)* **77**(6), 807-808 (1994).
9. Chantepie, M., Jacquet, E., Laulhe, C., Kazantsev, S.A., Petrashen, A.G., *Optics and Spectroscopy (USA)*, to be published in 1996.
10. Demkin, V.P., Kazantsev, S.A., *Optics and Spectroscopy (USA)* **78**(3), 337-352 (1995).

LINE SHAPE MEASUREMENTS OF VISIBLE LIGHT EMISSION FROM THE ALCATOR C-MOD TOKAMAK

B. L. Welch, H. R. Griem, and J. L. Weaver
Institute for Plasma Research, University of Maryland
College Park, Maryland 20742-3511

J. L. Terry, R. L. Boivin, B. Lipschultz, D. Lumma,
E. S. Marmar, G. McCracken, and J. C. Rost
Plasma Fusion Center, Massachusetts Institute of Technology
Cambridge, Massachusetts 02139

Measurements have been made of the line shapes of the Balmer series of hydrogen/deuterium in the edge and divertor regions of the Alcator C-Mod tokamak. The lower series members show mostly Zeeman splitting and Doppler broadening. The higher series members exhibit considerable Stark broadening. These measurements are being used to determine the magnetic field strength, ion temperature, and electron density in the edge and divertor regions. Examples of measurements from the various regions of the tokamak are presented.

I. INTRODUCTION

The Alcator C-Mod tokamak[1] is a compact (major radius = 68 cm, minor radius = 22 cm), high toroidal field ($B_T \leq 9T$) tokamak with the capability of producing various shaped (kappa ≤ 1.85), diverted plasmas.[2] The ion cyclotron radio frequency (ICRF) plasma heating uses up to 3.5 MW of RF power to produce electron temperatures of up to 6 keV and ion temperatures of up to 5 keV (for electron densities up to $2 \times 10^{20} m^{-3}$). C-Mod is unique in that it has molybdenum plasma facing components.

Visible and ultraviolet emission has been measured with a spectrograph viewing the plasma via quartz fibers positioned at various locations around the tokamak. Emission from the edge region of the main chamber was recorded from the side and above and emission from the divertor region was recorded from fibers looking into the lower divertor region. A drawing of the cross section of Alcator C-Mod, as well as the views, is shown in Fig. 1.

II. EXPERIMENTAL ARRANGEMENT

Up to 50 spectra are recorded throughout each discharge at 30 ms intervals. The spectrograph views a 200 Å region with a dispersion of 0.25 Å/pixel. Due to blurring by the microchannel plate intensifier, the instrumental width of the detector was determined to be 0.8 Å. The spectrograph can view the plasma through one of four quartz fibers. The fibers were positioned to view the plasma as described above and shown in Fig. 1.

III. MEASUREMENTS OF THE HYDROGEN AND DEUTERIUM ISOTOPE FRACTIONS

The first line of the Balmer series (n=3 to 2 transition) is the most prominent line in the spectrum and is used to determine the ratio of the hydrogen to deuterium concentrations in the discharge. This ratio is important for several of the ICRF heating scenarios used on

© 1996 American Institute of Physics

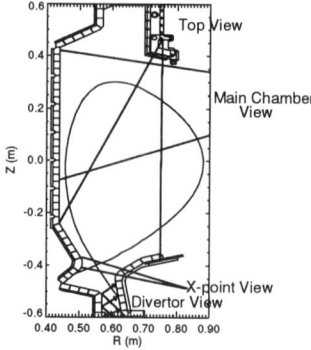

FIG. 1. Schematic drawing of the cross section of Alcator C-Mod showing four fiber views.

Alcator C-Mod. At 5.3 Tesla the RF power is resonant with the hydrogen ion cyclotron frequency in the core of the plasma and couples to the protons. This energy is then transferred to the electrons and deuterons by collisions, to heat the bulk of the plasma. A small hydrogen fraction is desired, and the ratio of the H_α to D_α intensity is used to determine this fraction. Hydrogen fractions on the order of 5 per cent are typical.

The interpretation of the H_α (6562.8 Å) and D_α spectra (6561.0 Å) is complicated by the fact that the transition is split into the three Zeeman components by the magnetic field. The isotope splitting is approximately equal to this splitting and therefore produces two overlapping Zeeman triplets. Examples of this are given in Fig. 2. In Fig. 2(a) the Zeeman triplet is evident (6559.5, 6561.0 and 6562.5 Å), with the discharge being primarily (more than 98 per cent) deuterium. In Fig. 2(b) the addition of the Zeeman triplet associated with hydrogen (6561.3, 6562.8, and 6564.3 Å) indicates a hydrogen fraction of 25 per cent. This assumes the population mechanisms for the two isotopes are the same and the isotope ratio is equal to the intensity ratio. Also shown in the figure are the fits to the data which were used to determine this ratio.

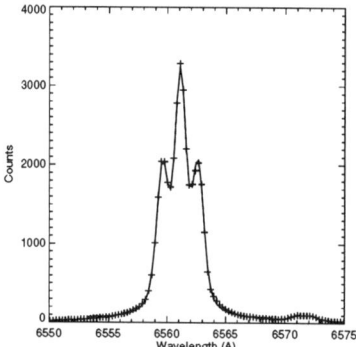 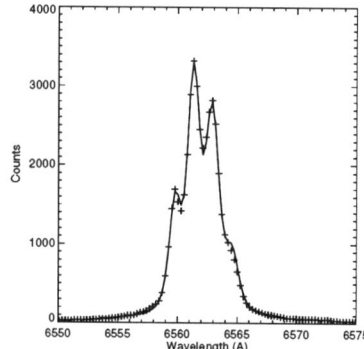

FIG. 2. Spectra of the first transition of the Balmer series with (a) nearly 100 percent deuterium and (b) approximately 25:75, hydrogen to deuterium.

An experiment has been done to determine the retention of the hydrogen isotopes in the molybdenum walls of the chamber. The fueling gas was changed from deuterium to hydrogen and the hydrogen fraction measured during a number of shots. The fueling gas was then changed back to deuterium and the hydrogen fraction again measured for a number of shots. Examples of the time history of the hydrogen fraction for two different discharges are given in Fig. 3. When the gas was first changed to hydrogen, the hydrogen fraction was seen to vary from 80 to 50 per cent during the discharge (see Fig. 3(a)). After ten shots the hydrogen fraction was seen to vary from 90 to 80 per cent (see Fig. 3(b)). The hydrogen fraction measured by the neutral particle analyzer (NPA) is shown in Fig.3 for comparison. The neutral particle analyzer measures the energy distribution of escaping neutral hydrogen and deuterium particles (simultaneously).[3] The isotope fraction is then obtained by taking the ratio of the distributions, corrected for isotope effects, e.g. in the charge-exchange and electron stripping cross-sections. The agreement between the NPA and the spectroscopic measurements is very good. The hydrogen fraction is seen to be higher earlier in the discharge, indicative of the pure hydrogen prefill. Later in the discharge the hydrogen fraction decreases, indicating that the discharge is being fueled from the deuterium in the walls.

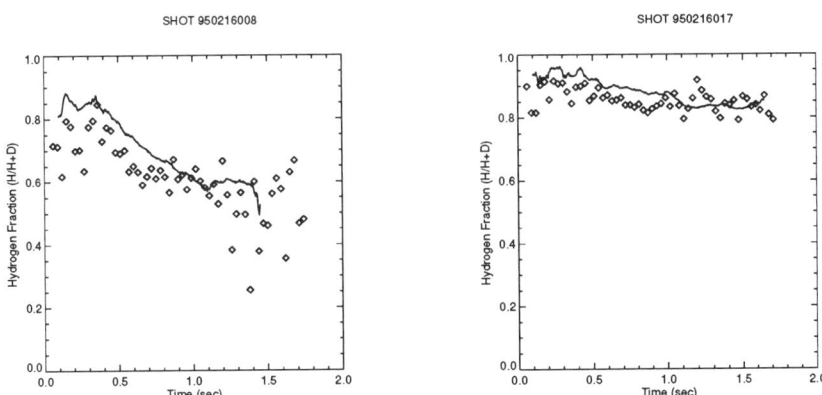

FIG. 3. Time histories of hydrogen fraction (a) immediately after switching from deuterium to hydrogen and (b) ten shots later. Diamonds are from spectroscopic measurements and the solid line is from the neutral particle analyzer.

A summary of the results of the isotope exchange experiment is given in Fig. 4. For shots 7 through 24 only hydrogen was puffed into the chamber. For shots 26 through 33 only deuterium was puffed. For the hydrogen puff, the hydrogen fraction is consistently higher early in the discharge. Conversely, after switching back to deuterium, the hydrogen fraction is lower early in the discharge. In both cases, this indicates a short term "memory" of the previous isotope abundance. An exponential fit to the late time (1.5 sec) data indicates an e-folding of 8-9 shots. This time is more relevant than the early time, since the RF pulse on Alcator is usually later in the discharge, during the current flat-top.

A higher resolution instrument has recently been put into operation which enables the determination of the width of the Balmer Alpha line of the hydrogen isotopes. An example is given in Fig. 5. Preliminary results indicate up to a factor of two larger line width observed in the main chamber than observed in the divertor. This difference may be caused by a temperature difference in the edges of the two regions. These widths would indicate a mean neutral energy of 2-7 eV.

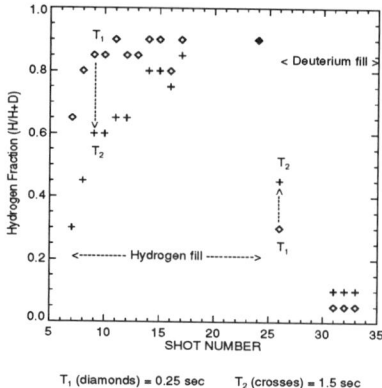

FIG. 4. Summary of isotope exchange experiment. Diamonds represent hydrogen fraction at $T_1=0.25$ sec and crosses represent hydrogen fraction at $T_2=1.5$ sec.

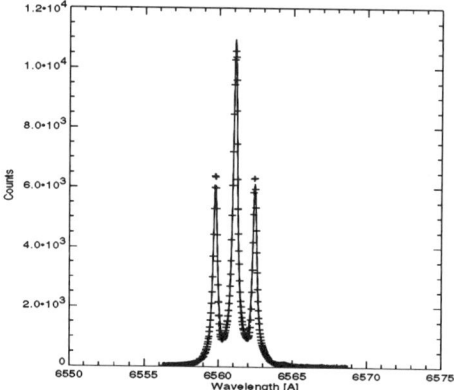

FIG. 5. High resolution spectra of the α transition of the Balmer series with nearly 100 percent deuterium.

IV. STARK BROADENING MEASUREMENTS

High-n members of the Balmer series were measured in the spectral region from 3700 to 3900 Å which indicate significant Stark broadening. This Stark broadening is being used to determine the electron density in the edge and divertor regions where the measurements are taken.[4] Examples of these spectra are given in Fig. 6. The spectrum in Fig. 6(a) indicates a low density plasma ($N_e = 0.3 \times 10^{20}\ m^{-3}$) and the spectrum in Fig. 6(b) indicates a high density plasma ($N_e = 12 \times 10^{20}\ m^{-3}$). Significant Stark broadening is evident in Fig. 6(b).

These spectra are fit with Voigt profiles and the resulting Lorentzian FWHM are used to calculate the electron density from Stark broadening calculations.[6] This electron density measurement can then be compared with other electron density measurements.

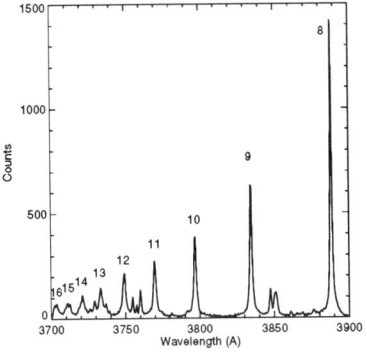 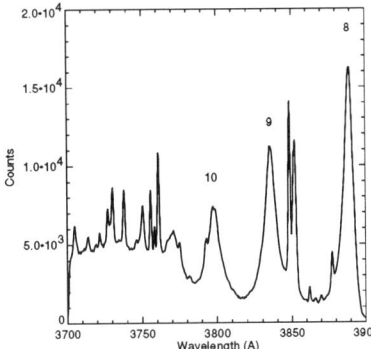

FIG. 6. Spectra from Alcator C-Mod showing Balmer series transitions up to $n = 16$. Spectrum taken (a) from divertor during typical (low density) discharge and (b) from main chamber during (high density) capillary gas puffing. Unlabeled lines are from oxygen and fluorine.

The electron density taken along a chord through the X-point has been compared with the average density in the main chamber, taken with the two-color interferometer system.[5] The results are presented in Fig. 7. As the main chamber density changes from 1 to 4 $\times 10^{20}$ m^{-3}, the density along the X-point view changes from 5 to 20 $\times 10^{20}$ m^{-3}. A linear fit to this relation shows a factor of five higher density measured along the chord through the X-point than the main chamber density. This is the density in the region of maximum Balmer series emission. These measurements were taken with no RF heating. Measurements with RF heating show a decrease in the density in the divertor region as the RF power is increased.

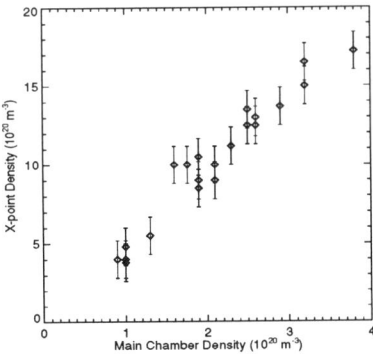

FIG. 7. Electron density in the X-point region (line averaged) versus the average main chamber electron density. The error bars are representative of the variation in the line width measurement.

An example of the time history of the electron density in the divertor, measured by Stark broadening, is given in Fig. 8, along with the electron density measured at the outer divertor plate with the domed Langmuir probes.[7] These time histories show excellent agreement, although exceptions to this have been seen for reasons which are still not clear. The wave shape of the RF heating pulse is shown for a timing comparison, and it can be seen that the electron density in the divertor decreases as the RF power is stepped up. A summary of this is given in Fig. 9. The ratio of the density in the divertor to the density in the main chamber is 3 to 4 with no RF heating. This is slightly below the value measured along the X-point chord as shown in Fig. 7. This value decreases to 1.5 to 2 as the RF heating power increases to 2 MW.

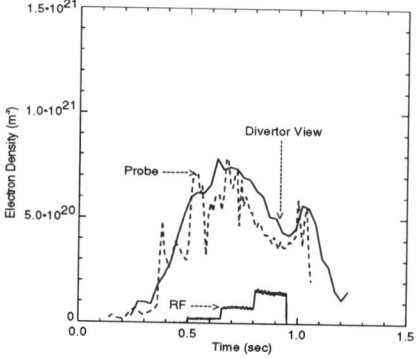

FIG. 8. Time histories of the electron density in the divertor region during RF heating. The electron density in the divertor was determined from the line widths of the 9-2 transition of the Balmer series (solid line) and the domed Langmuir probes in the divertor (dashed line). Also shown is the pulse shape of the RF pulse (RF power steps of 0.1, 0.4 and 0.8 MW).

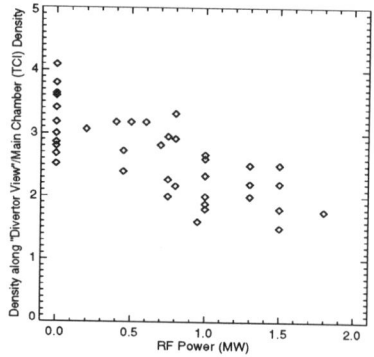

FIG. 9. Electron density in the divertor region (line averaged) versus the RF power input.

The electron density in a MARFE[8] phenomenon was also measured using this technique, along with a filtered diode array[9] which was used to identify the occurrence and location of the MARFE phenomenon. An array of diodes filtered to transmit the D_α (6561 Å) radiation views various chords of the plasma from beyond the outer scrape-off layer. The combination of these diagnostics made possible the localization (in the Z direction) of the spectrograph's spatially integrated measurement. It should be pointed out that the Zeeman pattern of the H_α/D_α line measured from this view, as previously discussed, corresponds to the magnetic field at the inner wall. This localizes the radiation in the radial dimension.

The line widths of Balmer transitions with upper levels from n=7 through n=11 have been measured simultaneously in pairs. The ratio of these widths should be independent of electron density and can therefore be compared with calculations. These ratios of adjacent line widths are shown in Fig. 10 along with the ratio of the calculated values. A scaling of $[(n+1) \times \lambda_{n+1}/(n \times \lambda_n)]^2$ is shown for comparison. The ratio of the calculated even to odd values is consistently higher than the experimental values. This may be due, in part, to the structure in the line shapes from the calculation. Examples of the line shapes of the 6-2 and 7-2 transitions are given in Fig. 11.[10] Notice the central peak in the odd-n transition and the central dip in the even-n transition.

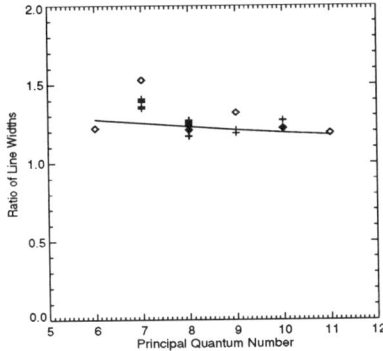

FIG. 10. Ratio of the widths of adjacent lines ($\Delta\lambda_{n+1}/\Delta\lambda_n$) versus the principal quantum number (n). The crosses are from experimental measurements from this paper. The diamonds represent data from Stark broadening calculations. The solid line is $[(n+1) \times \lambda_{n+1}/(n \times \lambda_n)]^2$ scaling.

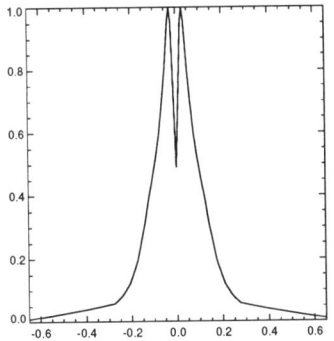 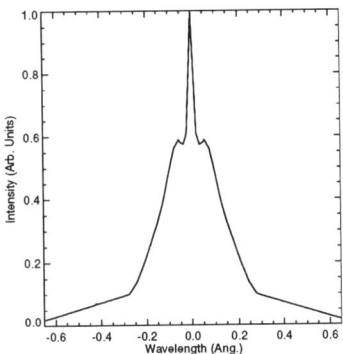

FIG. 11. Calculated Stark broadening line shapes of the (a) 6 to 2 and (b) 7 to 2 transitions of the Balmer series.

High-n transitions of the Paschen series of deuterium have recently been measured in the divertor region of the Alcator C-Mod tokamak. Spectra from a low density plasma indicated 13 Å Stark broadening of the n=9 to 3 transition whereas spectra from a high density plasma indicated 60 Å Stark broadening. The broadening of these lines indicate 3×10^{20} and $3 \times 10^{21} m^{-3}$, respectively.[11] These transitions can be used to extend this density measurement technique to much lower densities.

V. CONCLUSIONS

The line shapes of Balmer series transitions have been investigated in the Alcator C-Mod tokamak. The intensity of the D_α transition relative to the H_α transition allows the determination of the relative concentrations of the two isotopes. This ratio is important for the ICRF minority heating of the hydrogen. The wavelength separation of the Zeeman shifted components can be used to determine the magnetic field strength and an independent measure of the magnetic field localizes this emission measurement radially in the field. Chordal measurements of the D_α emission can localize the emission in the Z direction. High-n transitions can be used to determine the electron density from the Stark broadening of the transitions. Electron densities in the divertor and X-point regions of 4-5 times that in the main chamber have been measured. Stark broadening of the Paschen series transitions can extend this electron density measurement technique to lower densities, even with a modest resolution spectrograph.

VI. ACKNOWLEDGMENTS

The authors would like to acknowledge the help of the entire Alcator C-Mod staff. Without the continuing efforts of the entire group these results would not be possible. We would also like to acknowledge the discussions with P.C. Kepple concerning the theoretical Stark broadened line profiles. The present work was performed under the auspices of the U.S. Department of Energy by the University of Maryland under Contract No. DE-FG02-95ER54307 and by MIT under U.S. DOE Contract No. DE-AC02-78ET51013.

---

[1] I.H. Hutchinson, *Proceedings of the IEEE 13th Symposium on Fusion Engineering*, Knoxville, TN, edited by M. Lubell, M. Nestor, and S. Vaughan (Institute of Electrical and Electronic Engineers, New York, 1990), Vol. 1, p. 13.

[2] I. H. Hutchinson, R. Boivin, F. Bombarda, P. Bonoli, C. Fiore, J. Goetz, S. Golavato, R. Granetz, M. Greenwald, S. Horne, A. Hubbard, J. Irby, B. LaBombard, B. Lipschultz, E. Marmar, G. McCracken, M. Porkolab, J. Rice, J. Snipes, Y. Takase, J. Terry, S. Wolfe, C. Christensen, D. Garnier, M. Graf, T. Hsu, T. Luke, M. May, A. Niemczewski, G. Tinios, J. Schachter, and J. Urbahn, Phys. Plasmas **1**, 1511 (1995).

[3] C. Kurz and C. L. Fiore, Rev. Sci. Instrum. **61**, 3119 (1990).

[4] B. Welch, H. Griem, J. Terry, C. Kurz, B. LaBombard, B. Lipschultz, E. Marmar, and G. McCracken, Phys. Plasmas **2**, 4246 (1995).

[5] J. Irby, T. Luke, E. Marmar, S Wolfe, Rev. Sci. Instrum. **59**, 1568 (1988).

[6] R. D. Bengtson, J. D. Tannich, and P. Kepple, Phys. Rev. A **1**, 532 (1970).

[7] B. LaBombard, D. Jablonski, B. Lipschultz, G.M. McCracken, and J. Goetz, J. Nucl. Mater. **220-222**, 976 (1995).

[8] B. Lipschultz, B. LaBombard, E. S. Marmar, M. M. Pickrell, J. L. Terry, R. Watterson, and S. M. Wolfe, Nucl. Fusion **24**, 977 (1984).

[9] J.L. Terry, J.A. Snipes, and C. Kurz, Rev. Sci. Instrum. **66**, 555 (1995).

[10] P.C. Kepple, private communication (1994).

[11] G. Himmel and F. Pinnekamp, J. Quant. Spectrosc. Radiat. Transfer **13**, 555 (1973).

ATOMIC DATA AND DATABASES

Ion-atom collisions relevant to fusion plasmas

H. B. Gilbody

Department of Pure and Applied Physics,
The Queen's University of Belfast, Belfast, United Kingdom.

Abstract. Some of the wide variety of ion-atom collision processes relevant to heating, modelling and diagnostics of Tokamak plasmas are considered. In particular, results of recent experimental measurements on selected processes involving charge transfer and excitation (based on translational energy spectroscopy and photon emission spectroscopy) are discussed in relation to theoretical predictions.

INTRODUCTION

The important role of ion-atom collision processes in the design and operation of Tokamak devices is well known. Some of the cross section needs relating to current and next-step devices such as ITER are summarised in a recent report by Janev [1]. If we consider the processes of charge transfer, ionization and excitation we can easily identify the following three areas of continuing importance where reliable cross section data are required.

(a) Plasma heating by fast neutral beams of hydrogen or helium.
(b) Modelling of the plasma edge and the design of divertors for control and removal of impurities including the helium 'ash'.
(c) Diagnostics of plasma parameters by neutral beam probes.

Supplementary heating in the present large Tokamaks involves the injection of 40 - 80 keV amu^{-1} neutral beams of hydrogen isotopes. The processes of charge transfer

$$H^+ + H \rightarrow H + H^+ \quad (1)$$

and ionization

$$H^+ + H \rightarrow H^+ + H^+ + e \quad (2)$$

in collisions with plasma protons are important in this context. However, collisions with multiply charged impurities (many of which originate at the plasma facing walls)

$$X^{q+} + H \rightarrow X^{(q-1)+} + H^+ \quad (3)$$

$$X^{q+} + H \rightarrow X^{q+} + H^+ + e \quad (4)$$

can also result in serious modification of the energy deposition profiles. In (3) electron capture into high n states of $X^{(q-1)+}$ followed by spontaneous radiative decay can also result in serious plasma cooling. Cross sections for (1), (2), (3) and (4) are now quite well established for collisions of ground state H atoms over the energy range of interest. However, at energies above about 200 keV amu^{-1} (which are appropriate to next-step devices) where the radiative and collision times of the excited beam particles become comparable, cross sections for fast excited H atoms are required to determine the effect on beam attenuation. Recently, fast neutral helium beams have also been used for plasma heating. Here processes involving both one and two electron removal must be considered. The fast metastable content of the helium beams is an additional complicating feature.

In the comparatively cool plasma edge region, many different types of collision process are relevant. In the 1 - 500 eV energy range of primary interest, resonant or near resonant charge transfer processes or moderately exothermic charge transfer processes involving slow multiply charged ions may have high cross sections which can influence the particle, energy and ionization balance. Collisions involving hydrogen atoms, molecules and their ions together with helium atoms and ions are important as primary constituents of the plasma edge. A wide range of low and high Z impurities in various stages of ionization are of interest. Molecular impurities include H_2 and CO and a range of C_nH_m impurities may also be important. In addition, there is interest in introducing cold H_2 or rare gases into the divertor region to produce enhanced cooling or diagnostic information.

Diagnostics of plasma parameters has been carried out with both fast neutral heating beams and with dedicated beam probes. Charge exchange recombination spectroscopy has been used extensively with fast H atom beams [2] for the diagnostics of fully ionized plasma species such as He^{2+}, C^{6+} and O^{8+}. Observations are made of the decay of the H-like excited products formed in the highly state-selective electron capture process

$$X^{z+} + H \rightarrow X^{(z-1)+}(n,l) + H^+ \qquad (5)$$

Although the dominant excited product channels in (5) generally radiate in the VUV, observations of the much weaker non-dominant capture channels which radiate in the visible region can be advantageous because of the convenience of fibre optics coupling to the spectrometer and easier calibration. Measurements of this type have also been carried out using fast He and Li beams.

The technique of beam emission spectroscopy is based upon observations of the emissions from fast excited beam atoms formed in collisions with the plasma ions and electrons. Accurate cross sections for direct excitation of H or He atoms by H^+, He^{2+} or by particular X^{q+} impurity ions are needed in this approach. For example, studies of Balmer alpha emission emitted from fast H heating beams [3] have been used as a diagnostic of H^+ densities and other plasma parameters.

Reliable measurements of cross sections for processes involving H atoms, unlike stable gases such as H_2 or He, have required the development of special well characterised sources of atomic hydrogen and a number of different experimental approaches have evolved over the past few decades. Few of these measurements are absolute and most require the measured relative cross sections to be normalised to other data. The accuracy of early measurements [4] based on modulated crossed beam methods employing thermal energy beams of highly dissociated hydrogen were severely limited by poor signal to background ratios but the use of coincidence counting techniques and TOF spectroscopy in methods developed in this laboratory [5] now provide high sensitivity and accuracy. In other work [6], the use of intense ion beams together with hydrogen beams of high intensity from a Wood's tube discharge have allowed detailed studies of processes involving excitation by the use of photon emission spectroscopy (PES). Many studies of electron capture by ion impact have been based on the furnace target approach in which a tungsten tube furnace provides a target of highly dissociated hydrogen. Translational energy spectroscopy (TES), first used in this laboratory for studies of charge transfer in X^{q+}- H collisions [7], can provide detailed information on state selective electron capture.

RESULTS OF SELECTED MEASUREMENTS

Since the late 1970's a great deal of information on charge transfer and ionization processes involving collisions of multiply charged ions with ground state H, H_2 and He has been obtained, particularly at velocities v > 1 au corresponding to 25 keV amu^{-1}.

Fig 1 shows a typical set of experimental data for (3) and (4) which indicate the relative importance of charge transfer and ionization over a wide energy range. For q ≥ 3, charge transfer cross sections for a particular velocity can be seen to increase with q . At velocities beyond the peak values ionization cross sections also increase with q. High velocity cross sections for both charge transfer and ionization are found to be q rather than species dependent and, for q ≥ 3, scale according to simple relations of the form $\sigma = \sigma_o q^n$ where the scaling parameters σ_o and n vary slowly with velocity. These observations have provided the basis for the development of general q scaling relations which allow high velocity cross

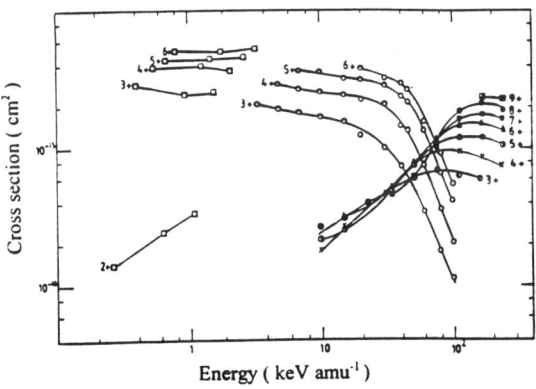

Fig 1. Cross sections for charge transfer and ionization in Ar^{q+} - H($1s$) collisions for q = 2 -9. □: charge transfer - furnace target method [8]; O: charge transfer - crossed beam coincidence method [9]. ◪, ●, ■, ▲, ◐, ×, ◓ : ionization - crossed beam coincidence method [9].

sections to be predicted within 20 - 30% for any high q primary ion in collisions with H, H_2 or He (see for example [10]). At intermediate velocities, calculations based on the Classical Trajectory Monte Carlo (CTMC) approach [11] provide a reasonable description of the experimental measurements.

At velocities v < 1 au, measured charge transfer cross sections cannot, in general, be described by simple q scaling relations. Exothermic electron capture may take place very effectively through avoided crossings of the adiabatic potential energy curves describing the initial and final molecular systems. These occur at internuclear separations $R_c \approx (q-1)/\Delta E$ where ΔE is the energy defect for the formation of product ions in specified excited states n,l. In many reactions the total electron capture cross section is dominated by a limited number of such crossings. The large and weakly energy dependent values for the exothermic $q \geq 3$ charge transfer processes in Fig. 1, are quite unlike the q = 2 cross sections, and indicative of a number of effective curve crossings. Accurate theoretical descriptions of state selective electron capture are difficult, and generally involve solution of the coupled channel equations with a multi-state expansion of the total electron wave function. However, a simple multi-channel Landau-Zener model is sometimes useful for predicting the main product channels. Experimental measurements based on the complementary techniques of TES and PES can, in principle, provide identification and relative cross sections for the excited product channels. The PES approach provides much greater energy resolution than TES but accurate optical calibration, particularly in the VUV is difficult. In TES, calibration with respect to known cross sections is simpler and the method (unlike PES) easily identifies product channels associated with any metastable ions present in the primary beam.

Both TES and PES measurements show that electron capture in collisions between slow fully stripped ions and H, He and H_2 is highly selective and, in most cases, dominated by a single avoided crossing. For example, in collisions of C^{6+}, N^{7+} and O^{8+} with H atoms, the dominant products are found [12], to be $C^{5+}(n=4)$, $N^{6+}(n=5)$ and $O^{7+}(n=5)$ respectively in accord with theoretical predictions. Hoever, for partially ionized species, where a number of avoided crossings may be important, the main product channels are less easy to predict theoretically.

Fig.2 shows a TES energy charge spectrum [13] observed for 15keV C^{3+}-H collisions in which the three main peaks can be identified with the $C^{2+}(1s^22s3s)^3S$, $C^{2+}(1s^22p^2)^1S$ and $C^{2+}(1s^22p^2)^1D$ product channels. This process has also been studied [14] using the PES approach and in Fig.2 are shown the cross sections derived from both theTES and PES methods together with theoretical predictions by Bienstock et al [14] and by Errea et al [15] based on close coupling methods; the latter most recent calculations are based on an expansion involving 22 molecular states. The limited degree of agreement between the TES and PES measurements and between the calculations in this case reflects the difficulty of carrying out both reliable measurements and accurate calculations.

In the case of electron capture in C^{4+}-H collisions there is much better accord between TES and PES measurements and close coupling calculations (see [16]). For many processes, no detailed calculations are available, and in these cases it is

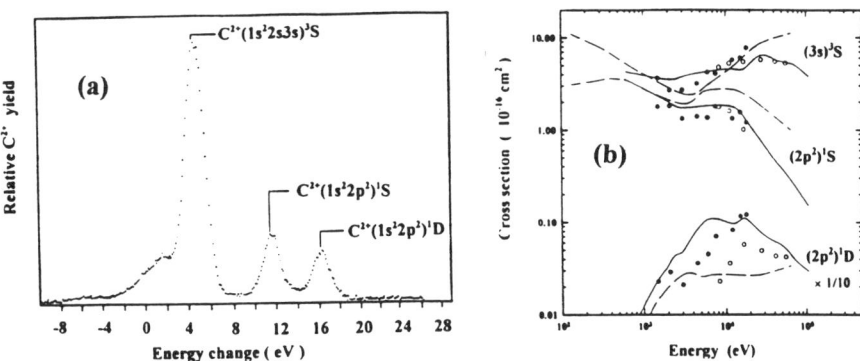

Fig.2. Electron capture in C^{3+}- H(1s) collisions showing (a) TES energy change spectrum [13] and (b) cross sections for $C^{2+}(n,l)$ formation. ●, TES measurements [13] ; O, PES measurements [6]; ——— , theory, Erea et al [15]; — — , theory, Bienstock et al [14].

of interest to investigate the extent to which MCLZ calculations can predict the main product channels. Fig.3 shows the energy change spectrum obtained for one

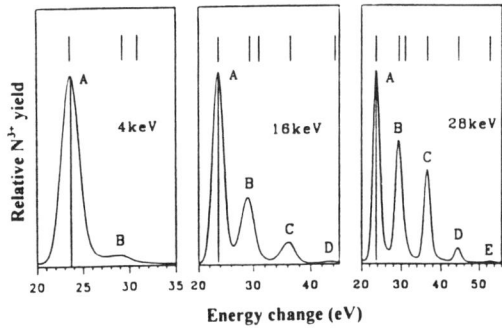

Fig.3. TES energy change spectra for one electron capture in N^{4+}-He collisions [16].

electron capture in N^{4+}-He collisions A number of well resolved peaks A, B, C, D and E can be identified with the $N^{3+}(1s^22p)2p^1S$, $N^{3+}(1s^22p)2p^1D,^3P$, $N^{3+}(1s^22s)2p^1P$, $N^{3+}(1s^22s)2p^3P$, $N^{3+}(1s^22s)2s^1S$ product channels respectively. The vertical lines are the results of our MCLZ calculations where, in order to facilitate comaparison with experiment, the largest calculated cross section has been normalised to the largest observed peak value. In this case the calculations only predict the main peak satisfactorily; peak B is grossly underestimated while peaks C, D and E are not predicted at all. In this and in other cases, MCLZ calculations seem to be only of limited value.

Many TES measurements reveal the presence of collision channels associated with metastable as well as ground state primary ions. In these cases quantitative analysis of the spectra is precluded since the metastable fraction present is unknown. For example, our TES studies of electron capture in C^{2+}-H collisions [7] show that at 4keV, product channels involving metastable $C^{2+}(2s2p)^3P^o$ primary ions account for about 60% of the total capture. Many partially ionized metallic species relevant to edge plasmas have metastable states of high statistical weight indicating their likely strong influence on measured cross sections.

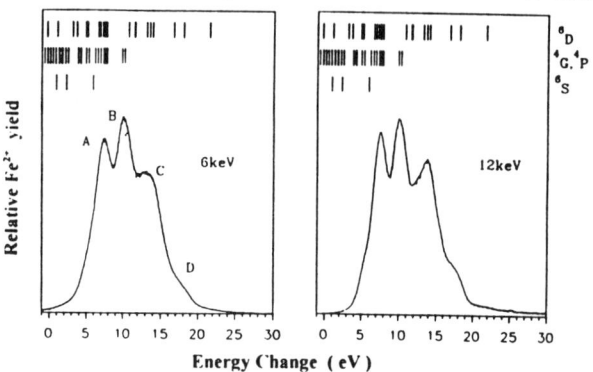

Fig.4. TES energy change spectra for one electron capture in Fe^{3+} - He collisions [17]

Fig. 4 shows TES spectra for Fe^{3+}-He collisions [17] where the positions of the possible product channels corresponding to ground state $Fe^{3+}(3d^5)^6S$ ions and the metastable primary $Fe^{3+}(3d^5)^4G,^4P$, and $(3d^4)4s^6D$ are shown. It can be seen that the peaks A, B, C and D are correlated with metastable rather than ground state ions. This is perhaps not surprising since the ground state has a statistical weight of only 4%. These observations indicate that many published cross sections based on measurements with unknown metastable fractions require cautious interpretation. In future work, it is essential to carry out TES measurements with state prepared beams. In this laboratory, plans to do this using double translational energy spectroscopy (DTES) are now well advanced. TES in a suitable gas target will be used to select product ions in known states which can then be used as primary ions in a second stage TES measurement.

As already noted, charge transfer collisions invoving molecules are of considerable interest since these may be large at low energies. For example, the one and two electron capture processes

$$He^{2+} + H_2 \rightarrow He^+ + H_2^+ \qquad (6)$$
$$He^{2+} + H_2 \rightarrow He + H^+ + H^+ \qquad (7)$$

with respective cross sections σ_{21} and σ_{20} have been extensively studied over a wide energy range. However, recent measurements by Okuno et al [19],which extend down to the very low energy of 0.3 eV amu^{-1}, show that σ_{20} is very large, even exceeding that for the corresponding resonant He^{2+}- He process, and is still increasing with decreasing energy. Shimakura et al [20] have explained this surprising result in terms of the avoided crossings between the various potential energy curves of the $(HeH_2)^{2+}$ system and shown that, at low energies, there is a very large probabilty for two electron capture into n = 3 states of He, while at higher energies, the one electron capture cross section σ_{21} begins to exceed σ_{20}.

Our recent TES measurements in the range 0.5 - 2.0 keV amu^{-1} [21] have provided a greater insight into the one electron capture process (6). At 0.5 keV amu^{-1}, one electron capture is dominated by the dissociative excitation channel

$$He^{2+} + H_2 \rightarrow He^+(1s) + H^+ + H(n=2) \qquad (8)$$

while the non-dissociative process (6), which is found to involve mainly He*(n=2,3) formation, accounts for only 1% of the total capture rising to 25% at 2.0 keV. This is the first direct evidence of the dominance of dissociative excitation in electron capture in this process at low energies.

Experimental studies of direct excitation processes of the type

$$X^{q+} + H(1s) \rightarrow X^{q+} + H(n.l) \qquad (9)$$

which have been based mainly on the PES approach using crossed beam configurations have still not been extensive. The recent data shown in Fig.5 for 2p excitation of H(1s) atoms by He^{2+} ion impact is an interesting case in which the observed [22,23,24,25] undulatory dependence on energy (which probably reflects competition between direct excitation and charge transfer channels), is generally confirmed by the predictions of close coupling calculations [24,26,] but not by calculations based on the hidden adiabatic crossings approach [27].

Balmer alpha n=3→n=2 emission cross sections (of relevance to beam emission spectroscopy) following direct excitation in 17 - 67 keV He^{2+}- H(1s) and 15 - 100 keV H$^+$- H(1s) collisions have been measured in this laboratory [28]. Close coupling calculations are in good accord with the He^{2+}- H measurements but in surprisingly poor agreement with the H$^+$- H data.

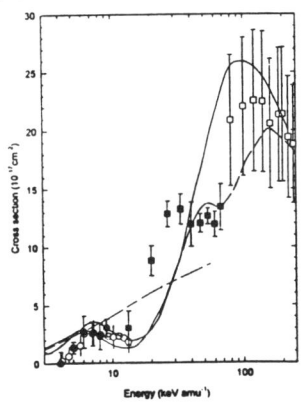

Fig.5. Cross sections for H(2p) excitation in He^{2+}- H(1s) collisions
Experiment:
●, Higgins et al [22]
■, Hughes et al [23]
O, Hoekstra & Beijers [see 24]
□, Detleffsen et al [25]
Theory:
— —, Fritsch et al [24]
———, Lundsgaard & Nielsen [26]
— - —, Krstic & Janev [27]

CONCLUSIONS

Measurements of total cross sections for electron capture and for ionization in collisions involving a wide range of multiply charged ions have been quite extensive particularly at velocities v > 1 au. At high velocities, the data have provided the basis for general q scaling relations which allow the prediction of approximate cross sections for collisions with ground state H and He atoms. There is still a need for measurements on excited H and He atoms. At velocities v < 1 au where simple charge scaling relations do not apply, both TES and PES measurements are providing a detailed insight into electron capture processes dominated by curve crossings which lead to selective population of a limited number of excited states. TES measurements have also indicated the important role of metastable ions and the need to develop techniques which will allow measurements with state prepared primary beams. PES studies of direct excitation of H by ion impact are still comparatively sparse and of limited accuracy. Further experimental and theoretical work is desirable.

REFERENCES

1. Janev, R.K., Summary report of IAEA Technical Committee Meeting on Atomic and Molecular Data for Fusion Reactor Technology, Vienna, *IAEA rept.* INCD(NDS)-277,(1993)
2. Fonck, R.J., Goldston, R.J., Kaita, R. and Post, D.E., *Appl.Phys.Lett.* **42**, 239, (1983).
3. Boileau, A., von Hellermann, M., Mandel, W., Summers, H.P., Weisen, H., and Zinoviev, A. *J. Phys.B.* **22**, L145, (1989).
4. Fite, W.L., Brackmann, T.R. and Snow, W.R., *Phys. Rev.* **112**, 1161 (1958).
5. Shah, M.B. and Gilbody, H.B., *J. Phys.B.* **14**, 2361, (1981).
6. Ciric,D.,Brazuk,A.,Dijkkamp, D., de Heer, F.J., and Winter, H., *J. Phys.B.* **18**, 3639 (1985).
7. McCullough R.W., Wilkie, F.G. and Gilbody, H.B. *J.Phys.B.* **17**, 1373, (1984).
8. Crandall, D.H., Phaneuf, R.A. and Meyer, F.W., *Phys. Rev. A* **22**, 379 (1980).
9. Shah, M.B. and Gilbody, H.B., *J. Phys. B.* **16**, 4395, (1983).
10. Janev, R.K. and Hvelplund, P., *Comments At. Mol. Phys.* **11**, 75, (1981).
11. Olson, R.E. and Salop, A., *Phys. Rev. A* **16**, 531 (1977).
12. Kimura, M., Kobayashi, N., Ohtani, S. and Tawara, H., *J.Phys.B.***20**,73 (1987).
13. Wilkie, F.G., McCullough, R.W. and Gilbody, H.B., *J.Phys.B.* **19**, 239, (1986).
14. Bienstock, S., Heil, T.G., and Dalgarno, A. *Phys.Rev.A* **25**,, 2850, (1982).
15. Errea,L.F.,Herrero,B.,Mendez,L. and Riera,A., *J.Phys.B.***24**,4061, (1991)
16. McLaughlin,T.K.,McCullough,R.W. & Gilbody,H.B.*Am.Inst.Phys.Conf.Ser.***274**, 202,(1993)
17. McLaughlin,T.K.,Tanuma,H.,Hodgkinson,J.,McCullough,R.W.,and Gilbody,H.B., J.Phys.B. **26**, 3871, (1993).
18. McLaughlin, T.K., Hodgkinson, J.M., Tawara, H., McCullough, R.W. and Gilbody H.B.,*J. Phys. B.* **26**, 3587 (1993).
19. Okuno, K., Soejima, K. and Kaneko, Y., *J. Phys.B.* **25**, L105, (1992).
20. Shimakura, N., Kimura, M. and Lane, N.F., *Phys. Rev. A* **47**, 709, (1993).
21. Hodgkinson,J.M.,McLaughlin,T.K.,McCullough,R.W.,Geddes,J.,and Gilbody,H.B., *J.Phys.B.* **28**,L393, (1995).
22. Higgins,D.P.,Geddes,J.,McCullough,R.W.and Gilbody,H.B.*Nuc.Inst.Meth.B* **103**,120,(1995)
23. Hughes, M.P.,Geddes, J.,and Gilbody, H.B.,*J.Phys.B.***27**,1143, (1994).
24. Fritsch,W.,Shingal.,and Lin,C.D.,*Phys.Rev.*A **44**,5686, (1991).
25. Detleffsen,D.,Anton,M.,Werner,A., and Schartner,K-H., *J.Phys.B.***27**,4195,(1994).
26. Lundsgaard, M.F.V.,and Nielsen, S.E., *Z.Phys.D.* **34**,97, (1995).
27. Krstic, P.S., and Janev, R.K. *Phys.Rev.*A.**47**, 3894, (1993).
28. Donnelly, A.,Geddes, J., and Gilbody, H.B. *J.Phys.B.* **24**,165, (1991).

The NIST Spectroscopic Database and Two New Critical Data Tables*

W. L. Wiese

National Institute of Standards and Technology (NIST)
Gaithersburg, MD 20899

The National Institute of Standards and Technology (NIST), formerly the National Bureau of Standards (NBS), operates three data centers on atomic spectroscopy. These centers critically compile the three principal quantities characterizing spectral lines:
  (1) Wavelengths and Energy Levels, including spectroscopic classifications,
  (2) Atomic Transition Probabilities, and
  (3) Line Shape and Shift Parameters.
Currently, two data centers are actively tabulating the spectroscopic quantities listed under (1) and (2), respectively, staffed in each case with two professionals. There is no activity, however, on the third area, line shape and shift parameters.

The data center activities comprise three distinctly different work areas:
  (a) Comprehensive, on-going literature searches for pertinent papers and maintenance of literature files and up-to-date bibliographies,
  (b) Critical evaluation and compilation of numerical data, and
  (c) Database development and data dissemination.

BIBLIOGRAPHICAL WORK

We search for pertinent papers by covering systematically the new issues of about 150 journals, mainly utilizing such abstracting services as *Current Contents* (1).

Until recently our (annotated) bibliographies have been published in book form, but are now, or will soon be, disseminated electronically on the World Wide Web (2) through the Internet. Their current status is as follows:

---

*Supported in part by the DOE Office of Fusion Energy under an Interagency Agreement.

(1) *Bibliography on Atomic Energy Levels and Spectra*, published as NBS Special Publications through 1987 (3,4); updates will soon be available on the World Wide Web (2).

(2) *Bibliography on Atomic Transition Probabilities*, published as NBS Special Publications through (1980) (5); pertinent references since 1977 are available on the NIST Physics Laboratory home page on the World Wide Web (WWW site: http://physics.nist.gov)

(3) *Bibliography on Atomic Line Shapes and Shifts*, published as NIST Special Publication 366 through March 1992 (6), planned to be made available with updates on the World Wide Web (for address, see above).

NUMERICAL DATA TABULATIONS

The main activity of the NIST spectroscopic data centers is the critical evaluation and compilation of all available literature data. A catalog of the references for the **most recent** major data compilations for wavelengths, energy levels and transition probabilities is presented in Table 1. The references are arranged by chemical element in the usual order with increasing atomic number Z. Most of the cited data compilations have been carried out at NIST (or, formerly, NBS) and this material is supplemented by recent tabulations from several other institutions. If more than one reference is listed for a specific quantity under one element, these data sources are complementary and may cover, for example, different stages of ionization, lines in different wavelength ranges, etc.

Further details on individual spectra, references to important reviews as well as some earlier compilations and a more extensive bibliography were given by Martin (7) in a 1992 review.

Table I. Catalog of major spectroscopic data compilations. The reference numbers for the most recent data tables are listed. Tables for heavier elements are usually quite incomplete, and often contain data on only a few stages of ionization.

Number (Z)	Element Symbol	Wavelengths	Energy Levels	Transition Probabilities
1	H	8	8	9
2	He	10,11,12	13	9
3	Li	10,11,12	14	9
4	Be	10,12	14	9

5	B	10,12,15	14,15	9
6	C	8,10,11,12	8	16
7	N	8	8	16
8	O	8,10,11,12	8	16
9	F	10,11,12	14	9
10	Ne	10,11,12	14	9
11	Na	10,11,12	17*	18
12	Mg	19*	19*,20*	18
13	Al	21*	21*,22*	18
14	Si	8,10,11,12	23*	18
15	P	10,12	24*	18
16	S	10,12,25*	25*,26*	18
17	Cl	10,11,12	14	18
18	Ar	10,11,12	14	18,27
19	K	10,11,12	28*	18
20	Ca	10,11,12	28*	18
21	Sc	29*	28*	30*
22	Ti	10,11,12,31	28*	30*
23	V	10,12,31*	28*	30*
24	Cr	10,12,31*	28*	30*
25	Mn	10,12,31*	28*	30*
26	Fe	10,11,12,31	28*	32*
27	Co	10,12,31*	28*	32*
28	Ni	10,12,31*	28*	32*
29	Cu	10,11,12,31*	33*	34
30	Zn	10,12	35*	34
31	Ga	10,12	14	34
32	Ge	10,12,36	37*	34
33	As	10,12,36	14	34
34	Se	10,12,36	14	-
35	Br	10,12	14	34

36	Kr	10,12,31*	31,38*	34
37	Rb	10	14	34
38	Sr	10,36	14	34
39	Y	10,36	14	34
40	Zr	10,36	14	-
41	Nb	10,36	14	-
42	Mo	10,31*	31,39*	34
43	Tc	10,36	14	-
44	Ru	10,36	14	-
45	Rh	10,36	14	34
46	Pd	10,36	14	-
47	Ag	10,36	14	(34)†
48	Cd	10,36	14	34
49	In	10,36	14	(34)
50	Sn	10,36	14	34
51	Sb	10,36	14	-
52	Te	10,36	14	-
53	I	10,36	14	(34)
54	Xe	10,36	14	34
55	Cs	10,36	14	34
56	Ba	10,36	14	34
57	La	10,36	40*	-
58	Ce	10,36	40*	-
59	Pr	10,36	40*	34
60	Nd	10,36	40*	34
61	Pm	10,36	40*	-
62	Sm	10,36	40*	-
63	Eu	10,36	40*	34
64	Gd	10,36	40*	-
65	Tb	10,36	40*	-
66	Dy	10,36	40*	34

67	Ho	10,36	40*	-
68	Er	10,36	40*	(34)
69	Tm	10,36	40*	34
70	Yb	10,36	40*	(34)
71	Lu	10,36	40*	(34)
72	Hf	10,36	14	-
73	Ta	10,36	14	34
74	W	10,36	14	34
75	Re	10,36	14	-
76	Os	10,36	14	-
77	Ir	10,36	14	34
78	Pt	10,36	14	-
79	Au	10,36	14	(34)
80	Hg	10,36	14	34
81	Tl	10,36	14	34
82	Pb	10,36	14	34
83	Bi	10,36	14	34
84	Po	10,36	14	-
85	At	10,36	-	-
86	Rn	10,36	14	-
87	Fr	10,36	-	-
88	Ra	10,36	14	-
89	Ac	10,36,41	41	-
90	Th	10,36,41	41	-
91	Pa	10,36,41	41	-
92	U	10,36,41	41	34
93	Np	10,36,41	41	-
94	Pu	10,36,41	41	-
95	Am	10,36,41	41	-
96	Cm	10,36,41	41	-
97	Bk	10,36,41	41	-

98	Cf	10,36,41	41	-
99	Es	10,36,41	41	-
100	Fm	-	41	-

†A reference number in parentheses indicates that less than 10 lines have been tabulated for this element.

*Asterisks indicate that the data from this reference are included in the NIST Physics Laboratory World Wide Web home page (2).

COMPUTERIZED SPECTROSCOPIC DATABASES

A. On-line databases

As part of the NIST Physics Laboratory home page on the World-Wide Web (2), as well as part of the NASA Astrophysical Data System (see also Ref. (2)), a general database on atomic spectroscopic data has been developed and assembled that includes a good part of the critically evaluated wavelength, energy level and transition probability data which are of recent vintage. Currently, this database includes energy levels for O II, for almost all spectra of the elements Na through S ($Z = 11 - 16$) and K through Zn ($Z = 19 - 30$), Ge ($Z = 32$), Kr ($Z = 36$), Mo ($Z = 42$) and the first several spectra of the rare earth elements La through Lu ($Z = 57 - 71$).

Recently compiled wavelength tables for all spectra of Mg, Al, S, Sc and for O II are included as well as those for all Ca-like to H-like spectra of the elements Ti through Cu, and Kr and Mo.

On transition probabilities, the recently evaluated data for about 18,000 lines of the spectra of the Fe-group elements are included.

Work is underway to expand the database with other existing NIST tables and to update it with new compilations as they become available.

B. PC Databases

Two databases are available on floppy disk for DOS operating systems with full search and select capabilities, the

Database for Atomic Spectroscopy (DAS), SRD 61 (42),

which contains wavelengths, energy levels and transition probabilities for practically all stages of ionization of the elements hydrogen through nickel ($Z = 1-28$),

Wavelengths of Atoms and Atomic Ions, SRD 38 (43),

which contains wavelengths and intensity estimates for prominent lines of the lower spectra, often up to the fourth stage of ionization, of all elements.

TWO NEW CRITICAL DATA TABLES

Two spectroscopic data tables of special interest for plasma applications have just been completed and are scheduled for publication in 1996. These are
Atomic Transition Probabilities of Carbon, Nitrogen and Oxygen (44),
and
Spectral Data for Highly Ionized Atoms: Ti, V, Cr, Mn, Fe, Co, Ni, Cu, Kr and Mo (31).
The first book contains about 12,500 transitions of C, N and O through all stages of ionization, which are arranged in order of multiplets and ascending quantum numbers so that it also serves as a multiplet table. The major part of the tabulation work has been the critical evaluation of the uncertainties of the various data sources. An earlier developed system of evaluation criteria was again applied to each data source (and also further refined). The four main criteria are:

    1. Consideration of the **critical factors** of a specific method by the authors. For example, one such factor is the treatment of electron correlation in atomic structure calculations by the use of multi-configuration approaches (discussed in detail in Refs. (16, 18));

    2. The degree of agreement among reliable data, based on extensive tabular and graphical comparisons;

    3. The authors' estimates of the uncertainty of their calculated or measured data;

    4. The fit of the data into systematic trends, e.g., along isoelectronic sequences, or deviations from them.

Most of the selected transition probability data are from recent sophisticated atomic structure calculations, and it has been estimated that they yield multiplet data accurate within $\pm 10\%$.

The second book, cited above, contains critically evaluated wavelengths, energy levels, ionization energies, and transition probabilities for Ca-like through H-like ions of the elements Ti through Cu, Kr and Mo, which are of special importance for fusion research. In this tabulation, the spectral data are arranged in order of decreasing wavelengths.

REFERENCES

1. *Current Contents on Diskette with Abstracts*, Inst. Sci. Inform. Philadelphia, PA (1996).
2. World Wide Web site: http://physics.nist.gov.
3. C. E. Moore, *Bibliography on the Analyses of Optical Atomic Spectra*, Nat. Bur. Stand. (U.S.) Spec. Publ. 306, Sec. 1 (1968); Secs. 2-4 (1967).
4. Hagan, L., Martin, W. C., Zalubas, R., and Musgrove, A., *Bibliography on Atomic Energy Levels and Spectra*, Nat. Bur. Stand. (U.S.) Spec. Publ. 363 (1972); Suppl. 1 (1977); Suppl. 2 (1980); Suppl. 3 (1985).
5. Fuhr, J. R., Miller, B. J., and Martin, G. A., *Bibliography on Atomic Transition Probabilities*, Nat. Bur. Stand. (U.S.) Spec. Publ. 505 (1978); Suppl. 1 (1980).
6. Fuhr, J. R., Wiese, W. L., Roszman, L. J., Martin, G. A., Miller, B. J., and Lesage, A., *Bibliography on Atomic Line Shapes and Shifts*, Nat. Bur. Stand. (U.S.) Spec. Publ. 366 (1972); Suppl. 1 (1974); Suppl. 2 (1975); Suppl. 3 (1978); Suppl. 4 (1993).
7. Martin, W. C., in *Atomic and Molecular Data for Space Astronomy*, Lecture Notes in Physics, Vol. 407, P. L. Smith and W. L. Wiese, Eds., Springer-Verlag, Berlin, (1992).
8. Moore, C. E., Selected Tables of Atomic Spectra, NSRDS-NBS 3, Sects. 1-11 (1965-1985). (All these separately published sections, except those on silicon, are reprinted in a single volume, *Tables of Spectra of Hydrogen, Carbon, Nitrogen and Oxygen Atoms and Ions*, ed. by J. W. Gallagher, CRC Press, Boca Raton, FL 1993.) The O II data from Martin, W. C., Kaufman, V., and Musgrove, A., J. Phys. Chem. Ref. Data **22**, 1179 (1993) are also included.
9. Wiese, W. L., Smith, M. W., and Glennon, B. M., *Atomic Transition Probabilities - Hydrogen through Neon*, NSRDS-NBS 4, U. S. Govt. Print. Office, Washington, DC. (1966).
10. Reader, J., and Corliss, C. H. Eds., in *CRC Handbook of Chemistry and Physics*, 76th Edition, D. R. Lide, Ed., CRC Press, Boca Raton, FL. (1995).
11. Striganov, A. R., and Sventitskii, N. S., *Tables of Spectral Lines of Neutral and Ionized Atoms*, IFI/Plenum Press, New York (1968).
12. Kelly, R. L., *Atomic and Ionic Spectrum Lines Below 2000 Angstroms: Hydrogen through Krypton*, J. Phys. Chem. Ref. Data **16**, Suppl. 1 (1987).
13. Martin, W. C., Phys. Rev. A **36**, 3575 (1987).
14. Moore, C. E., *Atomic Energy Levels*, NBS Circ. 467, Vol. I (1949); Vol. II (1952); Vol. III (1958); reprinted as NRDS-NBS 35 (1971). U. S. Govt. Print. Office, Washington, DC.
15. Odintzova, G. A., and Striganov, A. R., J. Phys. Chem. Ref. Data **8**, 63 (1979).
16. Wiese, W. L., Fuhr, J. R., and Deters, T. M., J. Phys. Chem. Ref. Data, Monograph 7 (1996).
17. Martin, W. C., and Zalubas, R., J. Phys. Chem. Ref. Data **10**, 153 (1981).
18. Wiese, W. L., Smith, M. W., and Miles, B. M., *Atomic Transition Probabilities - Sodium through Calcium*, NSRDS-NBS 22, U. S. Govt. Print. Office, Washington, DC (1969).
19. Kaufman, V., and Martin, W. C., J. Phys. Chem. Ref. Data **20**, 83 (1991).
20. Martin, W. C., and Zalubas, R., J. Phys. Chem. Ref. Data **9**, 1 (1980).
21. Kaufman, V., and Martin, W. C., J. Phys. Chem. Ref. Data **20**, 775 (1991).
22. Martin, W. C., and Zalubas, R., J. Phys. Chem. Ref. Data **8**, 817 (1979).
23. Martin, W. C., and Zalubas, R., J. Phys. Chem. Ref. Data **12**, 323 (1983).
24. Martin, W. C., Zalubas, R., and Musgrove, A., J. Phys. Chem. Ref. Data **14**, 751

(1985).
25. Kaufman, V., and Martin, W. C., J. Phys. Chem. Ref. Data **22**, 279 (1993).
26. Martin, W. C., Zalubas, R., and Musgrove, A., J. Phys. Chem. Ref. Data **19**, 821 (1990).
27. Vujnović, V., and Wiese, W. L., J. Phys. Chem. Ref. Data **21**, 919 (1992).
28. Sugar, J., and Corliss, C. H., *Atomic Energy Levels of the Iron-Period Elements: Potassium through Nickel*, J. Phys. Chem. Ref. Data **14**, Suppl. 2 (1985).
29. Kaufman, V., and Sugar, J., J. Phys. Chem. Ref. Data **17**, 1679 (1988).
30. Martin, G. A., Fuhr, J. R., and Wiese, W. L., *Atomic Transition Probabilities - Scandium through Manganese*, J. Phys. Chem. Ref. Data **17**, Suppl. 3 (1988). Also available as an interactive PC database from the Office of Standard Reference Data, Natl. Inst. Stds. and Tech., (Database 61, 1995).
31. Shirai, T., Sugar, J., and Wiese, W. L., *Spectral Data for Highly Ionized Atoms: Ti, V, Cr, Mn, Fe, Co, Ni, Cu, Kr and Mo*, J. Phys. Chem. Ref. Data, Monograph 8 (1996).
32. Fuhr, J. R., Martin, G. A., and Wiese, W. L., *Atomic Transition Probabilities - Iron through Nickel*, J. Phys. Chem. Ref. Data **17**, Suppl. 4 (1988). Also available as an interactive PC database from the Office of standard Reference Data, Natl. Inst. Stds. and Tech., (Database 61, 1995).
33. Sugar, J., and Musgrove, A., J. Phys. Chem. Ref. Data **19**, 527 (1990).
34. Fuhr, J. R., and Wiese, W. L., in *CRC Handbook of Chemistry and Physics*, 76th and earlier editions, D. R. Lide, Ed., CRC Press, Boca Raton, FL (1995).
35. Sugar, J., and Musgrove, A., J. Phys. Chem. Ref. Data **24**, 1803 (1995).
36. Meggers, W. F., Corliss, C. H., and Scribner, B. F., *Tables of Spectral Line Intensities*, NBS Monograph 145 (1975).
37. Sugar, J., and Musgrove, A., J. Phys. Chem. Ref. Data **22**, 1213 (1993).
38. Sugar, J., and Musgrove, A., J. Phys. Chem. Ref. Data **20**, 859 (1991).
39. Sugar, J., and Musgrove, A., J. Phys. Chem. Ref. Data **17**, 155 (1988).
40. Martin, W. C., Zalubas, R., and Hagan, L., *Atomic Energy Levels-The Rare-Earth Elements*, NSRDS-NBS 60, U. S. Govt. Print. Office, Washington, DC (1978).
41. Blaise, J., and Wyart, J.-F., *Energy Levels and Atomic Spectra of Actinides*, Internat. Tables of Selected Constants 20, Tables Internat. de Constantes, Paris (1992).
42. *Database for Atomic Spectroscopy (DAS)*, D. E. Kelleher, Ed., Standard Reference Database 61 (SRD), Natl. Inst. Stds. and Tech., Gaithersburg, MD (1995).
43. *Spectroscopic Properties of Atoms and Atomic Ions*, J. Reader and C. H. Corliss, Eds., Standard Reference Database 38 (SRD), Natl. Inst. Stds. and Tech., Gaithersburg, MD (1992).
44. Wiese, W. L., Fuhr, J. R., and Deters, T. M., *Atomic Transition Probabilities of Carbon, Nitrogen and Oxygen*, J. Phys. Chem. Ref. Data, Monograph 7 (1996).

Electron-Ion Collision Processes[§]

Gordon H. Dunn[§§]

JILA[***], *University of Colorado*
Boulder, CO 80309-0440

ABSTRACT Almost by definition, electron-ion collisions are critically important in neutral plasmas, and one must have a quantitative knowledge of these collisions to model and understand non-LTE plasmas. Experimental studies of such processes have been carried out for more than 30 years now, giving necessary input and feedback to theoretical studies so that ultimately plasmas ranging through those in astrophysical bodies, to those in controlled fusion devices, planetary atmospheres, flames, lasers, plasma generators and lighting devices can be modelled and understood. The challenge is a huge one for both the experimental and theoretical workers, since the number of processes is large: including elastic scattering, excitation, ionization and recombination for atomic ions. One has found that for each of these there are generally a number of mechanisms operative to lead to the final state. The processes are often referred to as "direct" and "indirect" processes. For molecular ions the multiplicities of mechanisms and parameters increases almost discouragingly; since the effects of internuclear motions can enter prominently into both the target and product states. It goes without saying that the number of species is essentially limitless. Thus, despite the long time of study and much progress, there remains much to be done to achieve the quest. The author in this brief space tries to give some indication of the status of a number of processes by describing some of the forefront experimental work presently going on and indicating some problems remaining to be overcome. There is no attempt to review all data and work in the literature.

[§] Author's portion of the work reported herein supported in part by the office of Fusion Energy, of the U.S. Department of Energy under Contract No. DE-A-105-86ER53237 with NIST.

[§§] Staff member in JILA of NIST.

[***] *A joint research institute of the National Institute of Standards and Technology and the University of Colorado*

© 1996 American Institute of Physics

INTRODUCTION

Fiercely optimistic activity in controlled thermonuclear fusion and heightened interest in astrophysics combined with advancing technology led in the early 1960's to the first successful electron-ion collision cross section measurements by Dolder, Harrison and Thoneman[1]. Since then, continuing pursuit of those plasmas and interest in many others such as lasers, planetary atmospheres, plasma etching and deposition, and lighting have supplied motivation to study electron-ion collisions in great breadth and detail. This is in addition to the strong incentives that come from the fact that electron-ion collisions are intrinsically very interesting. The experimental advances have been accompanied hand-in-hand with theoretical strides. There is clearly no intent nor hope of reviewing the 35 years of progress in this area in this brief space, but rather the reader is referred to detailed treatments elsewhere[2,3,4,5,6]. Here we only mention briefly some exciting recent experimental advances and techniques along with some remaining challenges. Important work using the Electron-Beam-Ion Trap (EBIT) or Electron-Beam-Ion-Source (EBIS) is not mentioned here, since it is covered elsewhere in this volume.

A feature of the Coulomb field that should be kept in mind is the infinitude of resonances that characterize the electron-ion system. Thus, the electron and ion may come together with the electron resonantly transferring its energy to a target electron so that a compound state is formed:

$$e + X^{n+} \rightarrow [X^{(n-1)+**}] \quad . \tag{1}$$

The compound state, $[X^{(n-1)+}]^{**}$ can decay via numerous routes. For example, the electron may simply leave again, and elastic scattering will have occurred with a "resonance" in the cross section; some of the excess energy may decay via radiation before the electron leaves again so that dielectronic recombination has occurred; if the core-excited electron was an inner-shell electron the compound state may have enough energy to eject *two* electrons so that resonant excitation double autoionization (REDA) has occurred, and the ionization cross section will have resonance features in it; if the core-excited electron jumps to a lower level not the ground state while the other electron is ejected, then resonant excitation will have occurred. Finally, if the target ion is molecular, the processes just mentioned may again occur with dissociation of the nuclei being another stabilization route - dissociative recombination is a particularly notable example of this. These resonances and their decay paths are particularly interesting, because the physics is fascinating and also because they can substantially affect the cross sections quantitatively and thus also affect the modelling of plasmas mentioned earlier. Indeed, much of today's challenges lie in more fully understanding resonances.

The internuclear distance in molecular ions may play a major role not only in leading to alternate decay routes of excited or compound states, but also through the fact that the transition matrix elements may be strongly dependent upon internuclear coordinates, which depend upon internal states associated with these coordinates (vibration and rotation). Thus, collisions with molecular ions take on an additional number of challenging qualities.

ATOMIC IONS

Ionization

Not only is electron-impact ionization of ions the first process for which cross section measurements were made[1], it is the process for which the most investigations have been made[2], and - arguably - it is the easiest to study. Experimental results are probably more accurate than those for electron-impact ionization of neutrals, since one need only measure electrical and geometrical quantities which are more accurately measured, normally, than neutral densities. The techniques have been perfected to the level that it may be considered almost a "made-to-order" task to measure absolute ionization cross sections of ions, and such measurements are primarily now made in crossed beams machines at the Strahlenzentrum at Giessen University[7] and Oak Ridge National Laboratory[8] for nearly any species. Indeed, the table of ions[9] for which measurements are needed for the edge plasmas of Tokamaks has now been essentially filled out.

The new and exciting development surrounding this process is the adaptation and implementation of the heavy-ion storage ring at Heidelberg to make measurements[6] on electron-impact ionization of ions. The storage ring provides a merged beams configuration with the associated kinematic narrowing of the relative energy distribution, a cold and dense electron target, and relatively intense beams of ions of nearly any species and charge state. This innovation is fairly recent so it has been used only to study the Li-like ions Si^{11+} and Cl^{14+} and the Na-like ions Cl^{6+}, Fe^{15+}, and Se^{23+}. The enhancement of the quality of data is dramatically illustrated in Fig. 1 containing cross sections for ionization of Fe^{15+}. It is particularly fitting to show these data[10], since Fe^{15+} is the ion for which REDA was first hypothesized[11]. Furthermore, when the crossed beams measurements of Gregory et al[12] were made only 9 years ago (three points only are shown in Fig. 1 for comparison), it was considered to be an important technological triumph. The distinction speaks for itself. Also, the precision of the new data in revealing the detailed structures of REDA compared with those calculated[10,13] emphasizes that experiment can now lead theoretical efforts to new progress as further systems are investigated with the new technology. One may anticipate that other storage rings operative for atomic physics will be retrofitted with detectors necessary for this class of investigation.

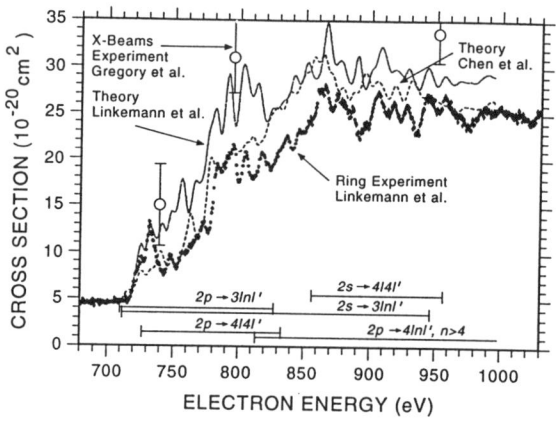

Figure 1. Cross section vs. interaction energy for electron-impact ionization of Fe^{15+}. Data are from Refs. 10, 12, 13.

Dielectronic Recombination

The study of dielectronic recombination (DR), so fundamental to ionization balance in plasmas, is also now committed almost entirely to ion storage rings. The quality of the data obtained using rings is emphatically higher than with any other technique - just as for ionization illustrated above. After the first measurements on DR[4] in the mid-1980's, a substantial number of measurements were made[4] in a single-pass merged beams configuration using the electron cooler being prepared for the Aarhus storage ring as the electron target. However, since then measurements on some 16 different ions have been performed on storage rings[6]. Using adiabatic expansion of the electron beam in the cooler as initiated by Danared[14], electron temperatures have been obtained in the range $kT_\perp \leq 0.010$ eV and $kT_\parallel \leq 0.0001$ eV with attendant resolution in observing resonances in DR of many ions. Theoretical calculations are generally good at reproducing or predicting the experimental results.

One notable issue remains outstanding. The experiments[4] at JILA on Mg^+ demonstrated a clear dependence on ambient electric fields in the collision region, and the agreement between theory and experiment could be considered very good. Also anomalous results in experiments[4] at ORNL could be interpreted in terms of similar electric field mixing; though the fields in the collision region could only be estimated. In these latter results, a number of inconsistencies arose - particularly for Li-like ions. Inclined beams experiments at Harvard[15] have attempted to resolve the inconsistencies, but the precision and scope of these experiments have left ambiguity. Fortunately, again the storage ring may come to the rescue. By imposing a small (100 - 400 μT) magnetic field in the electron cooler region of the Stockholm CRYRING, motional electric fields of up to 175

V/cm have been obtained[16], and very clear field effects at five different field values were observed in data on DR for the Li-like ion Si^{11+}. These new data - when fully analyzed - should help resolve the very important field-effect issue. Indeed, all plasmas have ambient fields whether they be plasma microfields or externally imposed ambient fields, and it is essential that this issue be resolved.

Excitation

In ionization and DR experiments, one typically detects the product ion/atom, and this can be done in beams configurations with nearly 100% efficiency. The excited product state in excitation is very transitory, but gives rise in many cases to photons which can instead be detected. The main problem is that this is done only with great loss in efficiency of detection - typically 10^{-4}-10^{-2}. Loss of signal with high backgrounds coupled with the need to do absolute radiometry with different sensitivity for each photon energy couple to make such experiments very difficult. Nevertheless, accurate and fairly precise measurments of absolute cross sections were performed for a number of ions and transitions and these have been reviewed in Ref. 3. Absolute cross sections were obtained for the multiply-charged ions Hg^{2+}, Al^{2+}, C^{3+}, and N^{4+} in addition to a fairly large number of singly-ionized ions.

To be able to measure excitation for a larger number of multiply-charged ions and for a larger variety of transitions, it became clear to us that an alternative to the fluorescence detection method which we had used so successfully (but with so much pain) would be required. A new technique was developed and implemented[17] which boasts the order of 100% detection sensitivity. The method, dubbed MEIBEL for Merged Electron Ion Beam Energy Loss, detects nearly all electrons which have lost the amount of energy characteristic of the transition being investigated. In the method, beams of electrons and ions are merged using crossed electric and magnetic fields, left to interact over some space, de-merged again using crossed fields, and the inelastically scattered electrons are detected on a position-sensitive detector. A similar technique is being implemented at Jet Propulsion Laboratory and a related technique is being implemented at Queen's University Belfast.

In addition to leading to measurements[17,18,19] of absolute excitation cross sections of some resonance transitions of multiply-charged ions, the technique led to the discovery[19] of strong backscattering of the electrons near threshold for resonant transitions. Also, it has been possible to make measurements of cross sections to non-radiating states, such as in intercombination transitions. An example[20] of the latter is illustrated in Fig. 2 for the excitation of a Kr^{6+} intercombination transition. Theoretical calculations for the process predicted strong resonance structure qualitatively resembling that observed and these calculations gave guidance to the experiments; but calculations duplicated the observed structure[21] only after numerous attempts and carrying the theoretical methods to some limits. It is speculated that resonance structure in excitation is more difficult to reproduce because interaction of resonances is direct rather than

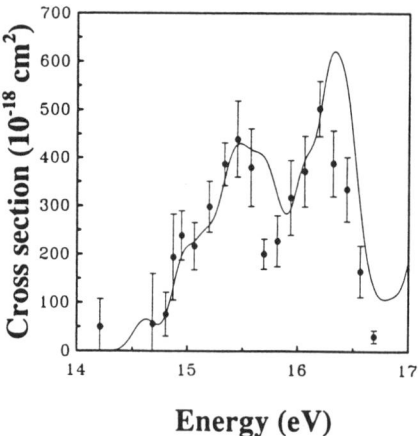

Figure 2. Cross section vs. interaction energy for the $4s^2(^1S_0) \rightarrow 4s4p(^3P_{0,1,2})$ electron impact excitation of Kr^{6+}. Experimental data and theoretical curve are from Ref. 20 and 21, respectively.

through the continuum as in the processes already discussed. Thus, though theory has been generally successful in predicting overall total absolute cross sections for excitation of resonance states, there is a challenge to experiments to further lead theory by more and better experiments showing resonances in excitation. Work on forbidden transitions with resonances would appear to command priority to guide further perfection of the theory.

No creative techniques have been suggested for implementing ion storage rings for measurements on atomic ion excitation.

MOLECULAR IONS

Perhaps counterintuitively, the physics of molecular ions has achieved a prominence in the physics of high temperature plasmas such as those in Tokamaks, due to their copious presence in the edge plasma[9]. More obviously, they play important roles in cooler plasmas such as atmospheres, flames, and plasma etching and depostion. As already alluded to, there is an additional richness to the possible physical processes not present with atomic ions due to effects of the added degrees of freedom brought by more than one nucleus. Energy deposited in a collision may go into electronic energy and/or vibrational and rotational energy. Excited states can radiate, be metastable, or dissociate, i.e. some electronic states are repulsive in the internuclear coordinates, and the molecule may fly apart with the excess energy coming as kinetic energy of the nuclei. Such dissociation usually occurs in times the order of $10^{-14} - 10^{-12}$ sec; so dissociative stabilization in such cases usually dominates over radiation with lifetimes the order of 10^{-9} sec. The dissociation products may have a characteristic angular distribution[22] following from the direction of the electron motion and the molecular state symmetries involved. The cross sections between electronic states may depend very strongly on internuclear states such as vibration. Perhaps the paradigm of this is encountered in the simplest of all molecules (and molecular ions) H_2^+. For this ion it was shown long ago both experimentally[23] and theoretically[24] that cross sections for electron-impact dissociation may vary more than a factor of 10^3 between the lowest and highest vibrational levels. Caution is needed, therefore in interpreting collision data for molecular ions.

Despite their major presence in plasmas of practical and intellectual interest and the vital role they play in the behavior and modelling of plasmas, there has been comparatively little work done on electron collisions with molecular ions. An exception to this may be the study of dissociative recombination which has been the focus of intense study over a period of more than 40 years. Perhaps the reason for a paucity of results in this area goes back to the complexity and the difficulty of obtaining "perfect" or well-defined conditions, i.e. single state targets and well-defined products - physicists often shun such situations as being too hard.

Ionization and Excitation

Ionization of molecular ions often results in dissociated product ions. It is experimentally difficult to determine whether one has dissociative excitation or dissociative ionization if only one of the resultant ions is detected. Thus, to measure ionization, it is often necessary to have a coincidence experiment to detect both product ions at once. The author is aware of the execution of this approach only once[25], that being again for the simplest of molecular ions, H_2^+. The only other measurements of ionization of molecular ions seem to be those on[26] CO_2^+ and on[27] H_3O^+. Clearly the field is wide open and needing attention.

Apparently, there is only one measurement[28] of electron-impact excitation of a molecular ion to a bound molecular excited state and that is for N_2^+. This employed the crossed-beams fluorescence technique and was for excitation of the first negative system. The situation is slightly better for dissociative excitation, i.e. excitation of a repulsive state which dissociates before it can radiate. Here there are a few targets[9] which have been studied in a definitive way: H_2^+, D_2^+, H_3^+, D_3^+, N_2^+, O_2^+, CO^+, H^3O^+. There are some other measurements on light hydrocarbons which have been mentioned in reports, but not with enough confidence for publication. These latter are needed for modelling fusion edge plasmas. Measurements on light hydrocarbons pose a special challenge. Since the hydrogen is so much lighter than the parent or other product ion, the various hydrogen ions resulting as products carry away essentially all of the dissociation energy. They thus come out with large angles and a broad laboratory energy spread making them difficult to detect with confidence. Ion storage rings may hold some of the answer for studies of some of these products and preliminary measurements[29] on some ions have been carried out. Other laboratory approaches have also yielded preliminary data[30]. The surface is only being scratched in terms of experimental investigation of excitation processes, and the theoretical situation is perhaps even more barren of results.

Dissociative Recombination

Dissociative recombination, generically repressented by $e + AB^+ \rightarrow [AB]^{**} \rightarrow A + B^*$, has been studied for over 40 years, and there is a very sizeable literature on it[5,31] and international conferences are held on this topic alone. This should not be taken as an indication that it is a "mature" subject. This is far from the situation, as issues of dependence upon internal states, final product states and branching, the effect of field in the interaction region, angle and energy distributions of final products and others have only been scratched on the surface. There are still controversies about orders of magnitude obtained with different techniques, and speculations about internal states and field effects enter intimately into the arguments. Indeed, it has been and continues to be a hotbed of activity because of its immense importance and intrinsic interest.

It should be pointed out that the implementation of ion storage rings has started a revolution in this area, and many of the issues may soon be better understood as more work is carried out on the rings. We call attention to two particular recent papers. Both deal with dissociative recombination of the simplest molecular ion, HD^+ (this isotopic variant is used since its vibrational levels relax in milliseconds in the ring, whereas H_2^+ vibrational levels have lifetimes the order of 10^7 s). High quality data for the absolute magnitude of the cross section and its energy dependence were obtained[32] at the Stockholm CRYRING, and agreement with the theory over relevant ranges is excellent. In another important breakthrough, two dimensional imaging at the TSR in Heidelberg was used[33] to determine the relative numbers in final electronic states of the product atom. The results were also used to make statements about the angular distributions of the product particles. Many of the issues that have plagued the study of dissociative recombination now seem tractable with ion storage rings provided that beam time on the limited number of machines can be made available for the work.

REFERENCES

1. Dolder, K.T., Harrison, M.F.A., and Thoneman, P.C., Proc. Roy. Soc. (London) **A264**, 367-378 (1961).
2. A. Müller, in *Physics of Ion Impact Phenomena*, Springer Series in Chemical Physics, Vol 54, ed. Deepak Mathur, Berlin: Springer, 1991, pp. 13-90.
3. Phaneuf, R.A., *AIP Conf. Proc.295, 18th Int. Conf. on the Physics of Electronic and Atomic Collisions*, eds. T. Andersen et al, New York, AIP, pp. 405-414.
4. *Recombination of Atomic Ions, Proc. of a NATO Advanced Research Workshop*, eds. W.G. Graham et al., New York, Plenum, 1992, 335 pages.
5. Mitchell, J.B.A., Phys. Rep. **186**, 217-248 (1990).

6. Müller, A., Atomic and Plasma-Material Interaction Data for Fusion, a supplement to the journal Nuclear Fusion, Vol. 6, 59-100 (1995).
7. Stenke, M. et al, J.Phys. B: At. Mol. Phys. **28**, 2711-2721 (1995); *ibid* 4853-4859.
8. Bannister, M.E. et al, Phys. Rev. A **52**, 414-419 (1995).
9. Dunn, G.H., At. and Plasma-Matl. Int. Data for Fusion 2, 25-39 (1992); Supplement to the journal Nuclear Fusion.
10. Linkemann, J. et al, Phys. Rev. Lett. **74**, 4173-4176 (1995).
11. LaGattuta, K.J. and Hahn, Y., Phys. Rev. A**24**, 2273-2276 (1981)
12. Gregory, D.C. et al, Phys. Rev. A**35**, 3256-3264 (1987).
13. Chen, M.H. et al, Phys. Rev. Lett. **64**, 1350-1353 (1990).
14. Danared, H. et al, Phys. Rev. Lett. **72**, 3775-3778 (1994).
15. Savin, D.W. et al, Phys. Rev. A **53**, 280-289 (1996).
16. Bartsch, T. et al, Private communication, December 1995.
17. Bell, E.W. et al, Phys. Rev. A **49**, 4585-4596 (1994).
18. Wåhlin, E.K. et al, Phys. Rev. Lett. **66**, 157-160 (1991).
19. Guo, X.Q. et al, Phys. Rev. A **47**, R9-R12 (1993).
20. Bannister, M.E. et al, Phys. Rev. Lett. **72**, 3336-3338 (1994).
21. Gorczyca, T. W. et al, Phys. Rev. A **51**, 488-493 (1995).
22. Dunn, G.H., Phys. Rev. Lett. **8**, 62-64 (1962).
23. Dunn, G.H., and Van Zyl, B., Phys. Rev. **154**, 40-51 (1967); Dance, et al, Proc. Phys. Soc. **92**, 577 (1967); Peart, B. and Dolder, K.T., J. Phys.B **4**, 1496-1505 (1971); ibid. **5**, 1554-1558 (1972).
24. Peek, J.M., Phys. Rev. **140**, A11 (1965); **154**, 52-56 (1967); Phys. Rev.A **10**, 539-549 (1974); Peek, J.A. and Green, T.A., Phys. Rev. **183**, 202-212 (1969); Collins, L.A. and Schneider, B.I., Phys. Rev.A **27**, 101-111 (1983); Kimura, M., Phys. Rev.A **35**, 4101-4107 (1987).
25. Peart, B. and Dolder, K.T., J. Phys. B **6**, 2409-2414 (1973).
26. Müller, A. et al, J. Phys. B **13**, L221-L223 (1980).
27. Stenke, M. et al, paper presented at Spring Meeting of the German Physical Society, 1990 (unpublished).
28. Crandall, D.H. et al, Phys. Rev. A **9**, 2545-2551 (1974).
29. Forck, P., PhD Thesis, Ruprecht-Karls-Universität, Heidelberg, 1994.
30. Djurić, N. et al, AIP Conference Proceedings 360, ICPEAC 1995, AIP, New York 1995, pp. 297-306.
31. *Dissociative Recombination: Theory, Experiment, and Applications,* (eds. Rowe, B.R. et al) New York, Plenum, 1993.
32. Strömholm, C. et al, Phys. Rev. A **52**, R4320-R4323 (1995).
33. Zajfman, D. et al, Phys. Rev. Lett. **75**, 814-817 (1995).

On-line Atomic Data Access

David R. Schultz

Physics Division, Oak Ridge National Laboratory, Oak Ridge, TN 37831-6373

Jeffrey K. Nash

Lawrence Livermore National Laboratory, Livermore, CA 94551

> The need for atomic data is one which continues to expand in a wide variety of applications including fusion energy, astrophysics, laser-produced plasma research, and plasma processing. Modern computer database and communications technology enables this data to be placed on-line and obtained by users over the INTERNET. Presented here is a summary of the observations and conclusions regarding such on-line atomic data access derived from a forum held at the *Tenth APS Topical Conference on Atomic Processes in Plasmas*.

I. INTRODUCTION

The use of data for a wide variety of applications has gone on practically since the earliest beginnings of atomic physics. In particular, present applications such as interpreting astrophysical environments (e.g. solar and stellar flares, supernovae, and the interstellar medium), developing fusion energy reactors (e.g. plasma composition and transport diagnostics, divertor chamber modeling), laser-produced plasma research (e.g. temperature and density modeling, x-ray production diagnosis), and the processing of material by plasmas (e.g. etching and deposition of semiconductor surfaces), require large databases of atomic collisional and spectroscopic data. Major new applied research projects are on the horizon, such as new laser facilities (e.g. the National Ignition Facility), the International Thermonuclear Experimental Reactor (ITER), various space missions (e.g. AXAF, SOHO), and many industrial programs utilizing atomic processes to aid in manufacturing.

With the advent of the INTERNET, the linking of computers world wide has recently provided the means to rapidly exchange data. Furthermore, the World Wide Web (WWW or simply the 'Web'), adoption of a standard for hypertext communication (HTML - *Hyper*Text Markup Language), and the widespread availability of so-called Web Browsers such as Mosaic and Netscape has allowed a very easy to use, yet powerful way for non-expert users to search for and retrieve data in a variety of formats. Thus, as a community of atomic data producers, collectors, and users, we have an unprecedented opportunity to preserve and make readily accessible needed atomic data.

© 1996 American Institute of Physics

II. FORUM

During the *Tenth APS Topical Conference on Atomic Processes in Plasmas* a one evening forum was held in order to

- Initiate discussion of issues relating to this opportunity to place atomic data on-line,

- Identify the consitutencies interested in this activity and discuss their roles, and

- Help to define some of the problems, open questions, and parameters for undertaking this endeavor.

The speakers included Douglass Post (ITER), Peter Smith (Harvard-Smithsonian CfA), Norman Bardsley (LLNL), Richard Lee (LLNL), Takako Kato (NIFS), Peter Mohr (NIST), and the present authors. Each speaker tried to address each of these three objectives especially as pertains to their own speciality. For example, Bardsley related his experience with forming an APS Forum on Low Temperature Plasma Science, and in particular the interaction of industry and the atomic physics community in linking producers and users of data in the semiconductor industry. Also, a number of the speakers who represent data centers which already have data available over the WWW provided demonstrations of these sites at the forum and, in fact, throughout the conference. Yuri Ralchenko (Weizmann Institute) has surveyed a large portion of the atomic data sites on the Web and has listed their WWW Uniform Resource Locator (URL) addresses along with a brief summary of their content. This manuscript in PostScript format may be viewed using a WWW browser helper such as "xdvi" at http://plasma-gate.weizmann.ac.il/~fnralch/app.dvi.

As an example of how to reach these resources, and how they are becoming linked, you can start by using a Web browser to connect to the Weizmann Institute site (http://plasma-gate.weizmann.ac.il) and from there navigate by following the hypertext links to a number of other sites. Similarly, one could begin at the home page for the Controlled Fusion Atomic Data Center at http://www-cfadc.phy.ornl.gov/ and follow the links page to the Weizmann Institute WWW page, or other atomic data sites such as the NIST Atomic Spectroscopic Database (http://aeldata.phy.nist.gov/nist_beta.html). Here we make no attempt to summarize the details of each speaker's presentation or the content of their Web site, rather we seek only to provide a synthesis of the ideas exchanged. Thus, the responsibility for the conclusions drawn here and any statements of fact rests with the present authors rather than the speakers.

III. GOALS AND METHODS FOR ON-LINE ACCESS

Storing atomic data electronically has many benefits. For example, data printed graphically or in tabular form in journals presents the end-user of that data with the need to locate the appropriate reference, obtain the particular volume, and to digitize or scan and recognize the graphs or tables, thereby converting them to a format usable in his application. Simply finding the appropriate reference is a step full of difficulty for a user who wishes not simply to find all references on a particular reaction but rather find a recommended cross section or rate coefficient valid over a wide range of energy. Another advantage of electronic storage of atomic data is the fact that many data sets can be stored and categorized in order that an expert can evaluate the data and make a recommendation regarding it. Finally, since the data, and hopefully the recommended data, are stored electronically, they can easily be transmitted to other sites over the INTERNET.

Thus, electronic storage of data facilitates

- locating needed data,

- preservation of data in a readily accessible format,

- collection of available data sets in centralized locations,

- the ability to compare and evaluate data sets so that a recommended set of values can be determined, and

- the means to share the fundamental and recommended data with others over the INTERNET.

Recognizing these benefits, it is next important to consider the means by which these data will be stored and disseminated. Typically, each research group or data center will have their own specific platforms and database products which they have judged to be most suitable for their applications or user groups. For example, atomic data may be stored on personal computers, mainframes, or UNIX workstations. Furthermore, it may be stored in a wide variety of formats, such as ASCII tables, relational databases (such as Lotus, ORACLE or INGRES), or some other specialized format such as ALADDIN. ALADDIN, for example, has been adopted by the International Atomic Energy Agency as a standard format for exchanging atomic data within the fusion energy research community. Since it does not seem practical to dictate the format used by individual groups, what is necessary is some means through which data can be exchanged.

Owing to its widespread use and the relative ease with which applications can be developed for it, the WWW appears to be the best choice for on-line atomic data access. Some experience exists with locally developed data interfaces written in C or FORTRAN, but performance of these tools over the INTERNET on a wide range of different accessing platforms lacks flexibility and often simply does not work on terminals or computers that they do not recognize. In particular, even within the framework of the X/Motif standard windowing environment, screen movements or character sizing are not controlled in a sufficiently uniform way across the different vendor's platforms to allow this approach to find much success. On the other hand, standards for Web browsers and the hypertext language they recognize are such that practically every computer on the INTERNET now has these tools. They alleviate the need of the research group or data center to develop the full interface to their data. Almost any database product, or FORTRAN, C, or ASCII file structure can be interfaced with the WWW in a relatively straightforward manner. The complexity and utility of these interfaces can range from simple file lists which allow ftp'ing of the data files, to graphical displays in GIF or PostScript format. Search engines are also easy to write so that facilities such as on-line bibliographies can be implemented. Thus, the WWW is the presently preferred medium for data exchange, and allows for the freedom of each group to decide on internal storage formats and data management products.

IV. ISSUES

A number of issues relating to the efforts to place data on-line were also identified at the forum. We comment on the following points in some detail:

- unrestricted data access,

- attribution of authorship,

- proper and effective data usage, and

- standardized data formats.

For example, it was the view of the large majority of the participants that access to atomic data should be open, on-line, and non-commercial. That is, data should be placed into the public domain rather than sold or distributed only to narrow interest groups. This reflects the general consensus of the scientific community regarding publication of their work in the open literature. Also, if data could be obtained on-line, rather than through, say receiving diskettes in the mail, or even worse in paper copies of tables or graphs, this would be the best way to efficiently disseminate data. On-line repositories of data would also allow a potential user to browse the available data, selecting for downloading only what they decided they needed after viewing it. Experience with bibliographic search requests often shows that the initial request of the users does not yield exactly what they are looking for, and that subsequent iterations are needed to refine the parameters of the search. Therefore, using mediums such as facsimile, electronic mail, or postal mail to transmit search results, sufficiently slows the overall process so as to compromise the utility of the search. Finally, it was generally agreed that users should not be required to purchase specific commercial products simply to obtain data. For example, only standard, commonly available, free Web browsers should be required in order to gain access to data.

Another important issue, as yet not completely resolved, is one of proper attribution. If a user downloads data from a Web site, it was agreed that there should be some mechanism to assure that the originator of that data is properly credited. Often the measure of a scientist's performance is judged by the citations that his or her work receives, and if placing data on the Web circumvents normal procedures for obtaining that attribution, then data producers might be quite reluctant to release their work. In addition, without attribution, the vital contacts between data users and producers breaks down.

Related to this issue is the perceived need to create mechanisms which would help assure that data is properly used. That is, all the information that a user would require about a particular data set needs to be included in the downloaded file. For example, is the data one set or a number of theoretical and experimental results, or is it evaluated or recommended data? What is the range of validity for the data if it is a theoretical result? What are the experimental uncertainties of the data? What precisely are the units and, of course, what is the full citation for the work (i.e. where can all the details be found, and who to contact for full information since this could differ from the lead author on a citation or since the data may not be formally published)?

Perhaps the most difficult issue to address is that of encouraging standard formats for on-line data. Some groups have already agreed within their community to distribute data in particular formats so as to minimize the difficulty in deciphering exchanged data. For example, one group might always fit their data using Chebyshev polynomials, whereas another always lists it as pairs of columns of raw numbers. The only way to resolve these issues seems to be by engaging in further discussions in future workshops and forums. Certainly, within a particular community, such as astrophysics, it might be possible to agree on some standards for atomic data exchange. As mentioned above, this has been done within the fusion community, but it is widely accepted that the current choice of an exchange format is not at all optimal.

Finally, it was emphasized that no single data access system seems appropriate for all needs. For example, some users require extremely large sets of data which could only be obtained by running large production-oriented codes, whereas other users require only a narrow range of data which must be of the highest accuracy possible. Thus, one user may seek to first remotely run a production code, then download hundreds of megabytes of spectroscopic line energies and transition probabilities, while another may want only a single number or bibliographic reference. Data repositories should also, ideally, allow for the interactive browsing and intercomparison of available datasets to facilitate choosing appropriate data. A distinction should also be made between facilities giving access to data archives versus those which include certain modeling or computing capabilities.

V. CONSTITUENCIES AND ROLES

The forum also helped to identify key constituencies each having specific roles to play in the quest to bring atomic data on-line. The most obvious identification sees individuals as either producers, users, or archivers of atomic data. Clearly, communication between users and producers should be improved so that producers know what users want, and users know what is available from producers. The Web sites which act as repositories of atomic data can facilitate this communication by posting information about data needs, for example. In fact, data centers play a very important role in that their missions most closely fit that goal of placing data on-line.

However, funding for data centers is particular difficult to come by. One of the diverse priorities of the communities wishing to either place data on-line or use it, should be to help increase the awareness of funding agencies of the importance of data centers. Such repositories form a kind of "community memory" by storing data. If allowed to shut-down, the hard work of collecting, managing, evaluating, and disseminating data is lost. In addition, traditional data center activities have focussed on collecting data sets which each individually consist typically of only a modest number of values, but often, researchers require very large, comprehensive data sets at whatever level of accuracy is possible. This highlights the prevailing focus on critically evaluated data, which might not best serve all users.

Related to this means of preserving information in data centers is the role that journals can play in on-line atomic data access. All journals should eventually be placed on-line, so as to realize the goal that the data contained in the articles can be electronically accessed. Much of the development along these lines also depends on the economies of electronic publishing. Archival journals such as Atomic Data and Nuclear Data Tables and the Journal of Physical and Chemical Reference Data are also important as repositories of needed data, often recommended or "best available" data, and should also be placed on-line.

Much of the business of coordinating data production, collection, and dissemination can be aided by networks and organizations such as the IAEA Atomic and Molecular Data Center Network, the Committee on Data for Science and Technology (CODATA), and the Atomic Data and Analysis Structure consortium. These networks are becoming quite aware of the need for on-line data access and are actively encouraging it. Similarly, meetings and workshops can provide forums for the discussion of issues relating to on-line data access. In particular, the interests present at the Atomic Processes in Plasmas Conference and the Gaseous Electronic Conference include particularly those influenced by these issues. It was noted that further workshops are already being organized specifically aimed at improving on-line atomic data access.

VI. CONCLUSIONS

Thus, the forum on on-line atomic data access showed that already much has been done to convert the interfaces to existing data archives to be compatible with the WWW, but that a number of key issues remain to be resolved. In particular, further coordination is needed to facilitate communicating data needs to producers, and the electronic distribution of results. Future workshops and conferences should also address issues regarding standardization of public data formats to ease exchange of data. It was noted that the focus on short-term results by both private and public sector funders is causing a great deal of pressure on data centers and individual data producers which are the source of the data needed for on-line access to be achieved. Particular difficulties which the community should seek to remediate include methods for easily locating existing data matching a user's needs, ways in which users can solicit new calculations and experiments, procedures for obtaining evaluated or recommended data in addition to or rather than primitive data, methods for assuring that data is used correctly and effectively, and that attribution is given to the originators of the data which is retrieved on-line.

ACKNOWLEDGMENTS

We wish to thank the speakers and the audience at the forum for their interest and contributions to the discussions. We are also grateful to the organizers of the *Tenth APS Topical Conference on Atomic Processes in Plasmas* for the opportunity to hold the forum and for their support of it. DRS wishes to express his thanks for support of the Controlled Fusion Atomic Data Center from the Office of Fusion Energy, US DOE through contract No. DE-AC05-96OR22464 with Oak Ridge National Laboratory, managed by Lockheed Martin Energy Research Corp. JKN acknowledges support provided under the auspices of the US Department of Energy by Lawrence Livermore National Laboratory under Contract No. W-7405-ENG-48.

ULTRA-SHORT PULSE LASER PLASMAS

Picosecond Time Resolved Spectroscopy of a Controlled Preformed Plasma Heated by an Intense sub-ps Laser Pulse

C.Y. Côté[+], Z. Jiang[+], J.C. Kieffer[+], A. Ikhlef[+], H. Pépin[+]
and O. Peyrusse[*]

+ INRS-Énergie et Matériaux, Université du Québec, Canada
and * CEA-Limeil, Villeneuve St-Georges, France

Abstract. The time history of the Al Li-like spectrum has been measured on the picosecond time scale in different regimes of interaction of a sub-ps laser pulse with controlled preformed plasmas. The a-d/k,j line intensity ratio is constant as a function of time, with a value close to the LTE value, for solid density plasmas obtained by radiation pressure confinement. For the cases of an interaction with a longer density gradient scale length the core excited lines are strongly enhanced by transient effects in thermal plasmas and by both transient and non-Maxwellian effects in non-thermal plasmas.

INTRODUCTION

The recent advent of compact high intensity subpicosecond lasers gives access to new regimes of laser matter interaction (1,2,3). Plasmas produced and heated by ultrashort laser pulses are characterized by extremely high energy density and highly transient and non-equilibrium states resulting from the very short temporal and spatial scales involved. The present experimental work focuses on the consequences of both the non-maxwellian electron energy distributions and of the rapid time variation on the atomic physics. Specifically we look at the time history of the K-shell emission of Al ions near 1.6 keV. Because the atomic structure data of Al are quite well known, this material is very convenient for experiments and the data can be used easily to benchmark numerical calculations.

Here we present new time resolved spectra obtained (with a 1.5 ps temporal resolution) when controlled gradient scale length preplasmas are heated by an intense 0.53µm high contrast pulse at 10^{18} W/cm^2. Results obtained for radiation confined plasmas, thermal plasmas and non-thermal plasmas are presented and compared. The picosecond time resolved spectra we previously presented were obtained at low intensity (10^{16} W/cm^2) for plasmas produced by a 1µm 500fs pulse (with a pedestal) (4) and by a high contrast (prepulse free) 0.53µm 400fs pulse (5). More recently we measured time resolved spectra for the case of the interaction of a very high contrast 0.53μm pulse with solids at 10^{18} W/cm^2 (6). A few other groups working with sub-picosecond lasers also repported very recently time resolved spectra of the keV emission (7,8) and of the soft x-ray (sub-keV) radiation (9).

EXPERIMENTAL

These experiments were carried out with the Table Top Terawatt laser system at Institut National de la Recherche Scientifique (INRS). A 550fs, 1.053 μm laser pulse with an energy up to 1J was provided by a Nd:glass system based on the chirped-pulse amplification technique (10). The infra-red beam was frequency doubled with a KDP type I crystal with a conversion efficiency of 50%. Near solid density plasmas were generated by focusing on solid targets the high contrast (10^{10}:1) green pulse with an off axis parabola (f/2.5) at 45° from the target normal with an intensity of 7×10^{17} W/cm^2. Longer gradient scale length plasmas were obtained with two beams having an adjustable time delay between them (0-200ps). The delay between the two beams allowed us to adjust the gradient scale length seen by the second pulse. A thermal plasma (plasma in which non-local effects are weak and hot electrons are negligeable) can be obtained with large delay when the intensity of the second interacting beam is relatively modest. This was realized by using a Michelson interferometer to generate two collinear 0.53 μm pulses which were focused with an f/6 spherical fused silica lens on thick solid planar targets (50 μm focal spot) giving intensity of 10^{16} W/cm^2 for each beams. Non-thermal plasmas (plasma in which hot electrons dominate) can be produced when the intensity of the beam which interacts with the long gradient preformed plasma is strongly increased. Such a higher intensity regime was obtained by focusing the unconverted 1 μm radiation by a lens (f/6) onto the target (at 10^{16} W/cm^2) to form a controlled preplasma with 100 μm in diameter while the secondary delayed green beam (p polarized) was focused at 45° incidence angle by means of an off axis parabola (f/3) to a 10 μm spot (7×10^{17} W/cm^2).

We study in particular the Li-like spectrum [1s2l2l'(l,l'=0,1)-1s^{2}2l(l=0,1)] which is an important diagnostic giving informations on the plasma density, on non-stationary effects and on non-Maxwellian effects (11,12). The time resolved spectroscopy is obtained with a streak camera having a 1.5 ps temporal resolution coupled to a Von-Hamos crystal spectrometer. The spectral resolution of the whole detection system was 5×10^{-4} nm at 0.8 nm, which allowed us to study the temporal behavior of the Li-like spectrum (6). The intensity calibration was verified by using time integrated spectra obtained on a SB5 Kodak film.

RADIATION CONFINED PLASMAS

At very high electron density the boltzman equilibrium is nearly established between the upper levels by the collisions, the energy difference between the doubly excited levels being small relative to the electron temperature. Such a high density (near solid density) regime was obtained with the high contrast green pulse incident at 7×10^{17} W/cm^2 (13). At this intensity the radiation pressure balance the thermal pressure, maintaining a very steep density gradient during the energy deposition and thus forcing the emission to be produced in a region where density is close to solid density (14). The Figure 1 below shows the time history of the a-d

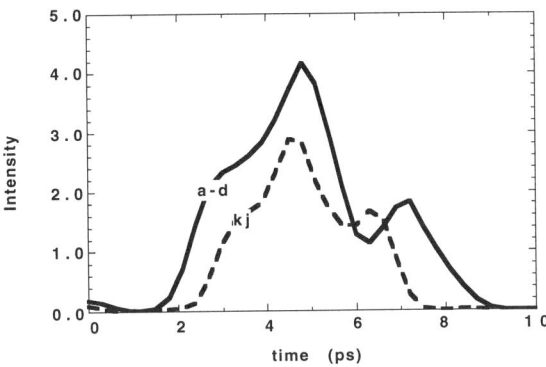

FIGURE 1. time history of the a-d (solid line) and kj (dashed line) Li-like lines emitted by a solid density plasma.

(upper levels mostly core excited) and k,j (upper levels populated by dielectronic capture) satelite Li-like lines. The lines peak at the same time as expected for level populations at the local thermodynamical equilibrium. It should be noted that in these conditions the line broadening starts to be significant and a good fit of the spectrum at the Li-like emission maximum was obtained with an electron density of 4×10^{23} cm^{-3} (the Al Li-like solid electron density is 6×10^{23} cm^{-3}).

THERMAL PLASMAS

Figure 2 presents the time history of the a-d and kj Li-like lines obtained with two collinear low intensity pulses separated by 100 ps. This very large delay was chosen in order that the second pulse interacted with a long gradient scale length plasma. In this case the laser intensity is relatively modest ($I\lambda^2 = 2.5 \times 10^{15}$ Wμm^2/cm^2), the gradient scale length L is large (approximately equal to 3.5 μm) and the beams are incident along the target normal. Thus non-linear processes (resonance absorption and parametric instabilities) are not important and the hot electron population is negligeable. Furthermore for large L/λ (where λ is the laser wavelength) the non-maxwellian effects related to non-local transport are weak and do not affect the Li-like satellites (15). We see that the collisional core excited a-d lines are produced before the k,j lines. The observed time history reveals the transient effects which enhance the inner-shell excited satellites due to a delayed appearance of the He-like ions limiting the dielectronic capture on the upper levels of the k,j transitions.

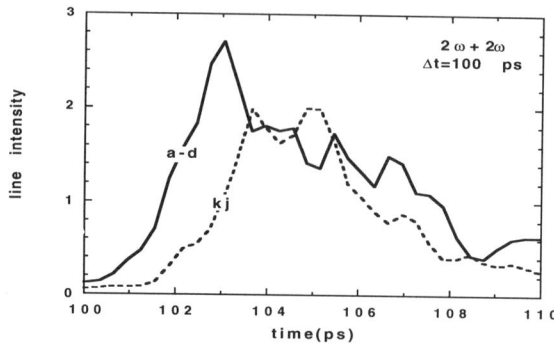

FIGURE 2. Time history of the a-d (solid line) and kj (dashed line) Li-like lines emitted by a thermal plasma which was produced by low intensity irradiation of a long gradient scale length preplasma.

NON-THERMAL PLASMAS

A non-thermal plasma is produced when the green pulse (p polarized) interacts at 7×10^{17} W/cm^2 with a long gradient scale length plasma preformed by the 1μm beam. The Figure 3 presents the Li-like time history obtained when the delay between the two pulses is 110 ps. In these conditions the fraction of laser energy in hot electrons is 9% and the hot electron temperature is 17 keV. Furthermore the backscattered light is red shifted relative to the fundamental frequency indicating that the ponderomotive pressure is pushing the critical density surface inward (16). In this regime both transient ionization and excitation and non-Maxwellian character of the electron distribution (due to non-local transport and collisionless processes) affect the early time history of the Li-like emission as observed in Figure3. At later time the Li-like emission is controlled by recombination and kj emission dominates.

As shown in Figure 4 the a-d/k,j line intensity ratio is constant as a function of time, with a value close to the LTE value, for solid density plasmas obtained by radiation pressure confinement. However for lower values of the electron density (less than 10^{23} cm^{-3}) the a-d/kj satellite line intensity ratio R can be roughly expressed as $R = N_{Li} \, g \, (1+fn_e) / N_{He} \, \alpha$ where N_{He} and N_{Li} are respectively the He- and Li-like populations, g and f are the collisional excitation rates for upper level of transition a-d from respectively the first excited state $1s^22p$ and from the upper level of kj transition $1s2p^2$, α the dielectronic rate for upper level of transition kj, and n_e the electron density. Transient and non-Maxwellian effects will respectively increase N_{Li} / N_{He} and g thus enhancing the ratio R. Such ratio enhancement is observed for the cases of an interaction with a longer density gradient scale length (Fig. 4). The core excited lines are strongly enhanced at early times both by non stationary and non-Maxwellian effects in non-thermal plasmas.

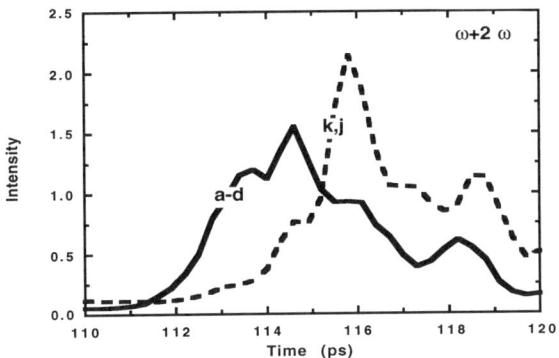

FIGURE 3. Time history of the a-d (solid line) and kj (dashed line) Li-like lines emitted by a non-thermal plasma which was produced by high intensity irradiation of a long gradient scale length preplasma.

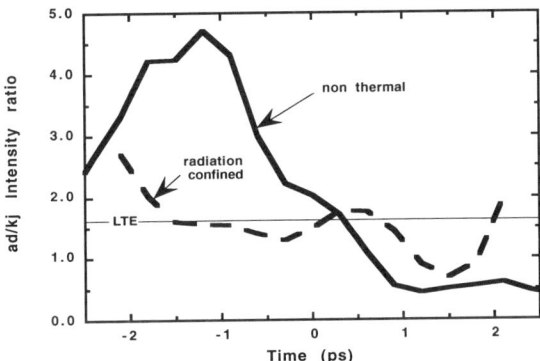

FIGURE 4. Time history of the line intensity ratio (a-d/kj) for a solid density plasma (radiation confined plasmas) and for non-thermal plasmas. The time 0 corresponds to the maximum of the a-d pulses.

CONCLUSIONS

We used time resolved spectroscopy to understand quantitatively the ionization and the plasma dynamics on the picosecond time scale. The high resolution spectroscopy of Li-like emission which has been used in the present work gives us the possibility to study various regimes of ultra-short laser matter interaction. The time history of these satellites appears to be particularly sensitive to non stationary and non-Maxwellian effects at early times in intermediate density plasmas. Thus by simply changing the delay between the prepulse and the interacting pulse or the intensity of the interacting beam it is possible to control the relative strength of these non-equilibrium effects and quantitatively study one of these effects in the appropriate interaction regime.

AKNOWLEDGEMENTS

This work was supported in part by NSERC of Canada and by Ministère de l'Éducation du Québec. The authors would like to thank J.P. Matte, S. Ethier, J.F. Pelletier and M. Chaker for many fruitful discussions.

REFERENCES

1. Milchberg, H. et al, Phys. Rev. Lett. 61, 2364 (1988)
2. Murnane, M. et al, Phys. Rev. Lett. 62, 155 (1989)
3. Kieffer, J. C. et al, Phys. Rev. Lett. 62, 760 (1989)
4. Kieffer, J. C. et al, in *X-ray laser 92*, Inst. Phys. Conf. Ser. 125, 201 (1992), Kieffer J.C. et al, Phys. Fluids B5, 2676 (1993), Kieffer, J.C. et al, OSA proceedings on short wavelength V, 17, 196 (1993)
5. Kieffer, J. C. et al, SPIE proceedings *short pulse high intensity lasers and applicationsII*, J.A.Paisner ed. 1860, 127 (1993)
6. Kieffer, J. C. et al, J. Opt. Soc Am. B13, 132 (1996)
7. Rouyer, C. et al, J. Opt. Soc Am. B13, 55 (1996)
8. Shepherd, R. private communication and Rev. Sci. Instrum. 66, 719 (1995)
9. Workman, J. et al, Phys. Rev. Lett.75, 2324 (1995)
10. Beaudoin, Y. et al, Opt. Lett. 17, 865 (1992)
11. Peyrusse, O. et al, J. Phys. B26, L511 (1993)
12. Mancini, R. et al, J. Phys. B27, 1671 (1994)
13. Jiang, Z. et al, Phys. Plasmas 2, 1702 (1995)
14. Peyrusse, O. et al, Phys. Rev. Lett. 75, 3862 (1995)
15. Matte, J. P. et al, Phys. Rev. Lett. 72, 1208 (1994)
16. Kalashnikov, M. P. et al Phys. Rev. Lett. 73, 260 (1994)

ATOMIC PHYSICS
IN DENSE PLASMAS

Population kinetics in dense plasmas

M. Schlanges[♯], Th. Bornath*, R. Prenzel* and D. Kremp*

*FB Physik, University of Rostock, D-18051 Rostock
♯FB Physik, Ernst-Moritz-Arndt-University Greifswald, D-17489 Greifswald

Abstract. Starting from quantum kinetic equations, rate equations for the number densities of the different atomic states and equations for the energy density are derived which are valid for dense nonideal plasmas. Statistical expressions are presented for the rate coefficients taking into account many–body effects as dynamical screening, lowering of the ionization energy and Pauli–blocking. Based on these generalized expressions, the coefficients of impact ionization, three–body recombination, excitation and deexcitation are calculated for nonideal hydrogen and carbon plasmas. As a result, higher ionization and recombination rates are obtained in the dense plasma region. The influence of the many–body effects on the population kinetics, including density and temperature relaxation, is shown then for a dense hydrogen plasma.

INTRODUCTION

The temporal relaxation of chemical composition and the possibility of population inversion in strongly coupled plasmas are of special interest in inertial confinement fusion and soft-X-ray laser research [1, 2, 3]. These properties can be studied using rate equations for the population densities of the different states of atoms and ions taking into account the relevant collisional and radiative processes.

Collisional rate coefficients describe processes such as impact excitation and ionization, deexcitation and three-body recombination. Usually, these rate coefficients are calculated in an elementary way from averaged cross section formulae neglecting dense plasma effects [4]. This leads to a restricted validity of the rate equations only for the ideal plasma state, because these rate coefficients are density independent. In thermodynamic equilibrium, there follows from the rate equation the well-known ideal Saha equation [5].

However, at high densities, there are strong interparticle interactions and the plasma becomes nonideal. Now one has to expect a dependence of the rate coefficients on density which can provide drastical changes in the population distribution in equilibrium as well as in nonequilibrium plasmas. This density dependence follows from many-body effects such as screening, self-energy, lowering of the ionization energy and Pauli blocking which are relevant for dense plasmas.

Of course, a systematic approach to describe the population kinetics of strongly coupled plasmas requires a strong foundation of rate equations on the basis of quantum statistical expressions for the rate coefficients taking into account many-body effects. In this paper we will consider the population kinetics of strongly coupled plasmas in such a way. We will discuss the influence of many-body effects on the rate coefficients and, finally, on the nonequilibrium level population for hydrogen and carbon plasmas.

RATE EQUATIONS FOR DENSE PLASMAS

The quantum statistical theory of the population kinetics in dense plasmas is based on kinetic equations for the distribution functions of the bound particles in different atomic or ionic states. Such equations were given in several papers [6, 7] basing on a real-time Green's functions approach. The various collision processes as elastic scattering, excitation, deexcitation, ionization and recombination are accounted for explicitly by collision integrals which are expressed in terms of in-medium T-matrices and distribution functions including plasma degeneracy effects.

Rate equations for the number densities of the bound particles can be obtained by integration of the kinetic equations with respect to the momenta. For a plasma of free electrons, singly charged ions and atoms with the densities n_e, n_i, n_A^j where j denotes the set of atomic quantum numbers, we get [8]

$$\frac{d}{dt}n_A^j = \sum_{\bar{j}}(n_e n_A^{\bar{j}} K_{\bar{j}j} - n_e n_A^j K_{j\bar{j}}) + (n_i n_e^2 \beta_j - n_e n_A^j \alpha_j). \quad (1)$$

The $K_{\bar{j}j}$ are the coefficients of collisional excitation (deexcitation) and the α_j, β_j are the impact ionization and three-body recombination coefficients, respectively. The theory now gives explicit statistical expressions for these rate coefficients valid for dense plasmas [7, 8].

The expressions are generalized ones because many-body effects are taken into account. Let us give a brief discussion of these effects. A first one is that the one-particle energies are energies of quasiparticles: $E_a(pt) = \frac{p^2}{2m_a} + Re\Sigma_a^R(p\omega t)|_{\hbar\omega = E_a(pt)}$.

Here, the influence of the surrounding plasma medium on the one-particle properties is condensed in an energy shift given by the real part of the retarded self energy function $\Sigma_a^R(p\omega t)$. A useful approximation is to replace the energy shift by a thermally averaged one, i.e.

$$E_a = \frac{p^2}{2m_a} + \Delta_a(t) \quad , \quad \Delta_a(t) = <Re\Sigma_a(pt)> \quad (2)$$

Starting from (2) in the so-called V^s-approximation for the self energy, the following simple expression can be found in lowest order [9]

$$\Delta_a(t) = -\frac{1}{2}Z_a^2 e^2 \kappa(t) \quad , \quad \kappa^2 = 4\pi \sum_a Z_a^2 e^2 \frac{d}{d\mu_a}n_a(\mu_a) \quad (3)$$

where κ is the inverse screening length and Z_a is the charge number.

A further important many-body effect is the lowering of the ionization energy. It results from the influence of the plasma medium on the two-particle properties. Consequently, the two-particle bound and scattering states have to be determined from an effective Schrödinger equation. In the dynamically screened ladder approximation it can be written as [9]

$$\left(\frac{p_e^2}{2m_e} + \frac{p_i^2}{2m_i} + \Delta_{ei}^{eff}(p_e p_i z) - z\right) \Psi_{ei}(p_e p_i z)$$
$$+ (1 - f_e - f_i) \int d^3q\, V_{ei}^{eff}(p_e p_i q z)\, \Psi_{ei}(p_e + q, p_i - q, z) = 0. \quad (4)$$

In comparison to the Schrödinger equation of an isolated pair of particles, there are some differences arising from the influence of the many-body effects: self energy correction Δ_{ei}^{eff} to the kinetic energy, Pauli-blocking factor $(1 - f_e - f_i)$ and effective dynamically screened potential V_{ei}^{eff}. Explicit expressions of these quantities and a discussion of (4) can be found in [9, 10].

A considerable simplification is possible, if V_{ei}^{eff} is approximated by the statically screened Coulomb potential and the self energy correction is assumed to be $\Delta_{ei}^{eff} = \Delta_e + \Delta_i$ with (3). This statically screened Schrödinger equation represents a useful and simple approximation to describe the main influence of plasma effects on the two-particle properties. Now, we have an effective ionization energy $I_j^{eff} = |E_j| + \Delta_e + \Delta_i - \Delta_j$ [8]. It is determined by the shift of the continuum edge $\Delta_e + \Delta_i$ and the bound state energy shift Δ_j.

IN–MEDIUM RATE COEFFICIENTS. EFFECT OF DYNAMICAL SCREENING.

For the calculation of the rate coefficients, the momentum distributions of the particles are assumed to be local equilibrium ones. If a nondegenerate plasma is considered, cross sections can be introduced formally. We will focus here on the ionization coefficient, results for the coefficients of (de)excitation in dense plasmas are given in [8]. The coefficient of ionization of the level j by electron impact can be written as

$$\alpha_j = \frac{8\pi m_e}{(2\pi m_e k_B T_e)^{3/2}} \int_{I_j^{eff}}^{\infty} dE\, E\, \sigma_j^{\text{ion}}(E)\, e^{-\beta E} \quad ; \quad E = \frac{p^2}{2m_e}. \quad (5)$$

In (5) the adiabatic approximation was applied because $(m_i \gg m_e)$. Instead of a full T–matrix calculation, we will consider the cross section in the weak coupling limit given by a modified first Born approximation

$$\sigma_j^{\text{ion}} = \frac{8\pi\hbar^2}{p_e^2 a_B^2} \frac{1}{4\pi} \int d\Omega_{p_e} \int_0^{\bar{p}_{max}} d\bar{p}\bar{p}^2 d\Omega_{\bar{p}} \int_{q_{min}}^{q_{max}} q\,dq \left|\frac{V_{ee}(q)}{\varepsilon(q, \frac{\bar{p}^2}{2m_e} + I_j^{eff})} P_{j\bar{p}}(q)\right|^2. \quad (6)$$

217

The limits of integration are determined by energy conservation in the considered ionization process [7, 8]. In Eq.(6) dynamical screening is included in the electron–electron potential with $V_{ee}(q)$ being the Fourier transform of the Coulomb potential and $\varepsilon(q,\omega)$ the dielectric function in random phase approximation (RPA) [9]. The atomic form factor is

$$P_{j\mathbf{p}}(q) = \int d^3r \Psi_j^*(\mathbf{r})\Psi_\mathbf{p}^+(\mathbf{r})e^{\frac{i}{\hbar}\mathbf{q}\cdot\mathbf{r}} \qquad (7)$$

where $\Psi_j(\mathbf{r})$ and $\Psi_\mathbf{p}^+(\mathbf{r})$ are the two–particle bound state and the scattering wave functions, respectively, which have to be determined from the effective Schrödinger equation discussed in the previous section.

In Fig.1 the cross section for ionization from the 2s atomic state of hydrogen is shown for different levels of approximation. The dashed curves are results from calculations where the screening in the electron–electron interaction as well as in the atomic form factor is accounted for in static approximation. There is a density and temperature dependent ionization threshold which moves to zero energy (Mott point) with increasing density. An other interesting in–medium effect is that higher lying bound states are shifted into the continuum due to screening in the form factor. This leads to resonances and, thereby, to an enhancement in the cross section [8, 11]. In contrast, the static screening in the electron–electron potential tends to lower the cross section.

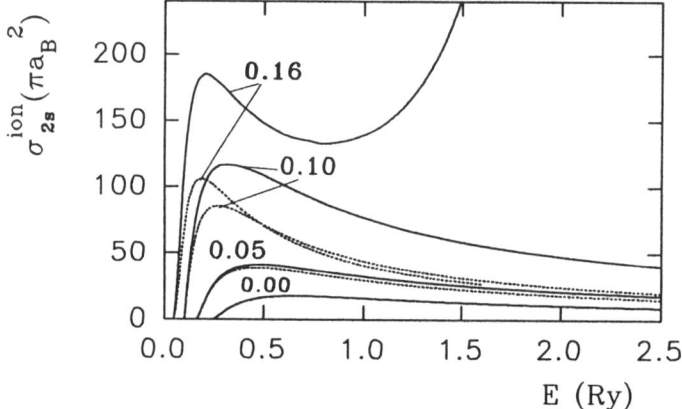

FIGURE 1. Total cross section for ionization of hydrogen 2s states versus impact energy for different sreening parameters $\lambda_D = \kappa a_B$. Two approximations are compared: dynamical screening in the e–e scattering (solid lines) and static screening (dashed).

The solid curves in Fig. 1 are results for dynamical screening included in the electron–electron potential. For small inverse screening length κ and low impact energies, the curves are similar to that for pure static screening. But at higher values of κ, there is a strong increase of the cross section for high impact energies, in contrast to static screening where the cross section is reduced. This behavior is connected with collective excitations in the strongly coupled plasma occuring for small q and for an energy argument in the dielectric function $\frac{\bar{p}^2}{2m_e} + I_j^{eff} \approx \omega_{pl}$. Therefore the concept of a cross section for impact ionization or excitation can not be applied in general (see also [12]). Although it is still possible to calculate the ionization coefficient according to Eq.(5), it is more appropriate to rewrite the basic statistical expressions for the rate coefficients. The ionization coefficient, e.g., is given by

$$\alpha_j = \frac{1}{n_e} \int \frac{d^3\bar{p}}{(2\pi)^3} \int d^3q V_{ee}(q) Im \varepsilon^{-1}(q, \frac{\bar{p}^2}{2m_e} + I_j^{eff}) n_B(\frac{\bar{p}^2}{2m_e} + I_j^{eff}) |P_{j\bar{p}}(q)|^2 \quad (8)$$

where $n_B(\omega)$ is the Bose function without a chemical potential. This expression is similar to that given in [13].

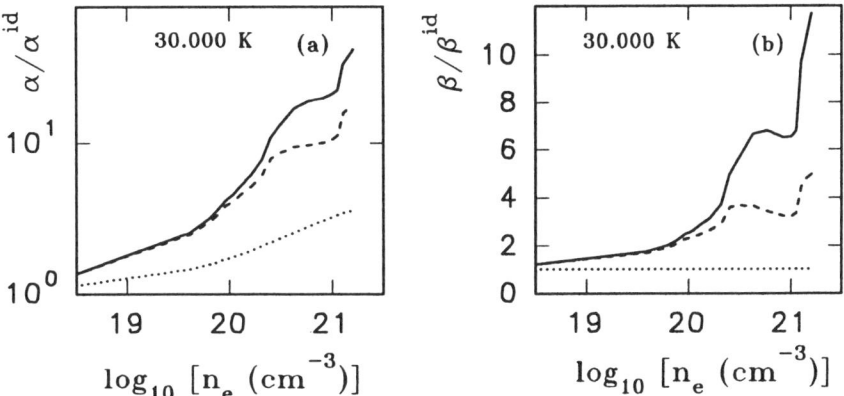

FIGURE 2. Ionization coefficient (a) and recombination coefficient (b) versus free electron density for hydrogen in the 2s state in different approximations as explained in the text: dynamical screening (solid lines), static screening (dashed) and the analytical formula (Eq.9) (dotted lines).

Fig. 2 shows the coefficients of ionization and recombination for the atomic 2s state of hydrogen as a function of the free electron density. Again, results for static and for dynamical screening are compared. Additionally, curves are given which correspond to the following analytical result

$$\alpha_j = \alpha_j^{ideal} e^{(\Delta_j - \Delta_e - \Delta_p)/k_B T} \quad (9)$$

with α_j^{ideal} being the value for the ideal plasma state ($n_e = 0$). This formula was obtained taking into account the lowering of the ionization energy only in a simple approximation [14, 15]. There is a strong influence of many-body effects on the ionization rate. Having calculated the ionization coefficients, it is rather simple to get the recombination coefficients from [7, 14]

$$\beta_j = \Lambda_e^3 \alpha_j e^{I_j^{eff}/k_B T}; \qquad \Lambda_e = (2\pi\hbar^2/m_e k_B T)^{\frac{1}{2}}. \tag{10}$$

The density dependence of the recombination coefficient is not as strong as that of the ionization coefficients. If the simple expression (9) is used, the recombination coefficient is independent of the density. This behavior changes drastically if many-body effects are included in a more systematic way as shown in this paper. Now there is an interesting density dependence which is determined by the competition between the screening in the electron-electron potential and that in the form factor.

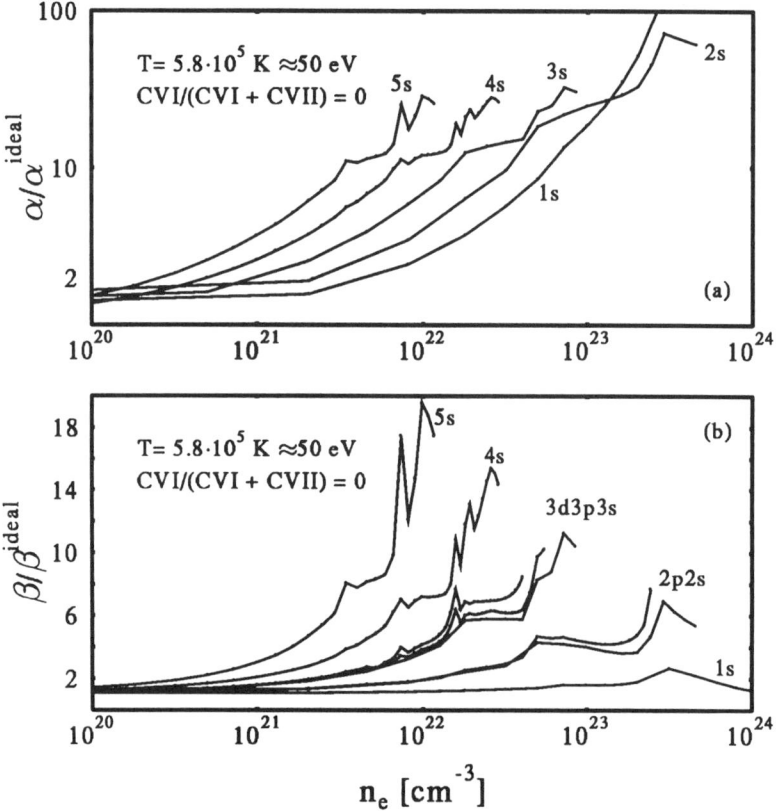

FIGURE 3. Ionization coefficient (a) and recombination coefficient (b) versus free electron density for carbon (CVI) in different states for a temperature $T_e = 50 eV$.

The calculations made for hydrogen can be applied to the important case of a dense carbon plasma consisting of free electrons, bare carbon nuclei and hydrogen-like ions in different states. The effective ionization energy is given now by [16]

$$I_j^{eff} = |E_j| - Z_{\text{CVII}} \kappa e^2. \qquad (11)$$

In Fig. 3, the ionization and recombination coefficients are shown for the ground state and some excited states. The screening was included here in static approximation.

NONEQUILIBRIUM LEVEL POPULATION

We consider a dense hydrogen plasma and determine the temporal evolution of the atomic population densities from the rate equations (1). For the electron temperature, the following balance equation was derived from the quantum kinetic equation [17]

$$\frac{dT_e}{dt} = \sum_j \frac{\frac{3}{2}k_B T_e - \frac{3}{4}\frac{e^2}{r_0} + |E_j|}{\frac{3}{2}n_e k_B + \frac{1}{4}\frac{e^2}{r_0}\frac{n_e}{T_e^2}\left(\frac{1}{T_e} + \frac{1}{T_i}\right)^{-1}} \frac{dn_H^j}{dt}. \qquad (12)$$

In the rate equation (1) as well as in (12), many-body effects are accounted for. In the temperature equation, the quasiparticle energy shifts are accounted for explicitely.

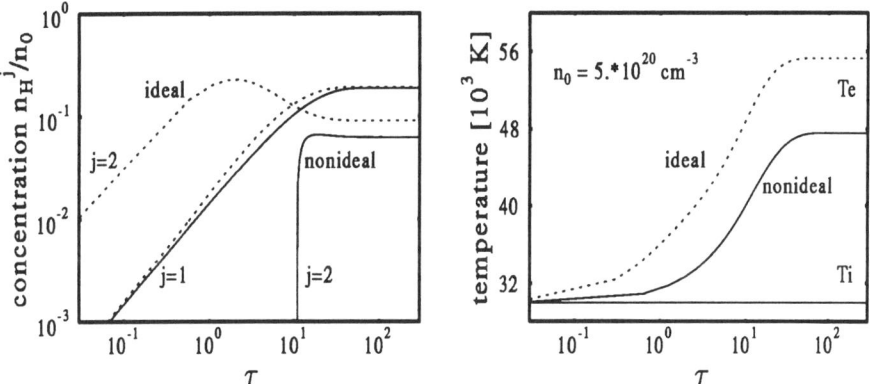

FIGURE 4. Time evolution of the occupation numbers and of the electron temperature for a hydrogen plasma:
$n_e^{tot} = 5 \cdot 10^{20} \text{cm}^{-3}, T_e(t=0) = 30000 K$. $\tau = t/t_0, t_0 = 3.685 \cdot 10^{-14} s$.

In Fig. 4 a density-temperature relaxation is shown for hydrogen which results from a numerical solution of the coupled system of the equations (1) and (12). The rate coefficients are used in the simple approximation where the lowering of the ionization energy is included only. The ionization coefficients are then given by (9). A fully ionized plasma is assumed in the initial state. There are drastical changes in the behaviour of the population densities due to the influence of many-body effects (nonideal plasma). An important effect can be observed from the behaviour of the excited level $j = 2$. If the ideal plasma theory is applied (dotted curve), this state is populated from the beginning of the relaxation process. But in the considered case of a strongly coupled plasma, the Mott condition must be taken into account. The consequence is that the nonideality of the plasma prevents the population of this excited level in the first period of the relaxation process. Thus, in the theoretical description of the nonequilibrium level population in dense plasmas, we are confronted with the interesting problem of the appearance and disappearance of bound states because of the Mott effect.

ACKNOWLEDGMENTS

This work has been supported by the Deutsche Forschungsgemeinschaft (Sonderforschungsbereich "Kinetik partiell ionisierter Plasmen").

REFERENCES

[1] Y. Leng, J. Goldhar, H.R. Griem, R.W. Lee, *Phys. Rev. E* **52**, 4328 (1995).
[2] D.J. Heading, G.R. Bennet, J.S. Wark, R.W. Lee, *Phys. Rev. Lett.* **74**, 3616 (1995).
[3] H.-J. Kunze, K.N. Koshelev, C. Steden, D. Uskov, H.T. Wieschebrink, *Phys. Lett. A* **193**, 183 (1994).
[4] L.M. Bibermann, V.S. Vorob'ev, I.T. Yakubov, *Kinetics of Nonequilibrium Low-Temperature Plasmas*, Consultants Bureau, New York, 1987
[5] Yu.L. Klimontovich, M. Schlanges, Th. Bornath, *Contrib. Plasma Phys.* **30**, 349 (1990).
[6] D. Kremp, M. Schlanges, Th. Bornath, "The Dynamics of Systems with Chemical Reactions", J.Popielawski (Ed.), World Scientific, Singapore, 1989, p.3
[7] M. Schlanges, Th. Bornath, *Physica A* **192**, 262 (1993).
[8] Th. Bornath, M. Schlanges, *Physica A* **196**, 427 (1993).
[9] W.D. Kraeft, D. Kremp, W. Ebeling, G. Röpke, *Quantum Statistics of Charged Particle Systems*, Plenum, London, New York, 1986.
[10] J. Seidel, S. Arndt, W.D. Kraeft, *Phys. Rev. E* **52**, 5387 (1995).
[11] Th. Bornath, Th. Ohde, M. Schlanges, *Physica A* **211**, 344 (1994).
[12] M.S. Murillo, J.C. Weisheit, this Volume.
[13] M.S. Murillo, J.C. Weisheit, in *Strongly Coupled Plasma Physics*, H.M. Van Horn, S. Ichimaru (Ed.), University of Rochester Press, 1993, p. 233
[14] M. Schlanges, Th. Bornath, D. Kremp, *Phys. Rev. A* **38**, 2174 (1988).
[15] W. Ebeling, I. Leike, U. Leonhardt, "Bound States and Ionization Kinetics in Dense Plasmas" in *AIP Conference Proceedings 257*, 1991, pp. 97–107
[16] K. Kilimann, W. Ebeling, *Z. Naturforsch.* **45a**, 613 (1990).
[17] Th. Ohde, M. Bonitz, Th. Bornath, D. Kremp, M. Schlanges, *Phys. Plas.* **2**, 3214 (1995).

Deexcitation and Recombination Processes Involving Doubly Excited Levels in High Density Plasma

Tetsuya Kawachi

The Institute of Physical and Chemical Research (RIKEN), Saitama, 351, Japan

Abstract. Enhancement of deexcitation and recombination from excited ions in dense plasma is considered for lithiumlike ions. The former process is caused by the thermal population in doubly excited levels of berylliumlike ions and their autoionization, and the latter is the collisional-radiative recombination which has been well established for the ground state ions. We have estimated the effective rate coefficients for these processes by adopting various approximations. We incorporate them into our collisional-radiative model for the lithiumlike aluminum recombining plasma. The population inversion and the gain of the $3d\ ^2D$-$5f\ ^2F$ laser transitions are calculated.

INTRODUCTION

The recent advent of high irradiance lasers makes it possible to produce dense and hot plasmas. Now we are able to study highly charged ions in dense plasma spectroscopically. For the purpose of applying these plasmas to inertial confinement fusion, soft x-ray laser, or high irradiance x-ray source, it is indispensable to make clear the population kinetics of the excited ions for the respective plasma conditions. Since the electron density (n_e) is of the order of 10^{24}-$10^{27}\mathrm{m}^{-3}$, the ions are substantially populated in doubly excited levels, and the atomic processes involving these levels may play an important role in the population kinetics of the ions. One of the atomic processes was first discussed by Fujimoto and Kato (1) for heliumlike ions in dense plasma: They proposed a process of virtual excitation of hydrogenlike ions via doubly excited heliumlike ions which are created from dielectronic capture of hydrogenlike ions. They also proposed the corresponding deexcitation process. They named these processes "dielectronic capture and ladderlike (DL) excitation and deexcitation", and showed that the process enhanced substantially the direct excitation or deexcitation rate coefficients in high density regions.

Recently, lithiumlike aluminum ions in the recombining plasma are intensively studied experimentally as a soft x-ray laser medium (2, 3). A serious discrepancy between the experiments and collisional-radiative (CR) calculations is pointed out for the gain of the $3d\ ^2D$-$5f\ ^2F$ transition (4, 5). One of the atomic processes neglected in these calculations is the DL processes. Other processes like recombination of excited ions may also operative in dense plasma. In the following, we discuss these processes and examine their effects on the excited level populations and the gain.

DL DEEXCITATION AND EXCITATION

DL deexcitation is explained as an autoionization process from doubly excited levels. Figure 1 shows a simple energy level diagram of lithiumlike ions. The k and i are levels of lithiumlike ions taken as an example, and we call them "parent levels". Each of these levels have a series of Rydberg levels accompaning it, and in the present case they are doubly excited levels of berylliumlike ions, (k, p) or (i, p), where k and i denote "the core electrons" and p is "the running electron". In high density plasmas, these doubly excited ions are substantially populated, and these populations may be lost by autoionization $(k, p) \rightarrow i + e$ $(i < k)$. Since the level (k, p) are strongly coupled with the parent level k and continuum electrons by collisional processes, the populations lost by this processes are supplied rapidly by recombination of the lithiumlike ions k with continuum electrons. As a result, a population flow from the level k, through the doubly excited level (k, p) into the level i is established, $i.e.$,

$$A^{q+}(k) + e \xleftrightarrow{\text{strongly coupled}} A^{(q-1)+}(k,p) \xrightarrow{\text{autoionization}} A^{q+}(i) + e \quad (1)$$

where e denotes a continuum electron. This population flow is nothing but deexcitation $k \rightarrow i$, which should be added to the ordinary direct deexcitation.

In low temperature dense plasmas, the high-lying doubly excited levels are approximately given by their LTE (local thermodynamics equilibrium) values with respect to the parent level and continuum electrons. Thus the DL deexcitation rate coefficient, F_{DL}, is approximately given as follows (1).

$$F_{\text{DL}}(k \rightarrow i) = \int_{p_C}^{\infty} A_a(k; p \rightarrow i) \frac{n(k,p)_{\text{LTE}}}{n_e n(k)} dp \quad (2)$$

where $A_a(k, p \rightarrow i)$ is the autoionization probability from level (k, p) to level i, $n(k,p)_{\text{LTE}}$ is the LTE population and p_C is the lower end of the LTE. p_C is given either by generalized Griem's level p_G (6) or Byron's level p_B (7), whichever the larger. For the doubly excited level lying above (k, p_G), the dominant depopulation process is the collisional excitation or deexcitation, while for levels lying below that the dominant process is autoionization or radiative decay. Figure 2 shows examples of the p_G for the levels (k, p) accompaning lithiumlike aluminum levels, $k=3p\ ^2P$ and $3d\ ^2D$. For the levels lying above (k, p_B) the dominant collisional depopulation is excitation, while for the levels below it is deexcitation. The autoionization probability can be derived from dielectronic capture rate coefficient by use of the principle of detailed balance. On the basis of the continuation properties of the dielectronic capture cross section across the excitation threshold to the excitation cross section, eq.(2) is approximated to

$$F_{DL}(k \to i) = \int_{E_C}^{E_k} (\frac{4Z_{eff}\varepsilon}{\pi a_0 h n^3}) \frac{g(k)}{g(k,p)} \sigma_a(i \to k; E') \frac{n(k,p)_{LTE}}{n_e n(k)} dE' \quad (3)$$

where E_C is the energy of the boundary level (k, p_C) which is measured from level i, $\sigma(i \to k; E')$ is the excitation cross section extrapolated below the excitation threshold $E(i, k)$. We treat the DL deexcitation for the $n = 2$-

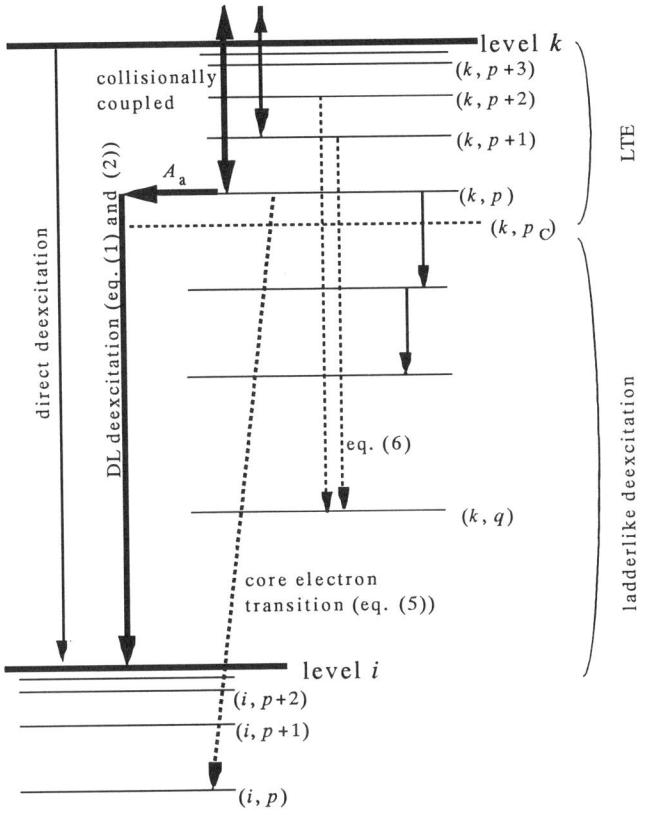

FIGURE 1. The schematic diagram of the DL deexcitation (thick solid arrows), radiative decay of core electron (thick dotted arrow), recombination of excited ions by collision (thin solid downward arrows) and by radiation (thin dotted downward arrows). A_a in the DL deexcitation means autoionization.

3, 2-4, 2-5, 3-5 transitions and the $2s\ ^2S - 2p\ ^2P$ transition of lithiumlike aluminum ions, where n is the principal quantum number. We ignore them for the n=3-4, 4-5 and Δn=0 (n=3, 4 and 5) transitions because the threshold energy of these transitions are so small that the interval of the integration in eq.(3) is small. The excitation cross section data are calculated by Zhang et al. (8) for the transitions from the ground state ($2s\ ^2S$) and the $2p\ ^2P$ level, and for other transitions, the cross section is calculated by the distorted wave method, the details of which is given in Clark et al. (9).

The DL excitation of $i \to k$ in lithiumlike ions may be expressed as a series of processes of the dielectronic capture of an electron by a lithiumlike ion in level i to form a berylliumlike ion in a doubly excited state (k, p), followed by the ladderlike excitation-ionization, resulting in lithiumlike level k. The effective rate coefficient $C_{DL}(i \to k)$ for $i \to k$ is approximately related to $F_{DL}(k \to i)$ by the principle of detailed balance (1).

$$C_{DL}(i \to k) = \frac{g(i)}{g(k)} F_{DL}(k \to i) \exp(-E(i,k)/T_e) \quad (4)$$

It should be noted that eq.(4) is exactly the same as the relation holding for the rate coefficients for the ordinary direct excitation and deexcitation.

In Fig. 2, the thick solid and dotted lines show the n_e-dependence of DL deexcitation rate coefficients for $2s\ ^2S\text{-}3p\ ^2P$ and $2s\ ^2S\text{-}3d\ ^2D$ transitions, respectively. The direct collisional deexcitation rate coefficients of these transitions are given by the horizontal lines. As the electron density increases, the Griem's boundary level comes down, and the DL deexcitation rate coefficient increases. For $n_e > 10^{25}$ m^{-3}, the rate coefficients of DL process become larger than those of the direct process. For $n_e > 10^{27}$ m^{-3}, DL deexcitation rate coefficients become constant, because the lower end of the integration in eq.(2) is determined by Byron's boundary which is determined by T_e.

CORE ELECTRON TRANSITION AND CR-RECOMBINATION OF EXCITED IONS

In addition to the DL processes, we have to consider another process involving doubly excited levels; The radiative or collisional decay of core electron, $(k, p) \to (i, p) + h\nu$ or $(k, p) + e \to (i, p) + e$ ($i<k$), from a level with $p > p_C$ may enhance the depopulation of level k. (See the thick dotted arrow in Fig.1.) The transition probability or the rate coefficient of this process is approximately expressed by use of the radiative decay probability or rate coefficient of the core electron $k \to i$, $A(k, i)$ and $F(k, i)$, as

$$A_{rad} = \int_{p_C}^{p_{limit}} \frac{n(k,p)_{LTE}}{n_e n(k)} dp\ A(k,i)$$
$$F_{rad} = \int_{p_C}^{p_{limit}} \frac{n(k,p)_{LTE}}{n_e n(k)} dp\ F(k,i) \quad (5)$$

where p_{limit} is determined by "the ionization lowering". For atomic hydrogen plasmas, Shimamura and Fujimoto (10) introduced an ion sphere model with the radius $r_i = (3/4\pi n_e)^{1/3}$, and derived that the upper limit and the number of the bound state were given by $n^* = (2.0 \times 10^{29}\ \text{m}^{-3}/n_e)^{1/6}$ and $N_{bnd} = (1.7 \times 10^{29}\ \text{m}^{-3}/n_e)^{1/2}$, respectively. We refer to the result and scale the radius r_i by n_e/z, i.e., p_{limit} is expressed as $(2.0 \times 10^{29} z \text{ m}^{-3}/n_e)^{1/6}$, where z is the effective nuclear charge of the berylliumlike aluminum ions. The integral in eq.(5) is identical to the sum of the LTE populations hanging below the level k of the lithiumlike ions. In the present case, the sum of the LTE populations is less than 10% of the parent level. It should be noted that, since the number of the bound state and the population of each level in LTE are propotional to $n_e^{-1/2}$ and n_e, respectively, the transition probability or rate coefficient is propotional to $n_e^{1/2}$.

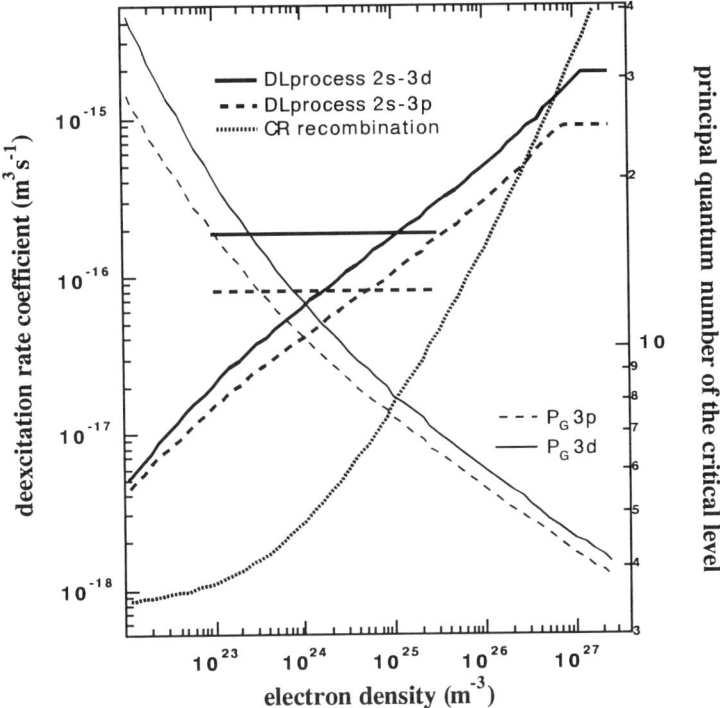

FIGURE 2. DL deexcitation for the $2s\ ^2S$-$3p\ ^2P$ (thick solid line) and the $2s\ ^2S$-$3d\ ^2D$ (thick dashed line) transitions. The thick dotted line: the recombination from the $3p\ ^2P$ level. The thin solid and dashed lines are generalized Griem's level for the $3p\ ^2P$ and $3d\ ^2D$ levels. The rate coefficients are referred to the left hand side ordinate, and the Griem's level is to the right hand ordinate.

The third process is the standard CR recombination process which is well established for the ground state ions (11). Its essential mechanism in the present context is that the LTE populations of the high lying levels (k, p) are lost by deexcitation of the running electron p. This process may be, in the case of $p_G > p_B$, the radiative decay $(k, p) \rightarrow (k, q) + h\nu\ (q < p_C < p)$

(the thin dotted arrows in Fig. 1), and, in the case of $p_G < p_B$, the ladderlike deexcitation $(k, p_B) \rightarrow (k, p_B\text{-}1) \rightarrow (k, p_B\text{-}2) \rightarrow \cdots$ (the thin solid arrows in Fig. 1). For both of the cases, the running electron in level q or $p_B\text{-}1, p_B\text{-}2\ldots$ finally reaches the 2s level by the ladderlike deexcitation to become berylliumlike ions in the level $1s^2 2sk$ or it may autoionize to become a lithiumlike ion. The effective rate coefficient for the former process, i.e., radiative decay process, can be calculated from,

$$\alpha_{rad} = \sum_{q=1}^{p_C-1} \int_{p_C}^{p_{limit}} \frac{n(k,p)_{LTE}}{n_e n(k)} A(k,p;k,q) \, dp \qquad (6)$$

where $A(k, p; k, q)$ is the radiative decay probability from doubly excited levels (k, p) to (k, q), and $q=1$ means the 2s state. $A(k, p; k, q)$ may be approximated to the spontaneous transition probability from p to q of hydrogenic ions. The rate coefficient of the latter process, i.e., ladderlike deexcitation process, is estimated from the collisional deexcitation rate coefficient of the $\Delta n=1$ transition from the lower end of LTE, $F(p_B+1, p_B)$.

In Fig. 2, the thick dotted line represents the effective rate coefficient for recombination of the $3p\ ^2P$ excited level, which is the sum of that due to radiation (eq.(6)) and that due to collisions.

We calculate the probability of the radiative decay of the core electron and the CR recombination rate coefficients for all the excited levels with $n \leq 5$.

CALCULATION OF THE 3D-5F GAIN OF Li-LIKE Al LASER

We take these three processes into our collisional-radiative model (5) for lithiumlike ions and calculate for the gains of lithiumlike aluminum laser. We assume that $T_e=30$eV which is a typical temperature for the lithiumlike aluminum recombination laser and that the plasma is in the recombining plasma scheme, i.e., the excited level populations are created only from heliumlike ions. Figure 3 show the populations of several excited levels without additional processes (a) and with these processes (b). The population inversion between $n=3$ and 4, $n=3$ and 5, and $n=4$ and 5 levels are generated for $n_e=10^{24} \sim 10^{25} \text{m}^{-3}$. In Fig. 3 (b), the excited level populations becomes smaller than those in Fig. 3 (a) for $n_e \geq 10^{26} \text{m}^{-3}$ which is mainly due to CR-recombination process from excited level.

Figure 4 (a), (b) show the amplification gains of several lines. The absorption profile of the laser lines is assumed to be thermal Gaussian. We have not included Stark broadening effect yet. The gain coefficients for the $n=3$-5 transitions are multiprized by factor of 10. As shown in these figures, DL excitation deexcitation and the recombination from an excited levels do not induce a significant change in the gains.

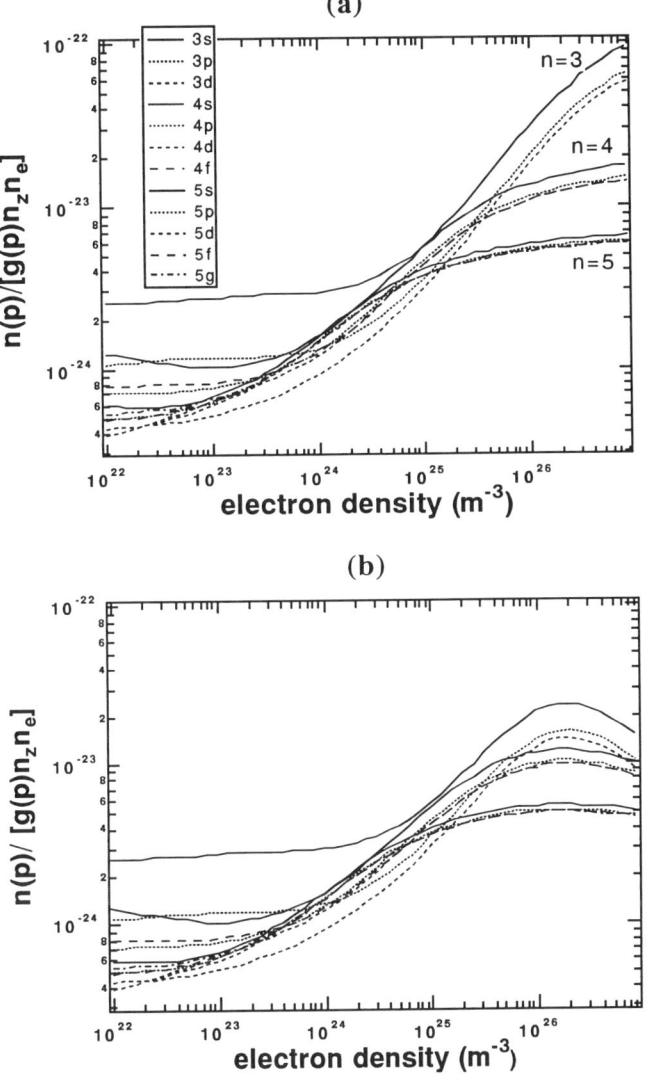

FIGURE 3. The populations of several excited levels of lithiumlike aluminum ions by use of our CR model. (a): additional atomic processes described in text are not included. (b): additional processes are included. In Fig. 3 (b), the decrease of the populations in $n_e > 10^{26} m^{-3}$ is due to the CR-recombination from excited levels.

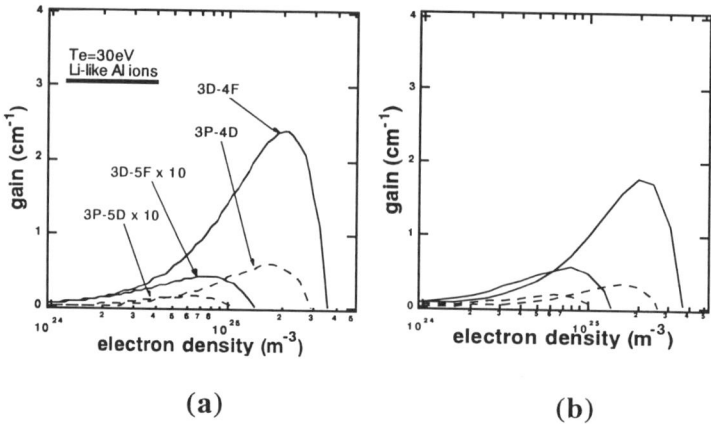

FIGURE 4. The amplification gain of several transitions of lithiumlike aluminum ions. (a): additional atomic processes described in text are not included. (b): additional processes are included.

ACKNOWLEDGEMENT

The author works in the institute (*RIKEN*) under the auspices of the special postdoctral researchers program.

REFERENCES

1. Fujimoto T. and Kato T., Phys. Rev. A **32** 1663 (1985).
2. Calliron A., Edwards M.J., Grande M., Henshaw M. J. de C., Jaegle P., Jamelot G., Key M. H., Kiehn G. P., Klisnick A., Lewis C. L. S., O'Neill D., Pert G. J., Ramsden S. A., Regan C. M. E., Rose S. J., Smith R. and Willi O., J. Phys. B **23**, 147 (1990).
3. Hara T., Ando K., Kusakabe N., Yashiro H. and Aoyagi Y., Jpn. J. Appl. Phys, **28**, L1010 (1989).
4. Klisnick A., Sureau A., Guennou, Moller H. C., and Virmont J., Appl. Phys. B**50** 153 (1990).
5. Kawachi T., Fujimoto T. and Csanak G., Phys. Rev. E **51**, 1428 (1995): Kawachi T. and Fujimoto T., Phys. Rev. E **51**, 1440 (1995).
6. Griem H. R.: Plasma Spectroscopy (McGraw-Hill, New York, 1964) p. 129.
7. Byron S., Stabler R. C. and Bortz P. I., Phys. Rev. Lett. **8**, 376 (1967).
8. Zhang H. L., Sampson D. H. and Fontes C. J., At. Data Nucl. Data Tables **44**, 31 (1989).
9. Clark R. E. H., Csanak G. and Abdallah J. Jr, Phys. Rev. A **44** 2874 (1991)
10. Shimamura I. and Fujimoto T., Phys. Rev. A **42**, 2346 (1990)
11. Fujimoto T., J. Phys. Soc. Jpn, **49**, 1561 (1980)

Plasma Density Effects on Ionization

Michael S. Murillo

Theoretical Division, Los Alamos National Laboratory
MS B212, Los Alamos, New Mexico 87545

Abstract. Traditionally, collisional rate coefficients for electron impact excitation or ionization of an ion in a plasma are obtained by first computing a binary collision cross section $\sigma(v)$ between a *single* plasma electron and an *isolated* ion. This quantity is subsequently averaged over the plasma electron flux to obtain the rate coefficient $\langle v\sigma(v)\rangle$. In a dense plasma this binary collision picture breaks down. A model is presented in which the collisional rate coefficent due to *many* interacting plasma electrons are computed directly, without reference to a binary cross section. Focus is directed toward dynamical screening in electron impact ionization of hydrogenic ions.

NONIDEAL VS. "ATOMICALLY DENSE" PLASMA REGIMES

Experiments involving short pulse (ns) and ultrashort pulse (sub-ps) lasers are capable of creating dense plasmas which are *not* in local thermodynamic equilibrium (LTE)[M95]. The theoretical description of such transient plasmas requires detailed knowledge of the the atomic population kinetics and therefore the rate coefficients of all relevant atomic processes[1]. Dense plasma modifications to the collisional rate coefficients has been an active subject of research in recent years[W88, MW93, SB93, J95a, J95b]. These modifications can be loosely grouped into two categories: nonideal plasma effects and "atomically dense" plasma effects. Although the present work is primarily concerned with the "atomically dense" regime, nonideality will briefly be described.

Nonideality in a plasma is defined by strong Coulomb coupling and/or Fermi degeneracy[I94]. Coulomb coupling, the ratio of the average potential energy to the average kinetic energy, for species s is described by the parameter

$$\Gamma_s = 2.321 \times 10^{-7} \frac{Z_s^2 n_s^{1/3}}{k_B T_s} \qquad (1)$$

where n_s is the particle density in cm^{-3} and $k_B T_s$ is the temperature in eV. The condition $\Gamma_s \geq 1$ identifies the strong coupling regime. Fermi degeneracy,

---

[1]This should be contrasted with most astrophysical dense plasmas (e.g. stellar interiors) which are in LTE. Population kinetics of these plasmas are well described with Saha relations.

the ratio of the Fermi energy to the temperature $k_B T$, is measured by the parameter

$$\chi = 3.64 \times 10^{-15} \frac{n^{2/3}}{k_B T}. \qquad (2)$$

The degenerate regime, where $\chi \geq 1$, involves the breakdown of classical statistics. This parameter is usually only relevant for the electrons. Although nonideal plasmas are theoretically interesting, most laboratory laser experiments take place at temperatures sufficiently high that both Γ and χ are less than unity for most of the duration of the experiment, i.e. the plasmas are ideal. Nevertheless, we still might expect to have some density effects resulting purely from the near solid densities characteristic of these experiments.

An "atomically dense" plasma refers to a plasma which is dense on time and length scales defined by the atomic environment of interest. Such plasmas may or may not be ideal. These scales can be defined by considering an atomic wavefunction

$$\psi_a(\mathbf{r}, t) = \phi_a(\mathbf{r}) \exp(-i E_a t / \hbar) \qquad (3)$$

which corresponds to some stationary state $|a\rangle$. We then define the time scale in terms of the frequency $E_a/\hbar$ or, in the context of transitions, $(E_b - E_a)/\hbar$ where $|b\rangle$ is some other stationary state. For example, consider an atomic transition with transition frequency $(E_b - E_a)/\hbar$. For some states $|b\rangle$ and $|a\rangle$ this frequency may be near the frequency ω_p associated with the electron plasma oscillation. The frequency ω_p in an inertial confinement fusion (ICF) experiment, for example, is 10^5 times that in a magnetic fusion experiment. Similar arguments hold for the spatial part of the wavefunction. For example, the energy E_a itself may be modified as a result of overlap of $\phi_a(\mathbf{r})$ with nearby ions and electrons. At high plasma density we have the possibility that the more deeply bound atomic levels, those that are usually not in LTE, are subjected to these types of perturbations. Note that the plasma may be "atomically dense" for certain atomic transitions and not for others.

Of the various collisional atomic processes, the focus here will be on electron impact ionization. The results, however, equally apply to impact excitation and the inverse processes of three-body recombination and de-excitation. The corresponding ion impact processes are not considered here. We begin with a brief review of the traditional approach to electron impact ionization rate coefficients. Simple modifications will be suggested that allow density effects to be included in this picture. Then, models more suitable to the dense plasma regime will be presented and compared with the results of the traditional approach.

ELECTRON IMPACT RATE COEFFICIENTS VIA CROSS SECTIONS

Collisional rate coefficients for collisional processes inside a plasma are typically obtained from cross sections. In general, a cross section $\sigma_{a\to b}(v)$ is obtained which is relevant for an atomic transition from state $|a\rangle$ to state $|b\rangle$ given that a free plasma electron impacts with initial velocity v. To obtain the *rate* w at which this reaction takes place we multiply the cross section by the flux $n_e v$ of the plasma electrons and average over the distribution of initial velocities, *viz.*

$$w = n_e \langle v \sigma_{a\to b}(v) \rangle. \tag{4}$$

Note that the plasma density enters here only as the factor n_e. It is clear that the cross section $\sigma_{a\to b}(v)$ contains all the information regarding the transition[2]. In this section a brief review of cross section calculations is given. A simple model will be used which treats an electron collision with a hydrogenic ion target in the first Born approximation (FBA). Then, modifications of the cross section for dense plasma effects will be discussed. The robustness of these modifications will be addressed in the following sections.

Consider a free plasma electron with velocity $\mathbf{v}$ impacting a hydrogenic ion initially in state $|a\rangle$. Later we find the atomic electron in some final free particle state $|b\rangle$ and the free plasma electron in a different free particle state $|\mathbf{v}_f\rangle$. Let us assume that we are interested in the total cross section for electron impact ionization. The cross section for this process can be written in the FBA as

$$\sigma_a(\mathbf{v}) = \frac{2\pi}{\hbar} \sum_{\mathbf{v}_f} \sum_b \left| \langle \mathbf{v}_f | \langle b | \hat{V} | \mathbf{v} \rangle | a \rangle \right|^2 \delta(E_f - E_i) \frac{1}{(v/\Omega)} \tag{5}$$

where Ω is the quantization volume, $\delta(x)$ is the Dirac delta function, E_f and E_i refer to the final and initial energies of the system, respectively, and $\hat{V}$ is the interaction potential between the atomic and plasma electron which, in vacuum, is the Coulomb interaction potential. The FBA neglects higher order contributions and exchange. A good review of these effects has been presented by Younger[Y85] and are not the main theme of the present article. It will be useful in what follows to Fourier transform the Coulomb potential such that the cross section may be written as

$$\sigma_a(\mathbf{v}) = \frac{2\pi}{v\Omega} \sum_b \sum_\mathbf{q} \frac{(4\pi e^2)^2}{q^4} \left| \langle b | e^{i\mathbf{q}\cdot\mathbf{r}} | a \rangle \right|^2 \delta\left(\frac{\hbar q^2}{2m} + \omega - \mathbf{v}\cdot\mathbf{q} \right) \tag{6}$$

where ω is the energy transferred to the atom.

---

[2] In highly non-equilibrium situations, deviations from a Maxwellian velocity distribution may also play an important role in accurately determining the rate coefficient.

Density effects can be incorporated into (6) by making intuitive modifications to the states of the target ion and/or the free plasma electron. In the former case, we view the ion as polarizing the surrounding plasma thereby producing an external potential which acts on the bound electron in addition to the Coulomb potential of the nucleus. Thus, the states $|a\rangle$ and $|b\rangle$ will have different forms from those used above. In an ideal plasma the total (external plus nuclear) potential is often taken to be a Debye-Hückel potential[R70]. To lowest order, this model predicts smaller ionization energies than that of the bare ion. Clearly this serves to enhance the ionization rate. The former case is treated by now taking the interaction of the plasma electron as also being screened. In this case the interaction $\hat{V}$ is also replaced by, say, a Debye-Hückel interaction. At high density this screening can cause a significant weaking of the electron-ion interaction and leads to a reduced ionization rate. These effects have been extensively explored by Jung in a semiclassical model for bound-bound transitions in both weakly coupled plasmas[J95a] and strongly coupled plasmas[J95b].

Formally we may describe the above dense plasma model via a Hamiltonian of the form

$$H = \frac{p_a^2}{2m} - \frac{Ze^2}{r_a}\phi_1(r_a) + \frac{p_p^2}{2m} + \frac{e^2}{|r_a - r_p|}\phi_2(|r_a - r_p|) \qquad (7)$$

where $\phi_1(r)$ and $\phi_2(r)$ are screening factors for the target-electron and the electron-electron interaction potentials and the subscripts a and p refer to the atomic electron and the plasma electron, respectively. Again, $\phi_1(r)$ and $\phi_2(r)$ are often taken to have the Debye-Hückel form[J95a]. The last term represents the perturbation that causes the transition. It is clear that such models are incapable of describing transitions caused by the plasma oscillation, for example. This is a result of the *ad hoc* assumption that the screening functions $\phi_1(r)$ and $\phi_2(r)$ can be taken to be static for all time scales. In the following sections we explore another type of model suitable for addressing this issue.

THE STOCHASTIC MODEL

Conceptually we may visualize an impact ionization event as a free plasma particle passing by an ion and thereby producing a time-dependent potential $V(\mathbf{r}, t)$. At high plasma densities it is then natural to imagine that this time-dependent potential is produced by *many* plasma particles simultaneously. If these plasma particles move independently of the ion, the surrounding plasma can be treated as an external time-dependent perturbation. This external perturbation can be treated as a stochastic potential at the ion which is described by its statistical properties. The Hamiltonian for such a model can be written in the form

$$H = \frac{p^2}{2m} - \frac{Ze^2}{r} + V(\mathbf{r}, t) \tag{8}$$

where $V(\mathbf{r}, t)$ represents the stochastic potential. This is the essense of the "Stochastic Model". Because there is no particular particle colliding with the atom, as there is in (7), the cross section referring to the scattering of a single electron with the target ion is not used and we resort to a direct calculation of the rate coefficient for the excitation of the ion *by the whole plasma*. In practice this is in no way a limitation since we are ultimately interested in the rate coefficient. In the low density limit this model should reduce to that of the previous section. This type of model was first used by Vinogradov and Shevel'ko[VS76] for bound-bound transitions of an atom located in a "random field", $V(\mathbf{r}, t)$. A similar model was developed independently by Weisheit[W88] which described the stochastic potential in terms of plasma density fluctuations of both electrons and ions. In this description the power spectrum of the stochastic potential is accurately described in terms of the plasma dynamic structure factor. That work was extended to bound-free transitions (ionization) by Murillo and Weisheit[MW93] and included level shift (continuum lowering) effects as well. A comprehensive reformulation of such models as a collision process between the entire interacting plasma and an ion was subsequently given by Murillo[M95]. The latter two approaches are described in the following sections.

The Semi-Classical Stochastic Model

The time-dependent perturbation $V(\mathbf{r}, t)$ clearly arises from the Coulomb interaction of all of the plasma particles with the ion simultaneously. If $n_e(\mathbf{r}, t)$ is the electron density and $n_i(\mathbf{r}, t)$ is the ion density we may write the interaction potential as

$$\begin{aligned}
V(\mathbf{r}, t) &= e^2 \int_\Omega d^3 r' \frac{n_e(\mathbf{r}', t)}{|\mathbf{r} - \mathbf{r}'|} - \overline{Z} e^2 \int_\Omega d^3 r' \frac{n_i(\mathbf{r}', t)}{|\mathbf{r} - \mathbf{r}'|} \\
&= \langle V(\mathbf{r}, t) \rangle + e^2 \int_\Omega d^3 r' \frac{\delta n_e(\mathbf{r}', t)}{|\mathbf{r} - \mathbf{r}'|}.
\end{aligned} \tag{9}$$

where $\overline{Z}$ is the average ion charge surrounding the target ion. In the second expression the average potential has been separated from the fluctuations which, in turn, have been assumed to arise solely from the electron density fluctuations $\delta n_e(\mathbf{r}, t)$. The time dependence of the density fluctuations will be specified statistically at the end of the calculation. It is instructive to combine the average potential of (9) with the Coulomb potential as

$$-\frac{Ze^2}{r}\phi(\mathbf{r}) = -\frac{Ze^2}{r} + \langle V(\mathbf{r}, t) \rangle \tag{10}$$

where a screening function $\phi(\mathbf{r})$ has been defined. This leads to a model Hamiltonian of the form

$$H = \frac{p^2}{2m} - \frac{Ze^2}{r}\phi(\mathbf{r}) + e^2 \int_\Omega d^3r' \frac{\delta n_e(\mathbf{r}',t)}{|\mathbf{r}-\mathbf{r}'|}. \tag{11}$$

Now we may compare this with the Hamiltonian of (7). Clearly the main difference in these expressions is that all of the electrons in (11) are treated on an equal footing whereas, in (7), one particular electron moves with the remaining electrons rigidly fixed to it.

If we treat the electron density fluctuations as a (weak) perturbation on the atomic system we may use the Fermi Golden Rule to obtain the ionization rate from some initial ionic state $|a\rangle$ to all continuum states $|b\rangle$. These states are assumed to be eigenstates of the first two terms in (11). After some manipulation, one obtains

$$w/n_e = \frac{2\pi}{n_e \hbar} \sum_b \sum_\mathbf{q} \left| \frac{4\pi e^2}{q^2 \epsilon(\mathbf{q},\omega)\Omega^2} \right|^2 \left| \langle b|e^{i\mathbf{q}\cdot\mathbf{r}}|a\rangle \right|^2 S_0(\mathbf{q},\omega) \tag{12}$$

for the total electron impact ionization rate coefficient. Here $\epsilon(\mathbf{q},\omega)$ is the plasma dielectric response function and $S_0(\mathbf{q},\omega)$ is the free particle dynamic structure factor. The relation

$$S(\mathbf{q},\omega) = \frac{S_0(\mathbf{q},\omega)}{|\epsilon(\mathbf{q},\omega)|^2} \tag{13}$$

has been used[I92]. The other terms are similar to those that appear in (6). The three factors can be interpreted as the effective interaction between the plasma and the ion, the atomic form factor, and the power spectrum of the thermal motion of the particles respectively. In this expression ω and $\mathbf{q}$ represent the energy and momentum transferred to the atom, respectively. A significant feature of this expression is the correct dynamical description of the effective interaction as a dynamically screened Coulomb potential. In this manner, effects such as ionization due to collective modes ($\epsilon(\mathbf{q},\omega) \approx 0$), such as the electron plasma oscillation, are automatically included. In fact, we may use this expression to categorize the effective interaction in one of three models: the no screening (NS) model ($\epsilon(\mathbf{q},\omega) \approx 1$), the static screening (SS) model ($\epsilon(\mathbf{q},\omega) \approx \epsilon(\mathbf{q},0)$), and the full dynamical screening (DS) model (no approximation for $\epsilon(\mathbf{q},\omega)$).

Numerical calculations based on (12) have been performed for He^+ ionization from the $n=3$ state in a $15eV$ plasma for each of the screening models. Level shifts have been suppressed to illustrate only effects arising from the various screening models. The results are shown in figure (1). The ionization rate coefficient is shown as a function of the electron density n_e. As expected,

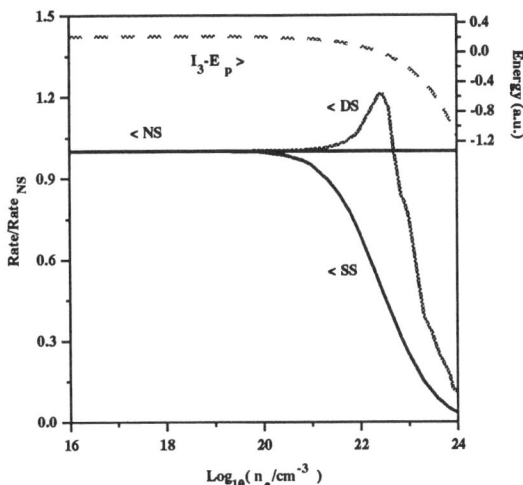

Figure 1: The total ionization rate of He^+ from the $n = 3$ state versus plasma density for various screening models. The plasma temperature is $15eV$. Specifically, the ratio of the rate for each screening model and the rate for the no screening case is shown (left ordinate) versus the logarithm of plasma density. Also shown is the difference in the ionization potential and the plasmon energy (right ordinate).

the NS model has no density dependence. This model is, of course, simply a different way of stating (4). This is illustrated in the next section. The SS predicts a significant decrease in the rate coefficient. This arises, as mentioned previously, due to the weakening of the effective interaction as the screening becomes more efficient. This model is essentially the one described by (7). The full DS shows qualitatively different behavior from the other two models. At electron densities below $\sim 10^{22} cm^{-3}$ there is little change from the NS model whereas the SS model predicts a $\sim 30\%$ reduction. At densities near $\sim 10^{23} cm^{-3}$, the DS model predicts an enhanced rate and which differs from that of the SS model by over a factor of 2. At very high plasma density the DS model shows behavior qualitatively similar to the SS model. The behavior of the DS model can be understood by comparing the ionization potential for this state with the plasma oscillation energy $\hbar\omega_p$. The difference in these two quantities is shown as a dashed curve in the figure. At low ($< 10^{22} cm^{-3}$) density, the ionization potential is larger than $\hbar\omega_p$ and therefore the values of ω in (12) all correspond to frequencies in the dielectric response function that exceed the plasma frequency. Hence the density fluctuations *are not screened*. Interestingly, for this particular transition and density/temperature regime, the model described by (7) is improved by taking $\phi_2(r) = 1$. Put another way, an electron moving with sufficient velocity to ionize this level cannot be screened by the surrounding plasma. The plasma simply cannot respond on this time scale. Models based on (7) assume that the plasma responds infinitely fast to all perturbations. At the peak of the DS ionization rate coefficient we see that the ionization energy is close to the energy $\hbar\omega_p$ and the transition is driven primarily by the plasma oscillation. This effect cannot be included in (7). At even higher electron density the transition energy is less than $\hbar\omega_p$. Fluctuation frequencies that drive the transition in this regime now appear slow to the plasma and can be screened in a manner similar to that described by (7).

The Quantal Stochastic Model

It is possible to gain more insight into the Stochastic Model by providing a purely quantum mechanical formulation. In doing so, we will see where its limitations are and thereby provide direction for future work. The starting point for the transition rate is

$$w = \frac{2\pi}{\hbar} \sum_f \sum_i F_i \left| \langle f | \hat{T} | i \rangle \right|^2 \delta(E_f - E_i) \tag{14}$$

where the states $|i\rangle$ and $|f\rangle$ are eigenstates of the entire system without the interaction between the target ion and the plasma, $\hat{T}$ is the T operator for the plasma-ion interaction, and F_i is the distribution of initial states (which includes one more bound electron than the final state). In this form, the rate (14) is more general than the Stochastic Model and we must make several approximations to recover it. We begin by writing the Hamiltonian as

$$\hat{H} = \hat{H}_{at} + \hat{H}_{pl} + \hat{H}_{at-pl} \tag{15}$$

where $\hat{H}_{at}$ and $\hat{H}_{pl}$ are the full Hamiltonians of the interacting ionic and plasma subsystems respectively. The final term represents the interactions between these subsystems. The exact eigenstates of the separate ionic and plasma Hamiltonians are given by

$$\begin{aligned} \hat{H}_{at}|a\rangle &= E_a|a\rangle \\ \hat{H}_{pl}|p\rangle &= E_p|p\rangle. \end{aligned} \tag{16}$$

In practice, of course, it is not usually possible to obtain these exact eigenstates. In the spirit of the Stochastic Model, we allow the plasma to remain separated from the ionic system and define the initial and final states of the combined system in terms of the factorized states of (16), viz.

$$\begin{aligned} |i\rangle &= |a\rangle|p_i\rangle \\ |f\rangle &= |b\rangle|p_f\rangle. \end{aligned} \tag{17}$$

With this approximation, exchange between the ionic electron(s) and the plasma electrons is neglected. However, we have formally included *all* interactions, including exchange, within each subsystem.

With these definitions in hand, it is now an easy matter to compute the first order transition rate. Once again simplifying to a hydrogenic atom, the interaction term can be written as

$$\hat{H}_{at-pl} = e^2 \int d^3r' \frac{\hat{n}_e(\mathbf{r}')}{|\mathbf{r} - \mathbf{r}'|} \tag{18}$$

where $\hat{n}_e(\mathbf{r})$ is the plasma density operator and acts only on the $|p\rangle$ states. The total electron impact ionization rate is then, to first order in the interaction (18),

$$w = \frac{2\pi}{\hbar} \sum_{b,p_f} \sum_{p_i} F(p_i) \left| \langle b|\langle p_f| e^2 \int d^3r' \frac{\hat{n}_e(\mathbf{r}')}{|\mathbf{r} - \mathbf{r}'|} |a\rangle|p_i\rangle \right|^2 \delta(E_f - E_i) \tag{19}$$

where E_f and E_i are the final and initial energies of the total system respectively and $F(p_i)$ is the distribution of initial plasma states. Again, the Coulomb interaction may be Fourier transformed and the matrix element squared. This yields the result

$$w = \frac{2\pi}{\hbar} \sum_b \sum_\mathbf{q} \left(\frac{4\pi e^2}{\Omega q^2}\right)^2 |\langle b|e^{i\mathbf{q}\cdot\mathbf{r}}|a\rangle|^2 \sum_{p_f,p_i} F(p_i) |\langle p_f|\hat{n}_e(\mathbf{q})|p_i\rangle|^2 \delta(E_f - E_i) \tag{20}$$

where all the plasma dependency has been collected to the right. The factor associated with the plasma dependency,

$$S(\mathbf{q},\omega) = \sum_{p_f,p_i} F(p_i) |\langle p_f|\hat{n}_e(\mathbf{q})|p_i\rangle|^2 \delta(E_f - E_i), \tag{21}$$

is the quantum mechanical version of the dynamic structure factor[P66] which, with (13), yields a rate similar to (12).

We may now introduce approximations to show that this form of the Stochastic Model reduces to the traditional approach in the high temperature and/or low density limit. Since the electrons do not interact with each other in the traditional approach, we must treat the dynamic structure factor in the ideal gas limit. The rate in this limit is

$$w = \frac{2\pi}{\hbar^2} \sum_b \sum_\mathbf{q} \left(\frac{4\pi e^2}{\Omega q^2}\right)^2 |\langle b|e^{i\mathbf{q}\cdot\mathbf{r}}|a\rangle|^2 \sum_\mathbf{Q} f(\mathbf{Q}) g(\mathbf{Q}-\mathbf{q}) \delta\left(\omega + \frac{\hbar q^2}{2m} - \frac{\hbar \mathbf{Q}\cdot\mathbf{q}}{m}\right) \tag{22}$$

where f is the Fermi-Dirac distribution function and $g = 1 - f$. If we define $\mathbf{v} \equiv \hbar\mathbf{Q}/m$ we see that this nearly takes the form of a velocity average over a quantity that is similar to a flux times a cross section. However, we see that this *cannot* be written exactly as an averaged flux times a cross section for a binary electron-ion collision process as a result of the factor $g(\mathbf{Q} - \mathbf{q})$ which couples the velocity averaging (the $\mathbf{Q}$ dependence) with the collision dynamics (the $\mathbf{q}$ dependence). This is just one form of the so-called "Pauli blocking" (see below). This form arises from a circumstance in which a plasma electron is blocked from making a transition to a state already occupied by another plasma electron. Thus we see that, even in the absence of interactions, degeneracy may prohibit a description in terms of binary cross sections. Of course, in the high temperature limit, the g factor becomes $g(\mathbf{Q} - \mathbf{q}) \approx 1$ and the traditional approach is recovered.

DISCUSSION

The basic result in this study of screening models is that the naive static screening model is quite frequently in error. In many laboratory experiments the highly stripped ions have large transition energies which can only be driven by fast, and therefore unscreened, electrons. That is, the plasma is not "dense" to these transitions. There may be states, however, that are close enough in energy that screened interactions might play a role. In a hydrogenic system, many of these states are eliminated by continuum lowering.

There are a large number of improvements that can be made to the existing Stochastic Model. A partial list is given here.

- MULTI-BODY IONIZATION. At very high plasma densities it may be possible for two or more dynamically screened plasma electrons to interact with the ion simultaneously. To see this, consider a hypothetical plasma of mean density n_e in which the electrons do not interact. Each plasma electron is then, on average, separated by a distance

$$D = 2.34 \times 10^8 \, n_e^{-1/3} \qquad (23)$$

where D is in Bohr and n_e is in cm^{-3}. At a density of $n_e \sim 10^{25} cm^{-3}$, conditions reachable in ICF experiments, the interparticle distance is about 1 Bohr, i.e. the electrons are separated by distances on the atomic scale and more than one electron interacts strongly with both the atom and the other electrons. Under such conditions, it is believed that a first order perturbation treatment is invalid. Furthermore, the electrons will no longer move independently of the atom since they cannot penetrate the atomic electron(s).

- NON-IDEAL EFFECTS. To get a complete theory of electron impact ionization in a dense plasma, nonideal plasma effects should also be included. At low temperatures, for example, the Pauli exclusion principle would prohibit *atomic* electrons from occupying filled *plasma* electron states. Such transitions are said to be "Pauli blocked". This is analogous to the Pauli blocking between plasma particles of (22). A complete description of both forms of Pauli blocking has been treated by Schlanges and Bornath[SB93].

- NON-THERMAL PLASMA. In a non-equilibrium plasma the electron velocity distribution may or may not be Maxwellian[L95]. Most calculations to date, however, have not used the true time-dependent distribution.

- ION MICROFIELD APPROACH. Although not discussed in detail here, the effects of level shifts are very important in the ionization rate. Most treatments of level shifts[R70, J95a, MW93, SB93] assume that the average potential $\langle V(\mathbf{r},t)\rangle$ has spherical symmetry. This is an inconsistent approach since, if the ions could move quickly enough to provide a spherical potential on the time scale of the electron impact ionization event, they would also participate dynamically in the ionization event. More likely, the ions do not move appreciably during the event and should be treated as a static, nonuniform distribution of ions that produces an microscopic electric field $\mathbf{E}$ at the ion. The final step in the calculation would then be to average over the distribution $P(\mathbf{E})$ of these microfields. This, of course, is what is done in line broadening theory[C85].

ACKNOWLEDGEMENTS

I gratefully acknowledge Jon C. Weisheit for numerous contributions and insightful suggestions related to this work. I also thank George Csanak for a careful reading of this manuscript accompanied with useful comments. Much of this work was performed whilst the author was at Rice University with NSF support through grants PHY-9321329 and PHY-9024397. Some of this work was performed under the auspices of the U.S. Department of Energy through the Theoretical Division of Los Alamos National Laboratory.

References

[M95] MANCINI, R. C., et al, *Atomic Processes in Plasmas*, ed. W. L. Rowan, AIP Conference Proceedings 322, 161 (1995).

[I94] ICHIMARU, S., *Statistical Plasma Physics, Volume II: Condensed Plasmas*, Addison Wesley (1994).

[Y85] YOUNGER, S. M., *Electron Impact Ionization*, eds. T. D. Märk and G. H. Dunn, Wien, Springer-Verlag (1985).

[R70] ROGERS. F., et al, *PRA* **1** (6), 1577 (1970).

[J95a] JUNG, Y.-D., *Phys. Plasmas* **2** (1), 332 (1995).

[J95b] JUNG, Y.-D., *Phys. Plasmas* **2** (5), 1775 (1995).

[VS76] VINOGRADOV, A. V. and SHEVELKO, V. P., *Sov. Phys. JETP* **4**, 167 (1976).

[W88] WEISHEIT, J. C., *Advances in Atomic and Molecular Physics*, Vol. 25, Academic Press, 101 (1988).

[MW93] MURILLO, M. S. and WEISHEIT, J. C., *Strongly Coupled Plasma Physics*, eds. H. M. Van Horn and S. Ichimaru, Rochester, University of Rochester Press, 233 (1993).

[M95] MURILLO, M. S., Ph.D. thesis, Rice University (1995).

[I92] ICHIMARU, S., *Statistical Plasma Physics, Volume I: Basic Principles*, Addison Wesley, 252 (1992).

[P66] PINES, D. and NOZIÉRES, P., *The Theory of Quantum Liquids*, New York, Addison Wesley (1966).

[SB93] SCHLANGES, M. and BORNATH, TH., Physica A 192, pp. 262-279 (1993) *and* BORNATH, TH. and SCHLANGES, M., Physica A 196, 427 (1993).

[L95] LAMOUREUX, M. ed. W. L. Rowan, *Atomic Processes in Plasmas*, AIP Conference Proceedings 322, 173 (1995).

[C85] IGLESIAS, C. A., et al, *PRA* **31** (3), 1698 (1985).

ATOMIC PHYSICS
IN STRONG FIELDS

Atomic Emission Spectroscopy in High Electric Fields

J.E. Bailey, A.B. Filuk, A.L. Carlson, D.J. Johnson, P. Lake, E.J. McGuire,
T.A. Mehlhorn, T.D. Pointon, T.J. Renk, and W.A. Stygar

Sandia National Laboratories, Albuquerque, N.M., 87185

and

Y. Maron and E. Stambulchik

Weizmann Institute of Science, Rehovot, Israel, 76100

ABSTRACT

Pulsed-power driven ion diodes generating quasi-static, ~10 MV/cm, 1-cm scale-length electric fields are used to accelerate lithium ion beams for inertial confinement fusion applications. Atomic emission spectroscopy measurements contribute to understanding the acceleration gap physics, in particular by combining time- and space-resolved measurements of the electric field with the Poisson equation to determine the charged particle distributions. This unique high-field configuration also offers the possibility to advance basic atomic physics, for example by testing calculations of the Stark-shifted emission pattern, by measuring field ionization rates for tightly-bound low-principal-quantum-number levels, and by measuring transition-probability quenching.

INTRODUCTION

The light-ion beam approach to inertial confinement fusion[1] proposes to achieve the required high energy density by accelerating a lithium ion beam to about 30 MeV using one or two acceleration stages, each driven with a high-power (~100 TW), ~30-nsec-duration pulse. Present experiments at the Particle Beam Fusion Accelerator II (PBFA II) facility routinely generate quasi-static, ~10 MV/cm, ~1 cm scale-length electric fields. This enables experiments studying the physics of the ion beam acceleration gap and these conditions also present a unique opportunity for extending experimental atomic physics into the 10 MV/cm regime. This paper describes our application of atomic spectroscopy to ion diode plasma physics issues, with an emphasis on the atomic physics required to understand the results. In addition, we describe preliminary experiments that illustrate the potential of using data from this device for basic atomic physics.

© 1996 American Institute of Physics

The primary motivation for atomic emission spectroscopy measurements in ion diodes is the need to understand and control the influence of the acceleration gap charged-particle dynamics on the ion beam brightness. Stark-shift measurements of the electric field distribution can be combined with the Poisson equation to determine key features of the charged-particle behavior[2,3]. Understanding the charged-particle distributions is of fundamental importance as we seek to increase the ion beam brightness because they largely control the enhancement of the ion current above the nominal Child-Langmuir space-charge limit[4,5] and because non-uniformities in the charge density can induce beam microdivergence or steering errors that reduce the focussed intensity[6-9]. Our results[3] show that theory and experiment are in reasonable agreement for the first 5 nsec of the ion beam pulse, but as the ion current grows significant discrepancies arise. The measurements provide evidence for new diode phenomena, including field-limited rather than space-charge-limited ion emission, a region with zero net-charge density near the anode, localized positive net-charge in the middle of the gap, and persistent azimuthal asymmetries. Our recent work has emphasized measurements of the small electric field component perpendicular to the direction of beam acceleration and on measurements of emission from Ba II dopants, both with the aim of improving understanding of ion beam divergence.

A new opportunity for atomic physics measurements is provided by the relatively long duration (tens of nsec) and large volume (~1000 cm^3) of the high-field region. Previous emission spectroscopy Stark effect investigations[2] were limited to electric fields below 1 MV/cm, about an order of magnitude below the fields in the PBFA II acceleration gap. Higher fields are certainly generated in short-pulse laser[10], plasma wakefield acceleration[11], and micro-needle experiments[12]. Also, the effect of high electric fields on hydrogen has been investigated[13] using the motional Stark effect to reach ~3 MV/cm. However, the short duration and/or small volume of the high-field regions has prevented acquisition of atomic spectra from these experiments. The conditions in present PBFA II experiments thus enable studies of the high-field Stark effect that were previously impossible. Comparison of independent calculations and results from different species supply confidence that theoretical predictions of the Stark pattern are accurate to within ± 5-10% at fields up 10 MV/cm. Preliminary results for transitions from the tightly-bound Li I 3d level are consistent with predictions of the electric field ionization threshold, and provide the first experimental evidence for transition probability quenching of these transitions.

EXPERIMENTAL METHODS

The experiments described in this paper were performed using a cylindrically-symmetric applied-magnetic-field ion diode[14-16]. PBFA II supplies a 20 TW,

~20 nsec, 10 MV power pulse through conical magnetically-insulated transmission lines connected to the top and bottom of the diode. The ion beam is accelerated radially inward from the inner surface of a cylindrical anode toward a target placed on the axis. An approximately 3 T magnetic field, applied parallel to the anode, insulates the anode-cathode (AK) gap against electron losses.

The visible spectroscopy diagnostic system[17] collects light from approximately-cylindrical 2-mm-diameter lines of sight aligned parallel to the anode. Multiple lines of sight are used to obtain radial and azimuthal spatial resolution, where the radially-resolved measurements provide a profile across the AK gap and the azimuthal measurements determine the degree of cylindrical uniformity. One configuration determines the global uniformity using lines of sight at opposing azimuths of the 15-cm-radius diode, while a second configuration with 2-10 mm azimuthal spacing measures the uniformity over a small azimuthal sector. The light is transported in fiber optics to remote streaked spectrographs for recording with ~ 1 nsec time resolution. A multiplexing technique enables time-resolved spectra with ~ 50 Å range and 1.5 Å resolution to be recorded simultaneously from 8 different spatial locations on a single spectrograph. Data from 18 locations were obtained in each experiment. The relative spacing of the lines of sight is accurate to $\pm$ 0.2 mm and the absolute accuracy of the line-of-sight bundle relative to the anode surface is $\pm$ 0.5 mm. The timing between spectra is accurate to $\pm$0.4 nsec and the timing accuracy relative to the electrically-recorded diode diagnostics is $\pm$ 2 nsec.

LI I ATOMIC PHYSICS IN HIGH ELECTRIC FIELDS

Most measurements to date have relied on Stark-shifted 2s-2p and 2p-3d emission from lithium neutral atoms. Lithium atoms are launched into the gap when a small fraction of the lithium ions charge exchange in a thin, dense, desorbed contaminant layer near the anode. The charge-exchange origin of the Li neutrals impacts the electric field measurements because the Li neutral atoms acquire a high velocity transverse to the beam acceleration direction, due to the ion beam source divergence. This high transverse velocity results in large Doppler broadening (typically 10-15 Å) that prevents resolution of the individual Stark/Zeeman line components. In addition, measurements on the cathode side of the gap are not possible until neutrals arrive there, typically ~10-15 nsec into the pulse. Nevertheless, the Li I charge-exchange neutral atom emission is convenient, it provides good accuracy without disturbing the normal diode operation, and measurements with neutral atoms provide brighter visible light emission intensities than ion emission measurements. We typically analyze 15-20 lineouts from each spectrum, averaging over 4 nsec intervals to improve the signal-to-noise ratio. A single Gaussian is fit to each spectral line, with a wavelength uncertainty determined using the fluctuation levels of the entire spectrum[18]. The shift is measured relative to the zero-field wavelength established using emission recorded after the power pulse (when the electric field is zero). The procedure determines the Stark shifts with a typical uncertainty ± 0.2 - 0.4 Å, compared to ~6 Å maximum shifts for the 2s-2p and 20-200 Å maximum shifts for the 2p-3d Li I transitions.

The spectral data provides the Stark shift as a function of time and space. The interpretation of this data requires calculations of the Stark / Zeeman line emission pattern as a function of E and B, for values of E that are higher than any previous terrestrial Stark-shift measurement. Independent calculations were performed at Sandia[19] and at the Weizmann Institute[20] in order to improve confidence in the results. Both calculations use direct diagonalization of the Hamiltonian without invoking a perturbation approach. The Zeeman and Stark effects are treated self-consistently and the effects of high-lying levels are included. The Weizmann Institute calculations also include level shifts due to interaction with continuum states. The calculations agree to ~1% with available published data[21,22] that extend up to ~ 0.4 MV/cm. A comparison of the two calculations for the Li I 2s-2p transition using B = 6 T and E = 0-11 MV/cm is shown in Figure 1, where the centroid of the emission pattern is averaged over all observation directions (in actual data analysis we take the directions of the electric and magnetic fields and the line of sight into account). Note that the shift is to the blue, so that larger shifts correspond to shorter wavelengths. The impact of the magnetic field strength on the interpretation of the experiments reported here is negligible because the high electric field dominates the centroid shift and the large Doppler broadening prevents observation of the individual line components. A shift assuming a quadratic dependence on E is also shown in Figure 1. The shift in both calculations is approximately quadratic in E for fields up to about 5 MV/cm, but at higher fields the calculated shifts deviate from the quadratic curve. The quadratic curve was used in the analysis presented below, since it lies between the two detailed calculations and the quadratic approximation is easy to use. At any given shift the fields determined from the two calculations agree to better than ± 5% for fields up to 10 MV/cm. This uncertainty is not included in the error bars presented below. The differences between the calculations at fields above 10 MV/cm are currently under investigation.

A streaked spectrum and a characteristic lineout from a PBFA II experiment are shown in Figure 2. In this experiment we first observe shifted Li I 2s-2p emission at ~56 nsec, simultaneous with the onset of ion current. The Li I 2p-3d emission is not observed until ~78 nsec and when it first appears it is split into red- and blue-shifted components. Interpretation of these data requires an understanding of the various population and de-population rates, as well as accurate calculations of the Stark/Zeeman patterns. The 3d level is initially populated by the same charge-

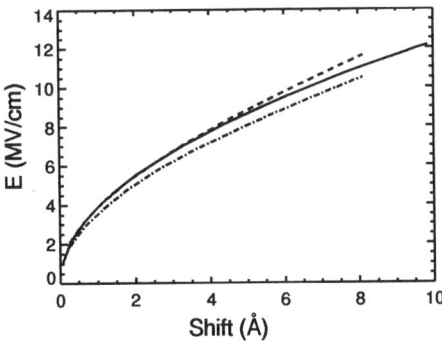

FIGURE 1. Electric field as a function of the Stark / Zeeman emission pattern centroid wavelength shift for the Li I 2s-2p transition at B = 6 T. The dashed curve is from Stambulchik and Maron[20] and the dot-dash curve is from McGuire[19]. The solid curve represents the quadratic relationship, shift (Å) ~ E(MV/cm)^2/15, illustrating the departure of the shift from the quadratic approximation at fields above about 5 MV/cm.

exchange events that populate the 2p level, as well as by ion impact excitation from the 2p level. The dominant de-excitation mechanisms are expected to be field ionization and radiative decay. Prior measurements of the state-selective charge-exchange cross section[23] indicate significant 3d population should be present early in the power pulse. However, the 2p-3d emission is not observed until near the end of pulse. This is almost certainly because the 3d electrons rapidly field ionize under the 5 - 10 MV/cm field. This is consistent with previous detailed calculations of the field ionization rate[24] and expectations based on semiclassical approximations.

As noted above, the 2p-3d is split into red-shifted and blue-shifted components when it first appears. The red-shifted component arises from the 3d m_L = 2 state and the blue-shifted component is a superposition of transitions originating from the 3d m_L = 0,1 states (throughout the text m_L refers to the absolute value of the orbital angular momentum quantum number). The m_L = 2 transition is red-shifted, and is less strongly perturbed, because the selection rules prohibit the 3p level from perturbing the 3d m_L = 2 state. The red-shifted feature appears first. The appearance of the blue-shifted feature is delayed because this component has a lower threshold for field ionization and it is more strongly shifted once it does appear. The larger Stark shift of the m_L = 0,1 states and the existence of even small electric field variations along the line of sight tends to smear the blue-shifted transitions into a broad line, making it more difficult to observe above the continuum. The field measured from the red-shifted feature is consistent with the field obtained from the 2s-2p transition (Figure 3; a more complete analysis including the shift of the m_L = 0,1 component and extraction of uncertainties from the data is in progress). While this agreement is satisfying, it does not represent a stringent test of the Stark pattern calculations because the uncertainty associated with measuring the field from the

2s-2p transition grows as the field shrinks below about 3 MV/cm. During the time of overlap the typical uncertainties for the fields measured from the 2s-2p and 2p-3d transitions are ± 50% and ± 5%, respectively. Note that the uncertainty for field measurements with the 2s-2p transition is typically much lower in experiments emphasizing the use of this line to measure higher fields (Reference 3).

The measurement of the 2p-3d emission enables observations of changes in the transition probability due to wavefunction mixing by the high electric field. While this effect is expected, we are unaware of previous experimental measurements for such low-principal-quantum-number levels. The spontaneous emission intensity per unit volume per steradian emitted in a spectral line is $P = n_{ij} A_{ij} h\nu_{ij}$, where n_i is the excited state population density, $h\nu_{ij}$ is the transition energy, and A_{ij} is the Einstein coefficient. The spontaneous transition probability A_{ij} from excited state i to final state j is proportional to the square of the dipole matrix element, $A_{ij} \sim <i|H|j>^2$, where H is the Hamiltonian. The electric field mixes the wavefunctions and A_{ij} depends on the extent of the mixing and, thus, on the field strength. Measurements of the line component intensities therefore reflect changes

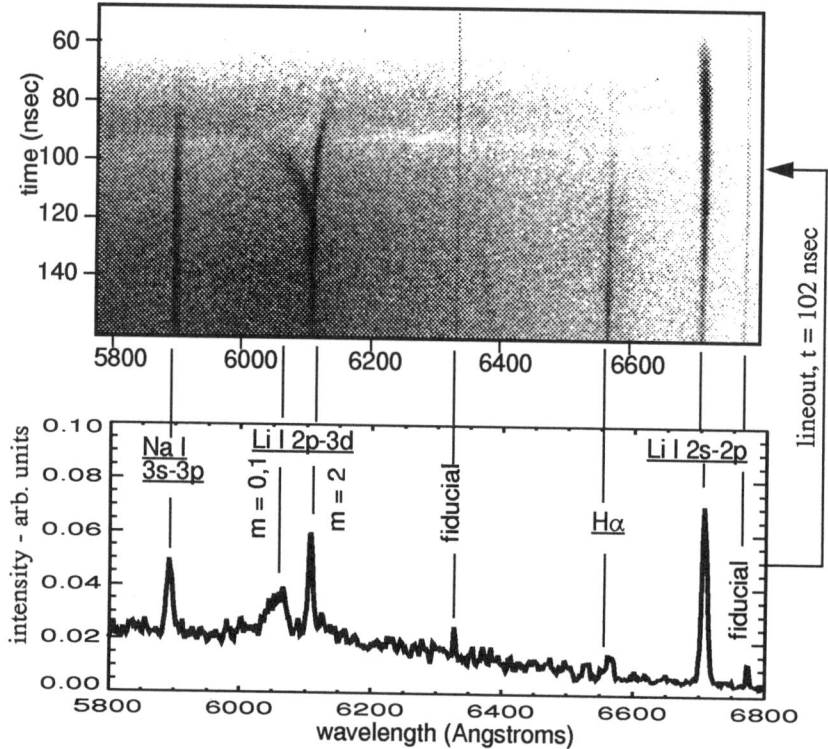

FIGURE 2. Top: Streaked spectrum measured 9 mm from the PBFA II anode surface. Bottom: Lineout averaging over 4 nsec, centered about t = 102 nsec.

in the transition probability caused by the electric field. Calculations of the change in the transition probability performed using the computer code described in Reference 20 show that the transition probability for the $m_L = 0,1$ component decreases by about 30% as the field increases from zero up to 3 MV/cm. The $m_L = 2$ component changes by only a few percent. The larger change for the $m_L = 0,1$ component is because these levels mix strongly with the nearby 3p level, while the $m_L = 2$ does not.

Measurements and calculations of the 3d $m_L = 0,1$ to 3d $m_L = 2$ intensity ratio are shown as a function of the electric field in Figure 4. The data and calculations are normalized so that the intensities from the two features are set equal at zero

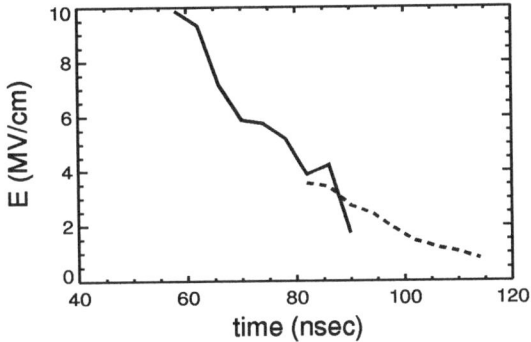

FIGURE 3. Electric field measured from the data shown in Figure 2. The solid curve is the result from Li I 2s-2p and the dashed curve is from Li I 2p - 3d ($m_L = 2$).

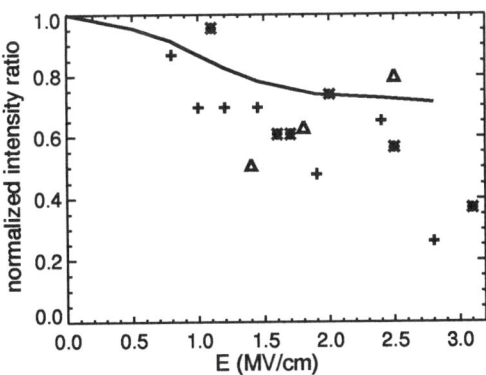

FIGURE 4. Intensity ratio of the Li I 2p-3d components $\{m_L = 0,1\} / \{m_L = 2\}$, normalized to zero electric field. The solid line is a calculation based on Reference 20. The plus, asterix, and triangle symbols are measurements from three separate PBFA II experiments. The experimental uncertainty in the ratio is roughly ± 25%.

electric field. The electric field displayed in this plot is measured from the Stark shift of the $m_L = 2$ (red-shifted) component. The data and calculations agree within the $\pm 25\%$ experimental uncertainty for fields up to about 2.5 MV/cm, although the theory systematically predicts a larger ratio. At higher fields the experimental ratio sharply decreases, unlike the theory predictions. This may be because the $m_L = 0$ sublevel is affected by field ionization (see below), an effect not included in the calculations. Note that in this analysis we assume other mechanisms that could affect the line intensity ratio are negligible. Plasma effects on the Einstein coefficient such as those discussed in References 25&26 are unimportant because the charged particle densities are[3] below about 5×10^{13} cm^{-3}. Also, the rates for collisional transfer of population between the $m_L = 2$ and $m_L = 0,1$ 3d levels were estimated and are far too low to affect the results. We therefore assumed that the ratio of the $m_L = 0,1$ to $m_L = 2$ excited state populations was defined by the statistical weights and was independent of the field strength. However, the population mechanism is uncertain, as we discuss below, and the population rate may in fact be sensitive to the field. For example, if the population is produced by ion-impact excitation (the most likely explanation at present) the rate depends on the dipole matrix element squared. Thus, for times short compared to the radiative decay time, the light intensity is expected to decrease quadratically with the dipole matrix element squared. This implies that the line intensity ratio should decrease more with increasing E than in the calculation shown in Figure 4, possibly improving the agreement with the data. The data should be regarded as confirming the expected qualitative trends in the theoretical predictions until a more detailed analysis is complete.

The $m_L = 2$ transition first appears at a field of ~ 3.7-3.9 MV/cm. This is roughly consistent with theoretical predictions that this state field ionizes in less than 10 nsec at E ~ 4.6 MV/cm. The $m_L = 0$ and $m_L = 1$ states are expected to field ionize at fields of 3.1 and 3.6 MV/cm, respectively, also consistent with the fields at which they appear. The transitions probably appear at field values lower than the ionization threshold because the field decreases in time and some additional time is required for the population to build up to an observable value after the field ionization stops depleting the level. However, we cannot rule out inaccuracies in the theoretical field ionization threshold until the population mechanism is better understood. At present, it appears that the earliest 3d emission we observe does not arise from charge-exchange neutral atoms created and excited into the 3d level at the anode. The velocity of the charge exchange neutrals during the early part of the pulse is determined in other experiments using measurements of the time-dependent Li I 2s-2p emission intensity on multiple lines of sight at different distances away from the anode. This velocity is typically ~ 50 cm/μsec, corresponding to a ~15-20 nsec time of flight between the anode and the spectroscopic line of sight located 9 mm away. This implies that if the origin of the Li I 2p-3d emission was charge exchange neutrals, they survived the high fields existing in the AK gap during the ~15-20 nsec prior to the initial observation of the 2p-3d emission (see Figures 3 & 5). Calculations of the field ionization[24] indicate that this is unlikely. Excitation from the Li I 2p state by Li ion impact excitation is also under investigation (electron impact excitation is small because of the relatively-high electron energies in the diode gap). The elapsed time between the field dropping below ~ 4.7 MV/cm (where fast field ionization is expected) and reaching the 3.7-3.9 MV/cm field (where the transition is observed) is about 3-5 nsec. The key question to evaluate for the ion impact excitation hypothesis is whether this is long enough to produce sufficient 3d m_L=2 population to account for the observed intensity, given the measured ion current density and 2p population density. A similar analysis can be applied to the 3d m_L= 0, 1 states. Resolution of these issues may enable testing of theoretical predictions for the field-ionization rates.

APPLICATIONS TO DIODE PLASMA PHYSICS

A temporal sequence of electric field profiles measured from Stark-shifted Li I 2s-2p emission in a single PBFA II experiment is shown in Figure 5. The AK gap in this experiment was 18 mm. The first measurement at 46 nsec corresponds to the onset of ion current. The diode physics aspects of these results are described in Reference 3. The field measured near the anode surface is 9-10 MV/cm, in contrast to the zero field expected for a space-charge-limited plasma ion source. This result is consistent with a recent theoretical hypothesis[27] that the LiF ion source produces Li ions via electron-assisted field desorption. The almost-flat electric field profile near the anode at peak ion power (62 nsec) implies that there is a ~ 2-3 mm thick, zero-net-charge region near the anode, in contrast to the positive net charge suggested by the simulations described in Reference 7. Note that according to the Poisson equation, the net charge density is proportional to the derivative of the field; hence, a field profile with zero slope implies zero net charge density. About 10 mm from the anode the electric field profile reverses its slope from negative to positive, implying a region with localized positive net charge in the middle of the gap. We emphasize, however, that these results are obtained by assuming uniform conditions along the line of sight. The effect of possible non-uniformities on the conclusions is under investigation. Azimuthal asymmetries between the measurements performed on opposite sides of the diode persist over much of the pulse, indicating that E x B drifting electrons are unable to cancel the asymmetries. These results suggest that improvements are required in both experiments and simulations in order to understand the power-coupling efficiency and divergence.

In the cylindrical barrel-diode geometry the electric field that accelerates the ions toward the target is primarily radial. However, non-radial field components can arise from electromagnetic instabilities and/or non-uniformities in the ion emission, cathode plasma, or virtual cathode electron cloud. These non-radial components deflect the ion beam and add to its divergence as it crosses the AK gap, effectively decreasing the beam power density irradiating the target. A method for measuring the field component in the azimuthal direction, E_ϕ (r,ϕ,t), is illustrated in Figure 6. We measure the magnitude of the electric field vector $|E| = [E_r^2 + E_\phi^2 + E_z^2]^{1/2}$ as a function of time on a rectangular array of spectroscopic lines of sight. Each line of sight is represented by a solid circle in Figure 6. The potential at each azimuth V_A or V_B as a function of radius is obtained by integration of the field, using the approximation that $E \sim E_r$. With this method we obtain the potential for adjacent azimuths as a function of radius and time. The azimuthal component of the electric field E_ϕ is then approximately the potential difference between two azimuths V_{AB}, divided by the azimuthal distance between them δ, using the fact that the line integral of E around the dotted path is zero (estimates for the dB/dt term in this equation indicate that this term is negligible). The challenge in applying this method is that the differences in $|E|$ are of order 10%, requiring that the uncertainty in $|E|$ be less than about 5% in order to arrive at statistically significant values for E_ϕ. The typical uncertainty in our present experiments ranges over $\pm$ 2-6% (at 1σ). This is adequate, but the accuracy of E_ϕ would clearly benefit from reduced uncertainties. Preliminary results for E_ϕ(r,t) are shown in Figure 7. The azimuthal field component grows with distance from the anode and it fluctuates on a 2-8 nsec time scale. If confirmed, the measured E_ϕ

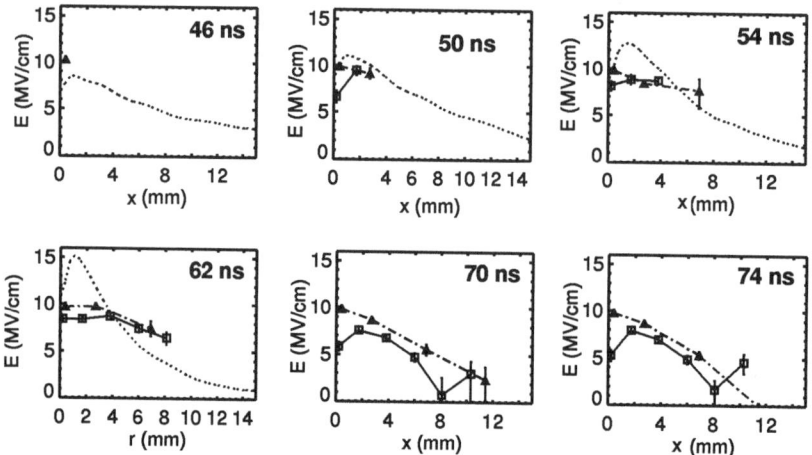

FIGURE 5. Electric field evolution as a function of x, the radial distance away from the anode. The squares and triangles are measurements from a PBFA II experiment at the 180° and 0° azimuths, respectively. The dashed curve is a QUICKSILVER simulation result (Reference 7).

within 5.5 mm of the anode is sufficient to cause ~25 mrad of beam deflection (averaged over the first 1/2-2/3 of the power pulse), comparable to the 25-35 mrad measured at the target on axis. Work is in progress to measure the radial and azimuthal extent of E_ϕ over larger radial and azimuthal distances in order to help identify its origin.

Additional aspects of both atomic physics and diode physics can be explored by adding impurity dopants to the anode. In one such experiment, BaF_2 and LiF were co-evaporated onto the anode in an approximately 1:1 molecular ratio. The primary motivations for adding BaF_2 to the anode were to provide a direct measurement of ion divergence and to enable Zeeman measurements of the magnetic field. Note that the lithium ion beam divergence is very difficult to measure directly because the

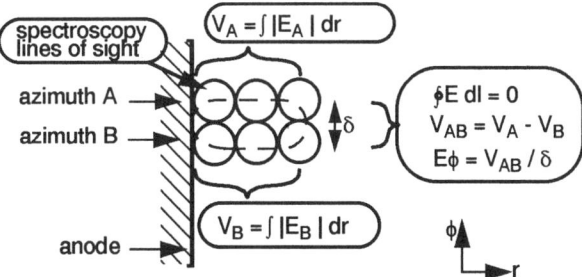

FIGURE 6. Top view of the diode AK gap illustrating the method used to determine the non-radial electric fields from the Stark shift measurements. The circles represent individual spectroscopic lines of sight that are approximately 2 mm in diameter. The dashed curve illustrates the path for the line integral. In actual experiments a larger array of points (up to eighteen total) is used to obtain more complete measurements of the radial and azimuthal dependence.

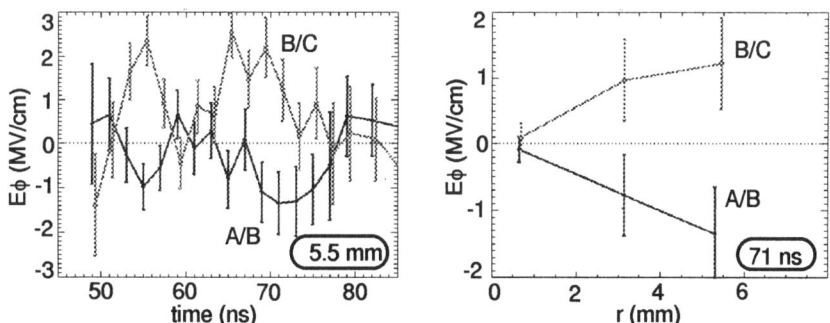

FIGURE 7. Preliminary measurements of the azimuthal component of the electric field obtained using a line of sight arrangement similar to that shown in Figure 6, but with an additional row of lines of sight at a third azimuth. The curves labeled A/B refer to one pair of azimuths and the curves labeled B/C refer to the other pair. Ion current onset in this experiment was at 41 nsec. The time history on the left corresponds to 5.5 mm from the anode surface and the spatial plot on the right corresponds to t = 71 nsec.

lithium ion excited states are not significantly populated. Measurements of Ba II Doppler broadening are straightfoward in comparison to Li II measurements, since the high Ba ion mass makes the Ba ion density relatively high and the easily-excited resonance transition lies in the visible regime. In addition, the Ba II Doppler broadening is expected to be smaller than for the Li I charge-exchange neutrals. This may enable measurements of the magnetic field profile from the Stark / Zeeman pattern.

These diode physics results will be discussed elsewhere. Here, we only point out that simultaneous measurements of the electric field from Ba II and Li I emission provide a useful cross-check of the Stark pattern calculations. A temporal sequence of lineouts measuring the Ba II 6s $^2S_{1/2}$-6p $^2P_{1/2}$ transition are shown in Figure 8. This transition is split into two components, one appearing at the zero-field wavelength and one that is Stark-shifted. The intensity of the unshifted component relative to the shifted feature grows in time. The appearance of a Stark-shifted component verifies that it is indeed possible to observe Ba II emission as the ions are accelerated across the gap. Interpretation of the fact that both shifted and unshifted light are observed simultaneously is in progress, but it appears to be consistent with acceleration of barium ions from both field-threshold-emission regions (as for LiF) and from plasmas. The time-dependent electric field measured from the Li I emission compared to the field determined using the shifted portion of the Ba II emission is shown in Figure 9. The fields agree to within ± 10%, although there is a systematic difference in the field determined form the Ba II 6s $^2S_{1/2}$-6p $^2P_{1/2}$ and 6s $^2S_{1/2}$-6p $^2P_{3/2}$ lines. This is probably due to inaccuracies in the oscillator strengths used in the Stark pattern calculations of Reference 20. Note that better agreement between the Li I and Ba II data cannot be expected, since in the

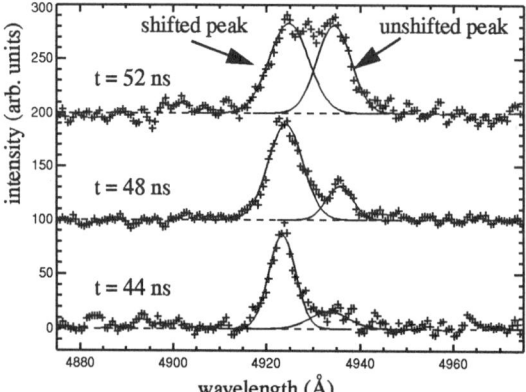

FIGURE 8. A temporal sequence of lineouts for the Ba II 6s $^2S_{1/2}$ - 6p $^2P_{1/2}$ transition, measured adjacent to the PBFA II anode. Each successive time step is displaced by 100 intensity units to facilitate viewing. Gaussian fits to the peaks (solid curves) obtained with the ROBFIT code are superimposed on the data (plus signs). The unshifted (zero field) wavelength is 4934.1 Å.

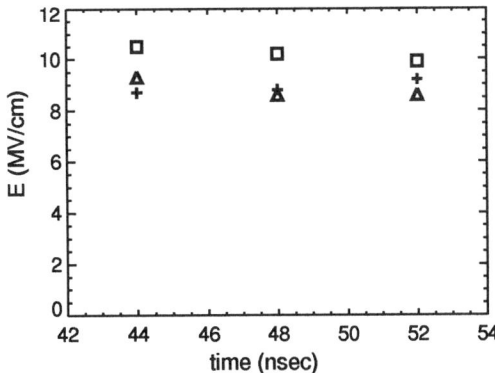

FIGURE 9. Electric field measured from Li I 2s-2p (triangle), Ba II 6s $^2S_{1/2}$-6p $^2P_{1/2}$,(squares), and 6s $^2S_{1/2}$-6p $^2P_{3/2}$ (plus sign) lines. The average deviation is less than 8%.

present experiment the Ba II and Li I emission were collected with lines of sight located at the same radial distance from the anode, but at adjacent azimuths separated by 2 mm. Data described above shows that differences of about 10% in |E| can exist over this azimuthal separation. We plan to perform a more stringent test of these Stark pattern calculations by collecting light in a single line of sight and splitting it into two streaked spectrographs. This should enable a cross-comparison of the Li I and Ba II results to within approximately ± 5%.

CONCLUSIONS

Measurements of Stark-shifted emission clearly provide new insight into the complex plasma physics in ion diode acceleration gaps. In particular, these measurements enable comparisons of the 3-D electromagnetic simulations that are used to predict diode operation with detailed measurements for the first time. Interpretation of the measurements requires accurate calculations of the line emission patterns under ~ 10 MV/cm electric fields and 3 - 6 T magnetic fields. In addition, the high field routinely produced on PBFA II invites basic atomic physics measurements for low-lying levels that were previously impossible. Measurements of the Li I 2p-3d transitions are consistent with calculations of field ionization thresholds, although stringent tests of the calculations require progress in understanding level population mechanisms. Preliminary measurements of the dependence of the line component intensities on the electric field are also consistent with predictions.

ACKNOWLEDGMENT

We would like to thank the PBFA II operations crew, P.M. Baca, and D.F. Wenger for technical assistance. We are also grateful to S.A. Slutz, T.R. Lockner, M.P. Desjarlais, and T. Nash for many useful discussions and to D.L. Cook, R.J. Leeper, and J.P. Quintenz for continuous support and encouragement. This work was supported by the U.S. Department of Energy under contract No. DE-AC04-94AL85000.

REFERENCES

1. J. P. VanDevender and D.L. Cook, *Science*, 232, 801 (1986).
2. Y. Maron, M.D. Coleman, D.A. Hammer, and H.S. Peng, *Phys. Rev. Lett.* 57, 699 (1986), *Phys. Rev. A.* 36, 2818 (1987).
3. J.E. Bailey et al., *Phys. Rev. Lett.* 74,1771 (1995).
4. S.A. Slutz, D.B. Seidel, and R.S. Coats, *J. Appl. Phys.* 59, 11 (1986).
5. M.P. Desjarlais, *Phys. Rev. Lett.* 59, 2295 (1987).
6. M.P. Desjarlais et.al., *Phys. Rev. Lett.* 67, 3094 (1991).
7. T.D. Pointon et.al., *Phys. Plasmas* 1, 429 (1994).
8. S.A. Slutz and W.A. Johnson, *Phys. Fluids* B4, 1349 (1992).
9. S.A. Slutz, *Phys. Fluids B* 4, 2645 (1992).
10. P.H. Bucksbaum, in *Atoms in Strong Fields*, Ed. by C.A. Nicolaides, C.W. Clark, and M.N. Nayfeh, Plenum Press, New York, 1990 (NATO ASI Series B: Physics Vol. 212) p. 381.
11. K. Nakajima et al., *Phys. Rev. Lett.* 74, 4428 (1995).
12. A.L. Pregenzer et al., *J. Appl. Phys.* 67, 7556 (1990).
13. T. Bergeman et al., *Phys. Rev. Lett.* 53, 775 (1984).
14. S. Humphries, Jr., *Nucl. Fusion* 20, 1549 (1980).
15. D.J. Johnson et. al., *J. Appl. Phys.* 53, 4579 (1982) and *Proc. 7th IEEE Pulsed Power Conf*, Monterey, CA, 1989, ed. by R. White and B.H. Bernstein, p. 944.
16. T.A. Mehlhorn, in *Proc. of 10th Int. Conf. on High Power Particle Beams*, San Diego, CA., NTIS # PB95-144317, p. 53 (1994).
17. J. Bailey, A.L. Carlson, R.L. Morrison, and Y. Maron, *Rev. Sci. Instrum.* 61, 3075 (1990) and J.E. Bailey, A.L. Carlson, and P. Lake, *Proc. 1994 IEEE Int. Conf. on Plasma Sci.*, Santa Fe, IEEE Cat. # 94CH3465-2, p. 133 (1994).
18. R.L. Coldwell and G.J. Bamford, *The Theory and Operation of Spectral Analysis Using Robfit*, AIP, New York (1991).
19. E.J. McGuire, (to be published) 1995.
20. E. Stambulchik and Y. Maron, Weizmann Institute of Science Report WIS-90/40/Sept.-PH (1994).
21. L.R. Hunter, D. Krause, Jr., D.J. Berkeland, and M.G. Boshier, *Phys. Rev. A* 44, 6140 (1991).
22. L. Windholz, M. Musso, G. Zerza, and H. Jager, *Phys. Rev. A* 46, 5812 (1992).
23. R. Odom et al., *Phys. Rev. A* 14, 965 (1976).
24. S.I. Themelis & C.A. Nicolaides *Phys. Rev. A* (1994)
25. H.R. Griem et al., *Phys. Fluids B* 3, 2430 (1991).
26. S. Suckewer, *Phys. Fluids B* 3, 2437 (1991).
27. T.A. Green, Sandia Report SAND95-1794 (available from NTIS, U.S. Dept. of Commerce, 5285 Port Royal Road, Springfield, VA. 22161) (1995), and R.W. Stinnett et.al, *Proc. 9th Int. Conf. on High Power Particle Beams*, Washington DC, NTIS PB92-206168, p. 788 (1992).

Laser beam propagation, filamentation and channel formation in laser-produced plasmas

P.E. Young, S.C. Wilks, W.L. Kruer, J.H. Hammer, G. Guethlein and M.E. Foord

Lawrence Livermore National Laboratory, P.O. Box 808, Livermore, CA 94551

Abstract. The understanding of laser beam propagation through underdense plasmas is of vital importance to laser-plasma interaction experiments, as well as being a fundamental physics issue. Formation of plasma channels has numerous applications including table-top x-ray lasers and laser-plasma-produced particle accelerators. The fast ignitor concept, for example, requires the formation of an evacuated channel through a large, underdense plasma. Scaled experiments have shown that the axial extent of a channel formed by a 100 ps pulse is limited by the onset of the filamentation instability. We have obtained quantitative comparison between filamentation theory and experiment. More recent experiments have shown that by increasing the duration of the channel-forming pulse, the filamentation instability is overcome and the channel extent is substantially increased. This result has important implications for the fast ignitor design and the understanding of time-dependent beam dynamics.

1. INTRODUCTION

The propagation of laser beams of moderate to high intensity through fully ionized plasmas is a topic that theoretical and experimental researchers are presently trying to understand. The behavior of the beam can be modified in surprising ways by the modification of the background electron density profile by the ponderomotive force of the laser pulse and by the onset of the filamentation instability. This topic has important practical implications for inertial confinement fusion targets in which the laser beam must propagate long distances through an underdense plasma. Recently, the fast ignitor concept has been proposed, in which a laser pulse is used to clear a channel through the coronal plasma around a compressed target to provide a low density path for a high intensity laser pulse which would ignite the target fuel [1].

In this article, we will discuss recent experiments in which we have investigated the propagation of > 100 ps laser pulses through preformed fully plasmas. We have observed the formation of density channels by the ponderomotive force which, for pulse lengths of 100 ps, terminate short of the peak density due to the onset of the filamentation instability. The axial extent of the channel can be substantially increased by increasing the laser pulse length.

© 1996 American Institute of Physics

In addition, we can measure the peak laser intensity in the filaments by detecting the energy of the ions ejected from the filament via the ponderomotive force.

2. EXPERIMENTAL DESCRIPTION

The experiments we will be discussing all had very similar geometry. They were performed with the Janus laser facility at Lawrence Livermore National Laboratory. The experiments used two opposing beams, one of which was used to explode a parylene (CH) foil which provided an underdense plasma for the second, 100 ps Gaussian beam to propagate through. A 0.35-μm wavelength, 50 ps probe was propagated through the target plane, perpendicular to the high energy beams, and was used for interferometry.

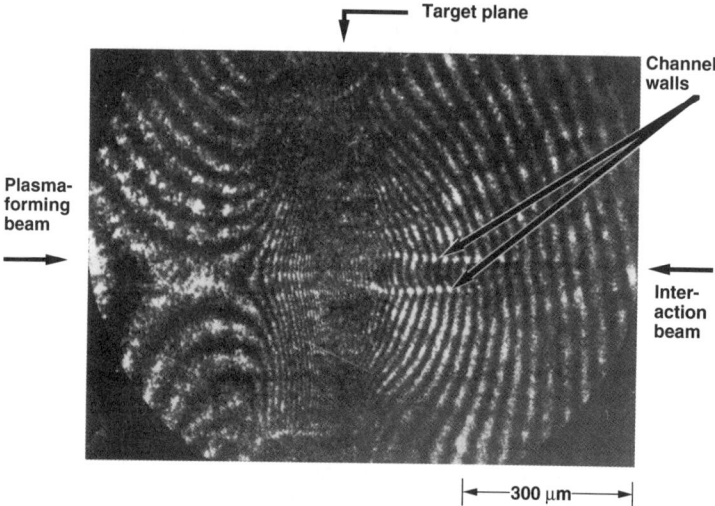

FIGURE 1. Sample interferogram which shows the density channel formed by a focused 100 ps beam propagating through a preformed plasma. Note that the channel is limited to the right hand side of the target plane where the peak density is located.

A sample interferogram is shown in Fig. 1. In the interferometer, the phase front of the reference beam is intentionally placed at an angle with respect to the phase front of the plasma probe beam; this produces, in the case of no plasma, a series of equally spaced fringes across the field of view. For our choice of angle between the plasma probe and reference beams, the presence of a plasma into the probe beam produces a phase shift with respect to the reference beam which shows up as a bending of the fringes to the right. On the left hand side of the target, it is possible for the plasma phase shift to nullify the background phase shift, producing an "X" fringe pattern such as is seen on the left hand side of Fig. 1. Between the "X" and the target, increasing density is indicated by fringes moving in the opposite direction.

3. 100 PS RESULTS

Our initial investigations at moderate intensity (1×10^{15} W/cm^2) immediately showed an interesting phenomenon: the channel formed by the 100 ps pulse had a limited axial extent [2]. In these foil produced plasmas, the peak density occurs at the foil plane and has a Gaussian axial distribution.

Simulations with a modified version of the F3D code [3] that includes nonlinear motion of the ions [4] suggested that the channels were being terminated by the onset of the filamentation instability [5]. Since the instability has a preferred wavelength, it could be detected by looking at the angular distribution of the transmitted light; the angle of deflection is given by

$$\sin^{-1}\theta = k_{\perp,max}/k_0 \qquad (1)$$

where [5,6]

$$k_{\perp,max} = (1/2)(v_0/v_e)(\omega_{pe}/\omega_0)(\omega_0/c) \qquad (2)$$

and v_0 is the electron oscillatory velocity in the laser field, v_e is the electron thermal velocity, ω_{pe} is the electron plasma frequency, and ω_0 (k_0) is the frequency (wavenumber) of the incident laser.

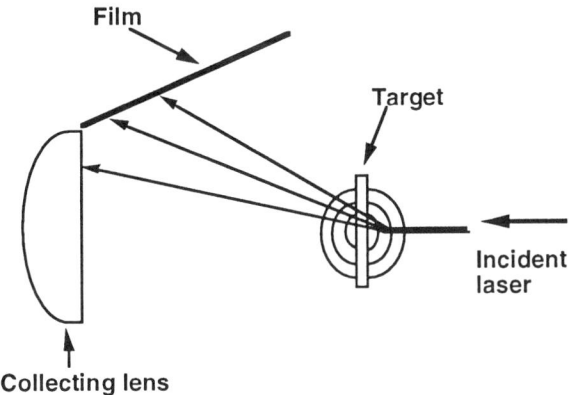

FIGURE 2. Light that is scattered by the filamentation was recorded by film which was placed outside of the collecting lens. The collecting lens was an f/2 aspheric lens which was identical to the focusing lens for the incident laser beam.

Experimentally, the deflection can be inferred by measuring the amount of the incident energy that is transmitted through the plasma within the original focusing solid angle. It can be measured directly by using film to record the pattern of the light outside the collecting lens (see Fig. 2).

The experiments confirmed the theoretical predictions. We observed a functional dependence of the deflection angle on laser intensity (see Fig. 3) and peak plasma density that agrees with Eq. (1). We also observe a corresponding decrease in the energy transmitted through the original focusing solid angle.

This observation sets a serious constraint on the fast ignitor model due to the difficulty in avoiding the filamentation instability in the coronal plasma around the ignitor target. We therefore next looked at the channel formation as a function of the input pulse width.

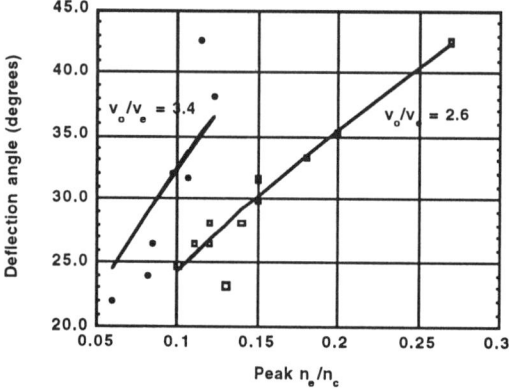

FIGURE 3. The measured deflection angle agrees well with the theoretical prediction based on the density grating set up by the filamentation instability.

4. TIME-DEPENDENT RESULTS

Figure 4 shows the transmitted energy through the collecting lens for different pulse widths [7]. For the highest allowed energies at pulse widths of 100 and 250 ps, the maximum transmission is less than 30%. If one drops the energy below the threshold, all the light goes through, although this is not compatible with the fast ignitor design. When the pulse width is increased to 500 ps, increasing the energy leads to a substantial increase in the transmitted energy. This is accompanied by an increase in the axial length of the channel.

Interferograms of the plasma taken at different times with respect to the channel forming pulse show quite clearly the evolution of the plasma. Figure 5 shows density profiles which were reconstructed from interferograms which show the plasma density distributions at times of 50 ps and 335 ps with respect to the leading edge of a 1 ns square pulse. At the earlier time, the channel is confined to the incident side of the plasma, but 300 ps later, the channel extends all the way through the plasma. This result is confirmed by LASNEX simulations and is understood in the following way. Although the laser light sprays out due to the density grating set up by the filamentation instability, there is still sufficient intensity upstream to begin to move plasma out of the beam. This takes longer, however, because the intensity is lower. Once the channel is formed, the density is too low for the filamentation instability to occur.

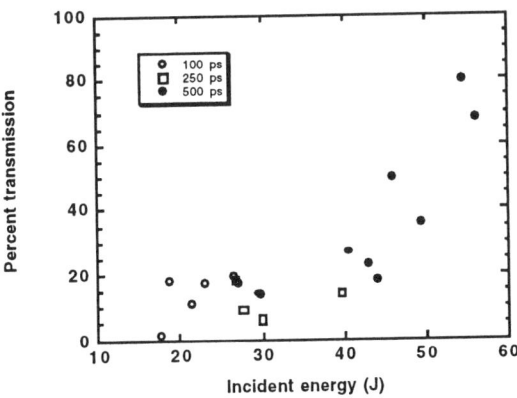

FIGURE 4. The percentage of the transmitted light collected by an f/2 lens is significantly increased when the pulse length is increased to 500 ps.

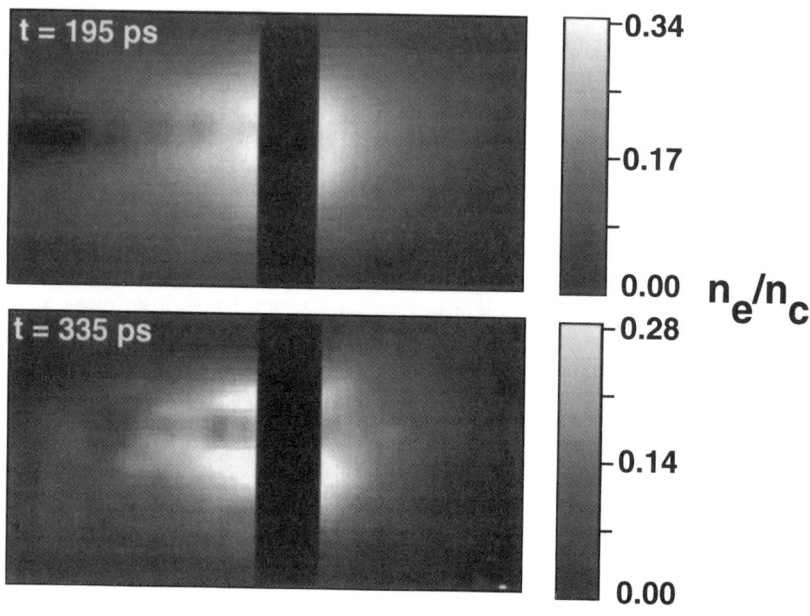

FIGURE 5. Density distributions calculated from Abel inversions of the experimental interograms. The interaction beam is incident from the left. At early times, the channel appears only on the left hand side of the target. At later times, the channel extends through the plasma.

5. FAST ION GENERATION BY FILAMENTS

The filaments that are produced by the 100 ps pulse described above can also accelerate ions to velocities which are much greater than the sound speed. A simple estimate shows the magnitude of the ion velocity that can be achieved. We equate the kinetic energy of the ion to the potential energy set up by the ponderomotive force produced by the laser pulse: $Mv^2_{max}/2 = Ze\phi = Zmv^2_{os}/4$ where M is the ion mass, m is the electron mass, ϕ is the ponderomotive potential and $v_{os} = eE/m\omega_0$ is the electron oscillatory velocity in a laser electric field E with a frequency ω_0. Solving for v_{max} gives:

$$v_{max} = (Zm/2M)^{1/2} v_{os} \qquad (3)$$

For hydrogen (Z = 1, m/M = 1/1836), and $I_L = 1.5 \times 10^{17}$ W/cm^2 (which results from the filamentation of the laser light in the plasma), we find $v_{max} =$

1.7×10^8 cm/sec which corresponds to an enrgy of 15 keV; for comparison, a typical sound speed and ion temperature for these plasmas is 3×10^7 cm/sec and 500 eV, respectively.

The ion veolcities were measured at 90 degrees to the incident laser beams by a time resolved ion spectrometer [8] which used a magnetic field to separate ions of different charge-to-mass ratios. The ions pass through a collimating slit and strike a microchannel plate so that an optical streak camera can be used to measure the ion time of flight. When the interaction beam is of sufficient intensity, a high energy proton tail forms with peak energies of the order of 20 keV. These ions are sufficiently energetic to have ion-ion mean free paths which are greater than the plasma diameter (400 μm).

From Fig. 6 we see that the increase in ion energy is coincident with the onset of the filamentation instability. The onset of the filamentation instability is determined from measurements of the light transmitted through the plasma. The light from the interaction beam that is transmitted through the plasma is collected with an f/2 lens and relayed to a full aperture energy calorimeter; since the interaction beam is focused with an f/2 lens, in the absence of deflections, all the incident laser light is collected by the calorimeter with the exception of backscattered light. As we have observed earlier, the beam breaks up into filaments; the density grating set up by the filaments deflects light out of the collection solid angle, leading to a decrease in the ratio of the measured transmitted energy to the incident energy.

As the laser energy is increased, the peak ion energy first increases with the laser energy, then reaches a constant value when the peak density is 0.25nc. If the peak density is reduced to 0.15nc, the input energy has to be increased before the ion energy increases, but at 30 J, we attain the same ion energy as at the higher density.

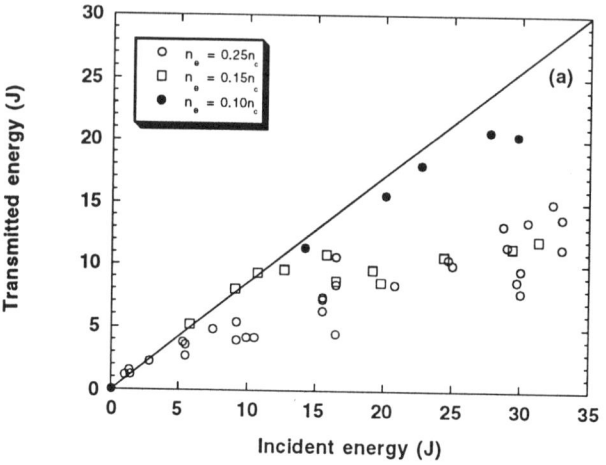

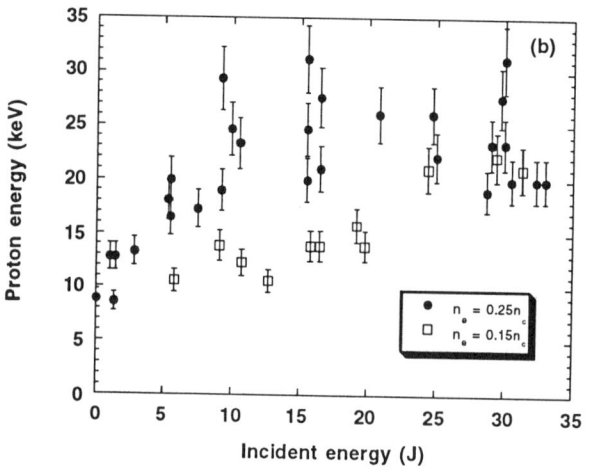

FIGURE 6. Comparison of (a) the onset of the scattering of the transmitted light due to filamentation, and (b) the increase in the peak proton energy as a function of incident energy.

In summary, we have established a versatile test bed for studying various aspects of laser propagation through fully ionized plasmas. Comparison to models, which are used to simulate the fast ignitor problem, have enabled us to validate those models and to point out the filamentation instability as an important concern. We have demonstrated that the laser pulse length is a very important parameter for propagating a laser beam through a relatively high density plasma, a result which is a necessary condition for the success of the fast ignitor concept. We have shown that the laser beam initially spreads due to filamentation, but is still of sufficient intensity to depress the plasma density and initiate a self-guiding process that, if the pulse is long enough, leads to the formation of a coherent channel.

ACKNOWLEDGMENTS

We would like to acknowledge useful conversations with M. Tabak. Important technical support was provided by W. Cowens, R. Gonzales, G. London, J. Bonlie, D. Chakedis, and A. Ellis. This work was performed under the auspices of the U.S. Dept. of Energy by Lawrence Livermore National Laboratory under contract W-7405-ENG-48.

REFERENCES

1. Tabak, M., J. Hammer, M.E. Glinsky, W.L. Kruer, S.C. Wilks, J. Woodworth, E.M. Campbell, and M.D. Perry, Phys. Plasmas **1**, 1627 (1994).
2. Young, P.E., H.A. Baldis, T.W. Johnston, W.L. Kruer, and K.G. Estabrook, Phys. Rev. Lett. **63**, 2812 (1989).
3. Berger, R.L., B.F. Lasinski, T.B. Kaiser, E.A. Williams, A.B. Langdon, and B.I. Cohen, Phys. Fluids B **5**, 2243 (1993).
4. Wilks, S.C., R.L. Berger, B.I. Cohen, W.L. Kruer, A.B. Langdon, B.F. Lasinski, D.S. Montgomery, and E.A. Williams, Bull. Am. Phys. Soc. **38**, 1935 (1993).
5. Wilks, S.C., P.E. Young, J. Hammer, M. Tabak, and W.L. Kruer, Phys. Rev. Lett. **73**, 2994 (1994).
6. Young, P.E., J.H. Hammer, S.C. Wilks, and W.L. Kruer, Phys. Plasmas **2**, 2825 (1995).
7. Young, P.E., M.E. Foord, J.H. Hammer, W.L. Kruer, M. Tabak, and S.C. Wilks, Phys. Rev. Lett. **75**, 1082 (1995).
8. Guethlein, G., et al., Rev. Sci. Instrum. **66**, 333 (1995).

RADIATIVE OPACITY

A Radiation Dependent Ionization Model for Laser Produced Plasmas

M.Busquet

CEA Limeil-Valenton, 94195 Villeneuve St Georges CEDEX, FRANCE

Abstract. RADIOM is a non-Local Thermodynamical Equilibrium Atomic Physics model, accounting for x-ray reabsorption. We present shortly the model, its introduction in hydrodynamic codes and a few application.

INTRODUCTION

Local Thermodynamical Equilibrium (LTE) is the situation of plasmas where all atomic radiative decay transitions are either negligible, either exactly balanced by the inverse photo-absorption transitions. In large zones of laser-created plasmas, LTE is not verified, even in indirect drive experiment where large radiation fields may induce significant amount of photo-absorption and may reduce deviation from LTE.

Then numerical simulation of laser-created plasmas requires a non-LTE atomic physics package for both equation of state (EOS), opacities (K) and emissivities (J). For high-Z material where radiation trapping may be significant, and in holhraums, atomic physics has to include the effect of the ambient radiation field. The atomic properties (EOS, K, J) can not be tabulated and has to be computed in-line. Solving the full rate equations system of a collisional-radiative model will be extremely CPU time consuming and even untractable for high Z material. A simplified model (XSN) has been proposed by R.M.More, W.A.Lokke and W.H.Grasberger in the early 80's, based on an average atom. Yet, it is a little too expensive and questions still remain on the validity of this average description.

I have proposed[1] a Radiation Dependant Ionization Model (RADIOM) based on the idea of an ionization temperature and taking profit of existing LTE atomic physics databases, thus allowing to use the best LTE datas when LTE conditions prevail.

A short description of non-LTE, then of RADIOM will be presented. Then I will present a few illustrations on non-LTE simulation and finally I will conclude.

NEED FOR NON LTE PROCESSING

Populations of the atomic levels result from the balance of updward and downward transition, both collisional and radiative. Collisionnal rates are balanced when electron distribution functions are close enough to Maxwellian distributions. Radiative decay rates are not balanced unless the radiation field is Planckian. However, they are negligible at high density or low temperature. Thus sufficient LTE condition is :

$$\beta = (N_e/10^{21} cm^{-3}) (T_e/100 eV)^{-7/2} < 0.3$$

which is called "collisionnaly dominated LTE."

Large zones of laser-created plasmas fall out of this domain, but let's underline that departure from LTE is sensitive to the radiation field.

Accounting for non-LTE will change the hydrodynamical and optical properties of the plasma. The ionic distribution is shifted toward lower charge state in non-LTE, thus internal energy, average charge, electron conductivity and laser absorption coefficient are lower than LTE properties. For the same reason, the opacity spectrum is shifted to lower photon energy. Moreover, as excitation is also reduced in non-LTE condition, the x-ray emissivity of the plasma is dramatically reduced.

A higher temperature results from the reduced heat capacity and radiative power losses for a same absorbed energy .

The general idea to obtain non-LTE microscopic properties is to compute the balance of all transitions between all levels. The skill of atomic physicists is to neglect scarcely populated levels, group together others,... This approach leads to the so-called "collisional-radiative" models. However, there are so many levels and transitions that these models can not be made general, and become rapidly untractable, especially for high atomic number, or when accounting for reabsorption. Dramatic simplification is required for reasonnable CPU time and memory, in order to use them inside an hydrodynamical code.

BUILDING A SIMPLE MODEL

The statistical average atom model

The first approach to simplify the atomic structure is the "statistical average atom" approach, constructed on a screened hydrogenic atomic structure (XSN). It has been developed first by Lokke, Grasberger and More,[2] and recently revisited by Decoster[3] and Pollak.[4] In these approach, balance of "average transition rates" between "average occupation number" of shells of an "average atom" is performed, and the properties derived from this procedure are assumed to be representative. In constructing the RADIOM model, we have shown that the atomic properties result from some "convolution" between the atomic structure (here XSN) and a "non-LTE state." This gives some credit to this assumption. However my experience of such models is that this model is still time consuming, which I believe comes from poor convergence properties[1] due to highly non linear structure.

The RADIOM model

To build this radiation dependent model, I start from the concept of an ionization temperature Tz, which is by definition is the temperature at which LTE distribution of charge state populations is centered at the same value than actual, non-LTE, distribution. It is easy to understand that LTE datas (average Z, internal energy,...) obtained at Tz mimic the non LTE-values. It

---

[1] despite major efforts of A.Decoster who implement different relaxation coefficients for each shells (private communication).

has been confirmed that the opacities are also well reproduced. An "extended Saha equation", which in first approximation does not depend on the atomic structure, but only on density, temperature and x-ray radiation field, allows to obtained direectly Tz without computing the exact non-LTE structure :

$$\frac{N_{Z+1,c}/\xi_{Z+1,c}}{N_{Z,c}/\xi_{Z,c}} = \omega_e \exp(-\chi/kTe) \frac{1+\beta \overline{h\nu/kT}^3 \overline{I_\beta/B_\nu}}{1+\beta \overline{h\nu/kT}^3},$$

$$\overline{h\nu/kT}^3 \overline{I_\beta/B_\nu} = \frac{\sum f^{rad} h\nu^2 \exp(-h\nu/kTe) \int I_\nu \phi_\nu h\nu^{-4} d\nu \Big/ \int B_\nu \phi_\nu h\nu^{-4} d\nu}{\sum f^{coll} h\nu^{-1} \exp(-h\nu/kTe)}$$

where ϕ_ν is the line or continuum profile,

$\omega_e = 6 \; 10^{21} \; T_{eV}^{3/2} \; Ne^{-1}$,

and $\beta = 0.13 \; (Te/100eV)^{7/2} \; (Ne/10^{21})^{-1}$

This equation also lead to the non-LTE source function

$$S_\nu = j_\nu/k_\nu = B_\nu \times \frac{1+\beta \overline{h\nu/kT}^3 \overline{I_\beta/B_\nu}}{1+\beta \overline{h\nu/kT}^3}$$

One other important effect of non LTE, first pointed out by Mihalas[5] but only for isolated lines, is a strong reduction of the coupling between radiation and matter. Let us write the energy exchange rate between radiation and matter :

$$W_\nu = k_\nu (S_\nu - I_\nu) = k_\nu \frac{1}{1+\beta \overline{h\nu/kT}^3} (B_\nu - I_\nu)$$

It results from this equation an apparent reduction of the coupling, which can be understood as follow: enhanced excitation follows absorbed photons, which in turn imply increase of emission, counterbalancing the absorption. Thus the net result is that most of the absorption can be converted in angular diffusion (and also in some frequency diffusion).

A FEW APPLICATIONS

In order to check precision of the model, two sets of comparisons have been made. We first compare Tz given by its definition, and Tz* given by the direct calculation mention above. In Fig.1 is presented contour lines of this ratio.[2] The largest discrepancies appear in the K-shell region, where the atomic properties varies slowly with temperature. Difference of about 10% are found in the domain close to LTE, where excited populations are important. These levels are not actually used to compute Tz. (see ref 1) Accounting for them will reduce the difference. When a good approximation of Tz is obtained, we havec checked it gives very good results for "hydrodynamic" properties, and rather good (better than expected !) agreement for opacities (see Fig.10 of ref 1)

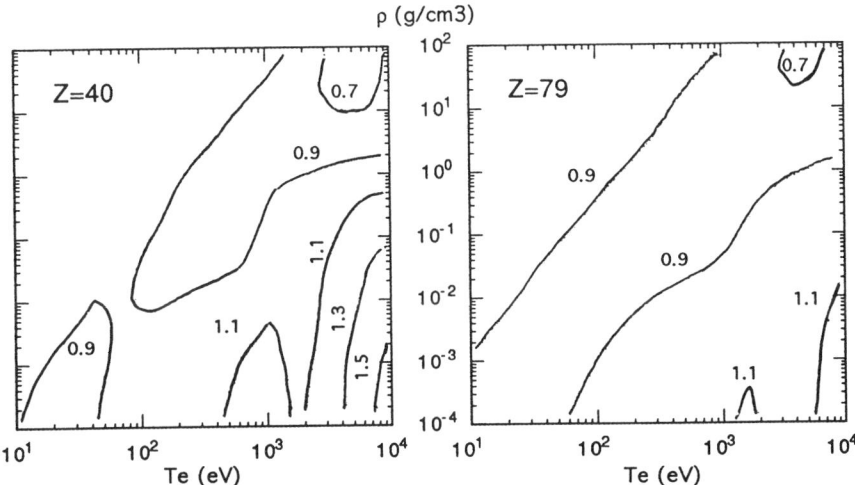

FIGURE 1. ratio of Tz given by its definition over Tz* computed directly from Te, Ne and {Ev}

Implementation of RADIOM in hydro.codes is relatively straigthforward: given density, temperature and radiative

---

[2] work performed by G.Lachèze-Murel and F.Garaude.

energy density, Tz is computed. LTE datas at Tz are then used and a few corrections are added to obtain non-LTE properties. Thus the hydro.code, converted from LTE to non-LTE in this easy way has not been really slowered. Compared to full LTE run, computation time of a simulation performed with RADIOM in a simulation is increased typically by 0% to 50%, which results from time spent to compute the ionization temperature and source function, but mainly time lost or gained (depending of the local conditions) in iterative convergence procedures and time step regulation: generally, time step can be larger in non LTE runs than in LTE runs.[3]

Large departure from LTE is found in Indirect Drive FCI

The modified code are then fast enough to be used routinely in all simulation of laser-created plasmas. A run performed by S.Laffite of a gold holraum heated by a Megajoule laser show non-LTE in most of the x-ray conversion zone. (see Fig. 2)

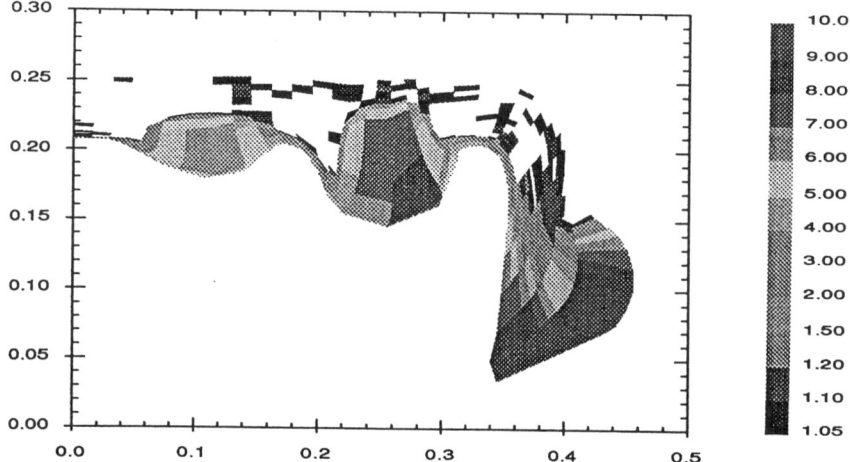

FIGURE 2. ratio of Te over Tz from a 2D calculation of an indirect drive hohlraum. The ratio peaks to more than 10, which is strong non-LTE.

---

[3] I have no experience of XSN, but runs performed with NOHEL, a "XSN-type" module written by A.Decoster, have shown computation time multiplied by a factor from 5 to more than 100, when coupled to radiation. In the optically thin approximation, when values can be tabulated in advance, computation times are essentially equal.

Radiation dependance increases X-ray conversion efficiency by 10%

In order to check the need of the radiation dependance in the atomic code, we compare x-ray conversion efficency computed in the optically thin approximation and in the fully coupled mode. A difference of 10% for a typical gold target is found (see Fig. 3). The effective coupling between radiation and matter (which vary with the energy of the photons) present reduction which may be more than 2 orders of magnitude (see Fig.4). Consequently, relaxation times and relaxation lengths are increased.

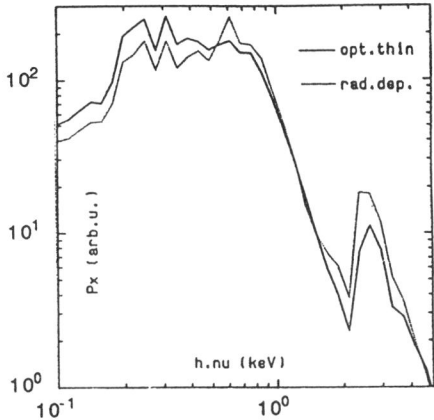

FIGURE 3. X-ray emitted spectrum of a plane gold target with a "PS22" pulse shape, in the optically thin approximation and a radiation dependant model.

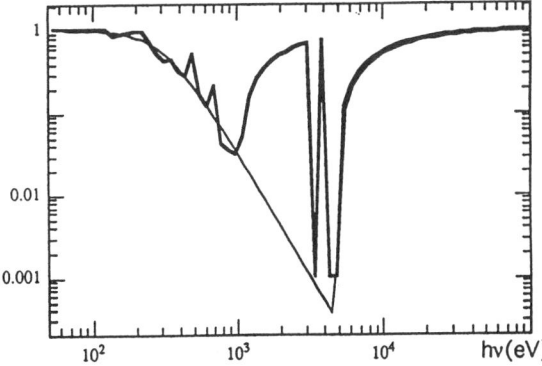

FIGURE 4. coupling coefficient between radiation and matter, relative to its LTE value, versus photon energy

In conclusion, one should use **RADIOM**

Detailed studies of results, and comparisons with other ionization models, including XSN type models, has shown that the non-LTE atomic description can be understood as some kind of convolution of the atomic structure and the non-LTE ionization state. Use of an ionization temperature and of an average atom description appear to be rather equivalent when the same atomic database is used, but evidently not so costly.

Differences between radiation dependent model and optically thin model has been illustrated.

The RADIOM models presents a 10% precision (less for Tz, more for Sv), is easily implemented in any hydro.code, can use the best LTE data base available, is fast and finally shows some insight on basic phenomenas.

ACKNOWLEDGEMENTS

Implementation of RADIOM in the Limeil's code FCI1 and FCI2 has been perforemed by G. Schurtz and F. Wagon. It has also been imlemented and tested in the NRL hydro-codes, thanks to D. Garren, M. Klapisch, D. Colombant, J.P. Dahlburg, J.H. Gardner and A.J. Schmitt. I want to mention the collaboration with D.Vanderhaegen on Mihalas's "equivalent diffusion".

REFERENCES

1. Busquet, M., *Phys. Fluids* **B 5,** 4191 (1993).
2. Lokke, W.A., Grasberger, W.H., More, R.M., unpublished LLNL report, 1977
3. Decoster, A., private communication, see also S.Bayle, Thesis, Paris 1991 (unpublished)
4. Pollak, G., unpublished LANL report, 1990
5. Mihalas, D., *Stellar Atmospheres*, San Fransisco,W.H.Freeman & co, 1978

New Methods For Probing The Opacity and Optical Properties of Dense Low-Temperature Plasmas

A. N. Mostovych, L. Y. Chan,[†] and K. J. Kearney[*]

Laser Plasma Branch, Plasma Physics Division, U.S. Naval Research Laboratory, Washington, D.C. 20375

We have demonstrated new techniques by which well characterized dense (10^{19}-10^{20} cm^{-3}), cold (1-15 eV), and strongly coupled ($\Gamma \sim 1$) plasmas are produced by laser vaporization and ionization of thin metallic films. By limiting the plasmas to small (r~ 100-500µm) but unconfined volumes it is possible to create fully accessible plasmas for diagnostic inquiry. The plasmas are very absorptive but, as a result of their small size, their optical depths are typically less than one. Laser interferometry, absorption probing, and spectroscopy are used to characterize the plasmas. Detailed measurements of the plasma density, temperature, opacity and spectra are compared to theory in the regime were the photon energies are of the same order as the average inter-particle energies of the plasma.

INTRODUCTION

The physics of high density, low temperature plasmas is in a regime where the plasmas are typically strongly coupled and theoretical modeling and experimental measurements are difficult (1). Transport properties, opacities, and the equation-of-state of these plasmas are important issues in astrophysics, laser-fusion, shock wave research, as well as in the physics of high current discharges.

At low temperatures, the electrons and ions of dense plasmas interact strongly to produce highly non-ideal gas systems. The interaction to thermal energy ratio Γ is no longer small as in "weakly coupled" ideal gas plasmas. As a result, the description of these plasmas requires new approaches. There are many new theoretical and numerical investigations but relatively few experiments. The measurement of optical properties of such plasmas is of particular interest because the optical properties are sensitive to the state of the plasma through their strong dependence on the equation of state, particle collisions, plasma microfields, and atomic line shapes in the plasma. The largest effects are expected for photons with

---

[†] NRC-NRL Post Doctoral Fellow
[*] Present Address: LLE, University of Rochester, Rochester, NY 14623

energies comparable to the plasma interaction energies, i.e., hv/kT ≤ Γ. At these energies the photon-plasma interaction is primarily determined by electron-ion collisions and transitions between high-lying atomic levels. Both of which are subject to strong perturbations from the plasma. This parameter regime is also one of the most difficult to investigate because the very high opacities of these plasmas prevent the use of optical diagnostics at low photon energies.

We have demonstrated techniques which overcome these problems (2). Our method succeeds by producing small optically-transparent plasmas by laser vaporization and heating of thin metallic films. These plasmas are well characterized, with measured temperatures and densities, making them useful for quantitative studies of plasma opacity in the dense, low-temperature regime. In this review, we discuss our plasma techniques, present measurements of plasma emission and opacity, and provide comparisons to current opacity and line broadening theory.

PLASMA SOURCE

A laser-produced plasma is used as the source for these experiments; see Figure 1. A glass substrate is coated with an Al film and a laser beam vaporizes this film by irradiation through the glass. The expanding supersonic vapor, with a diameter comparable to the laser focal spot (d~1.0mm), flows through a slit ($\Delta x \approx 100 \mu m$) which limits its transverse extent and total mass. After the vapor slab has reached the appropriate expansion length ($l \approx 1000 \mu m$ at $t \approx 100 ns$) and desired vapor density, a second laser ($\lambda = 1.054 \ \mu m$; 10^{10} - 10^{11} W/cm^2) is used to ionize the Al vapor. The heating occurs over a time of about 5-8 ns and produces a fully ionized plasma (roughly 1000μm by 1000μm) with an average ionization $\langle Z \rangle \approx 1$-4. With this technique, the density and temperature of the plasma can be chosen semi-independently. The width of the slit and the expansion time before heating determine the density whereas the laser heating level controls the plasma temperature and degree of ionization.

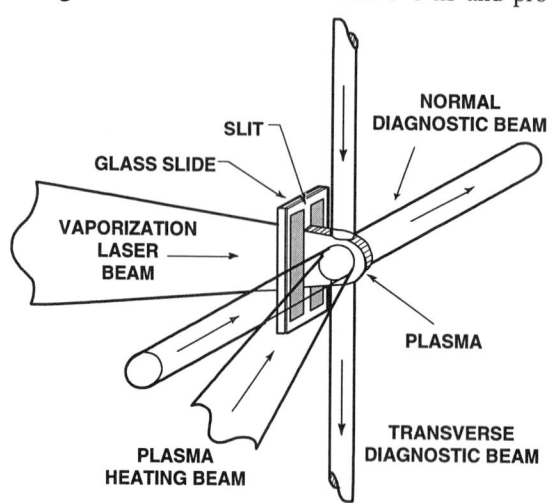

Figure 1. Schematic of Experimental Setup.

Typical diagnostics of the experiment are shown in Figure 2. The experiment is designed to simultaneously measure the plasma density, temperature, opacity, and emission spectra with temporal and spatial resolution. Three short pulse (700ps) probe beams (1.054, 0.527, and 0.351 μm) measure the transmission through the plasma normal to the slab geometry. The beams are focused to a small spot (d≈100μm) in the center of the plasma to provide spatial resolution and to insure an uniform density in the focal plane. The degree of transmission is measured with a set of fast ($\tau \approx$ 350ps) incident and transmitted photodiodes. The transmission probes are much brighter than the plasma and the transmission measurements are not affected by plasma emission. Two unfocused interferometry beams, one normal, one transverse, measure the line integrated density from the two orthogonal orientations. These measurements are unfolded to give the density profile along the line-of-sight of the transmission measurement. The emission of the plasma, in the focal volume of the trans-

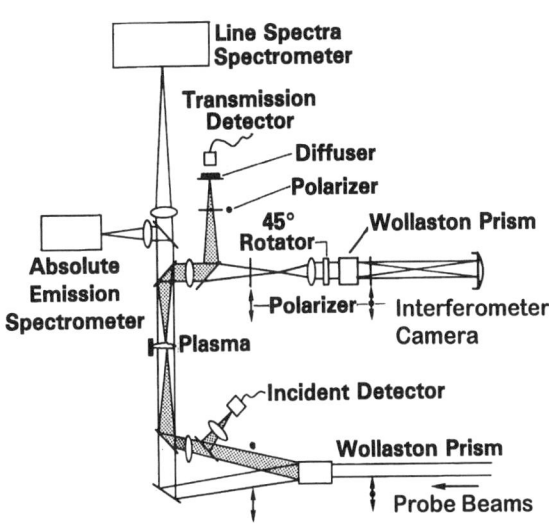

Figure 2. Diagnostic configuration. Density, temperature, opacity, and emission spectra are all measured simultaneously with temporal and spatial resolution.

mission probes, is measured with an absolutely calibrated 0.5m monochrometer with a temporal resolution of 0.5ns. The plasma temperature is determined from the absolute emission and the degree of transmission at 0.527 μm. For plasmas in local thermodynamic equilibrium (LTE) these quantities are related to the average temperature through the radiation transfer equation and Kirchoff's law (3), i.e., $I(v,T)=I_p(v,T)(1-I/I_0)$, where $I_p(v,T)$ is the Planck distribution, I/I_0 is the transmission fraction and v is the frequency of the absorbing radiation. The plasma emission spectra (2000-8000 Å) are recorded with a 0.33m spectrometer coupled to a streak camera or gated micro-channel plate detector.

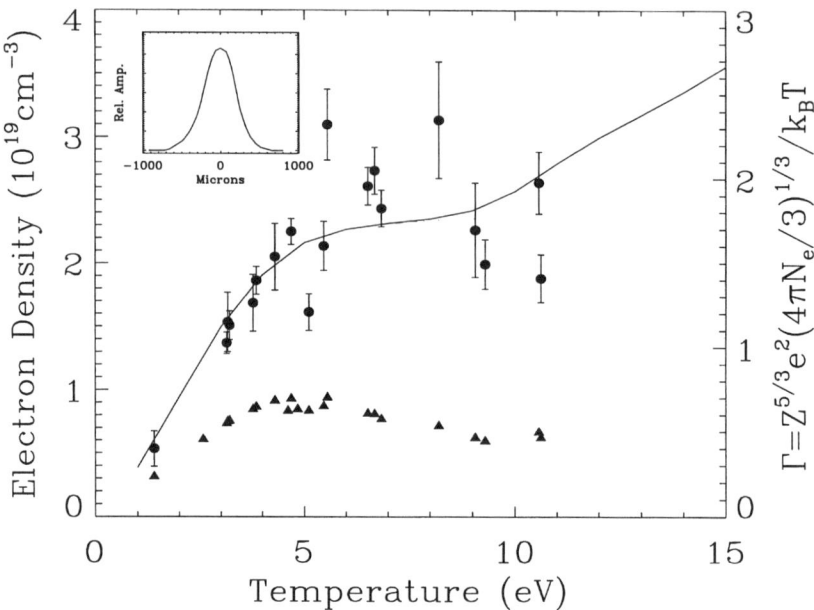

Figure 3. Measured plasma properties. Solid curve due to Saha equation calculation for an aluminum mass density of $\rho_m = 3.6 \times 10^{-4}$ g/cm^3. Triangular data points give the strength of the ion coupling parameter. Curve in insert is a typical transverse density profile of the plasma.

The initial plasma geometry is slab-like; however, the density profile of the plasma along the transmission-probe line-of-sight evolves into a gaussian-like profile as a result of heating and expansion; see insert in Figure 3. The profiles are symmetric as long as the vapor areal-mass-density remains below some critical value. In these experiments, the maximum Al mass density is restricted to approximately (4.5×10^{-4} g/cm^3; $N_{ion} \sim 10^{19}$ cm^{-3}) to insure symmetry in the density profile and to insure full fringe visibility in the highly absorbed transverse interferometer beam. The average mass density, i.e., ion density, is inferred from the measured electron density and temperature and the degree of ionization given by a Saha equation calculation for Al. The Saha equation and the equations of state in the OPAL (4) and STA (5) codes give ion densities which are within 2-3%. Typical electron densities in the experiment, averaged over the line-of-sight profile, are plotted in the solid circles of Figure 3. The uncertainty in the temperature for these data is typically 10-20%. The solid curve through the data is a Saha calculation for an average mass density of 3.6×10^{-4} g/cm^3. The spread in the data is due to shot-to-shot variations in the mass density of the aluminum vapor, resulting from variations in the film thickness, laser power, beam alignments, and slit width. The solid triangles in Figure 3 correspond to the strength of the ion coupling parameter [$\Gamma = e^2 Z^{5/3} (4\pi N_e/3)^{1/3}/T_e$]. Coupling between ions is the strongest ($\Gamma \approx 0.75$) at temperatures (4-5 eV) where the population of triply ionized aluminum peaks.

OPACITY MEASUREMENTS

The transmission of the laser probes (1.054, 0.527, and 0.351 μm) is measured at the peak of the heating pulse and in temporal and spatial synchronization with the interferometry and absolute emission measurements. The transmission T through the plasma is related to the absorption coefficient κ through T=exp(-∫κ(x)dx). For comparison between experiment and theory it is convenient to define an average absorption coefficient $\langle \kappa \rangle$ = -ln(T)/L where L is an average plasma thickness calculated from the measured density profile $n_e(x)$ such that $L \int n_e^2(x)dx = \left\{ \int n_e(x)dx \right\}^2$. With this choice, processes which depend on $n_e(x)^2$ are well described by average quantities even though the plasma profile is not flat. This model is useful for our conditions because the STA and OPAL opacity codes and the data show a $n_e(x)^2$ dependence for the absorption coefficient. In practice, L is also used to calculate the average electron densities. L is comparable to the thickness containing 90% of the plasma mass (0.06-0.1cm) and is about 1.5 times the full width half maximum (FWHM) of

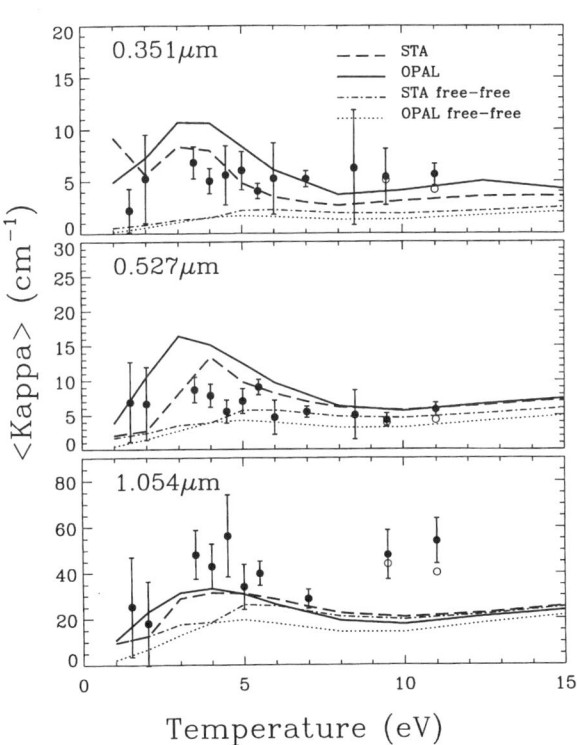

Figure 4. Measured absorption coefficient compared to OPAL and STA model calculations.

the density profile.

The 0.351, 0.527, and 1.054μm absorption coefficients are plotted in Figure 4. These opacity data are a subset corresponding to only those measurements having an initial mass density of $\rho_0 = 3.6 \pm 1.8 \times 10^{-4}$ g/cm^3. In addition, the data are scaled by $(\rho_0/\rho_m)^2$ to account for the finite distribution of actual mass densities ρ_m and the n^2 dependence of the opacity. Error in the data is minimized by averaging over multiple data in temperature steps of 0.5 eV. The error bars include instrumental, alignment, and calibration uncertainties of the transmission measurements, as well as the effects of the uncertainties in density and temperature on the absorption coefficient.

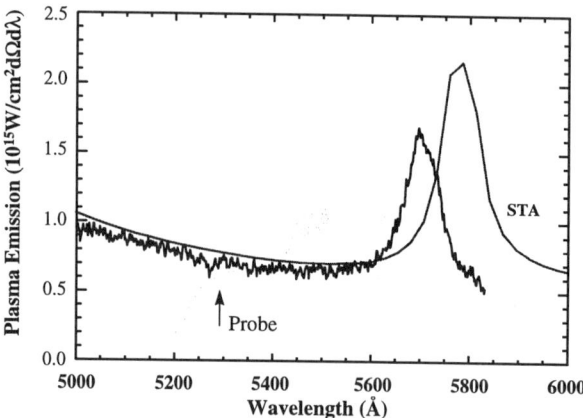

Figure 5. Emission in the vicinity of the 0.527mm probe. Curve due to STA calculation at 3.5 eV and 7x10^{18}cm^{-3}.

The data are compared to predictions of the STA and OPAL opacity models for the averaged density and temperatures measured in the experiment. The calculations represent the complete absorption coefficient $\kappa = \kappa_{ff} + \Sigma \kappa_{bf} + \Sigma \kappa_{bb}$, including free-free, bound-free, and bound-bound contributions. These are LTE models that include contributions from all ionization stages in the plasma as well as from all LTE populated bound and free states of each ionization stage.

Both STA and OPAL predict enhancement of bound state contributions in the 2 to 6 eV temperature range, where the Al III ionization stage and its large set of spectral lines in the UV-visible-IR range are well populated. Even though the probing wavelengths do not coincide with aluminum absorption lines, the wings of many spectral lines contribute to the absorption as a result of spectral line broadening by the dense plasma. The measured opacity, while greater than the free-free opacity, is nearly constant and does not show significant enhancement in this temperature range. The 0.351μm and 0.527μm data indicate that bound states contribute less than is calculated by STA and OPAL. The sensitivity of the opacity to line broadening issues is illustrated by the difference in STA and OPAL opacities. The OPAL bound state contributions are larger, primarily as a result of broader (2x) line profiles in OPAL calculations.

EMISSION SPECTROSCOPY

Spectral measurements were used to better characterize the conditions of the opacity measurements and to investigate line broadening in these plasmas. Typically, as in Figure 5, the absorption probes were found in smooth regions of the emission spectra. The emission lines were broad and did not fall directly on top of the opacity probes. The broadening of these lines was investigated in detail at somewhat lower densities. This permitted comparison of the line widths without the complication of optical depth issues. Sample data of the Al III (3d-4p) doublet and an underlying Al II line are displayed in Figure 6. Under conditions of the experiment the instrument, Doppler, and Stark ion broadening widths are insignificant and the lines should be Lorentzian as is expected for electron Stark broadening.

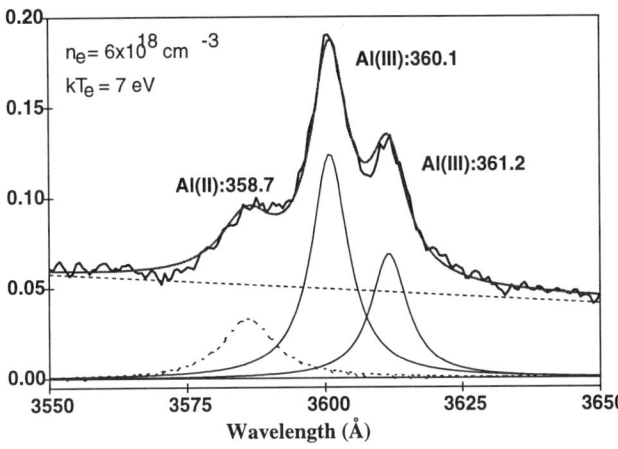

Figure 6. Al linewidth measurements.

The data are fit with Lorentzian profiles to obtain the experimental line widths. The line widths are found to scale linearly with density (Figure 7) and are in very good agreement with the calculations of semi-empirical theory

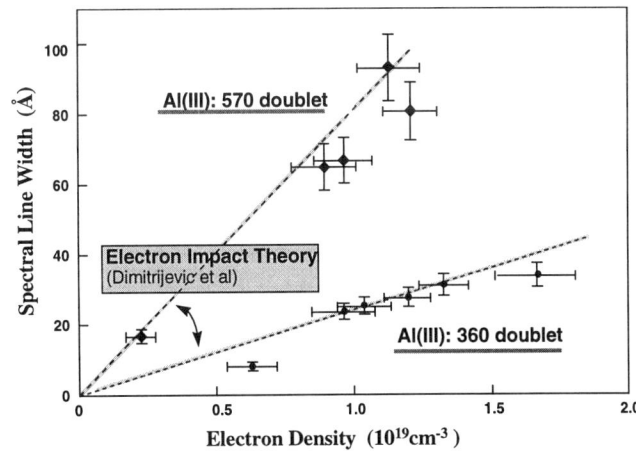

Figure 7. Comparison of line broadening calculations to theory.

of Dimitrijevic (6). This is significant because this theory is typically used in opacity calculations, as is the case for the OPAL and STA codes.

CONCLUSIONS

We have demonstrated techniques by which dense, cold, and strongly coupled plasmas can be produced under controlled and well characterized conditions. The opacity and emission spectra of these plasmas have been measured and compared to theory. Typically, the opacity of these plasmas is overestimated by the calculations whereas the measured line widths are in agreement with semi-empirical line broadening theory.

ACKNOWLEDGEMENTS

We gratefully acknowledge the contributions by C. A. Iglesias and F. J. Rogers in supplying OPAL calculations and D. Garren and M. Klapisch in supplying STA calculations in the course of this work. This work was supported by the U.S. Office of Naval Research.

REFERENCES

1. Good reviews are found in S. Ichimaru, H. Iyetomi, and s. Tanaka, Phys. Rep. **149**, 91 (1987); S. Ichimaru, *Statistical Plasma Physics* (Addison-Wesley,1992); *Strongly Coupled Plasma Physics*, ed H. M. Van Horn and S. Ichimaru (Univ of Rochester, 1993), and references within.
2. A. N. Mostovych *et al.*, Phys. Rev. Lett. **75**, 1530 (1995); A. N. Mostovych *et al.*, Phys. Rev. Lett. **66**, 612 (1991).
3. Ya. B. Zel'dovich and Yu. P. Raizer, *Physics of Shock Waves and High-Temperature Hydrodynamic Phenomena*, (Academic Press, New York, 1966).
4. F. J. Rogers and C. A. Iglesias, Astrophys. J. Suppl. Ser. **79**, 507 (1992).
5. A. Bar-Shalom *et al.*, Phys. Rev **A40**, 3183 (1989).
6. M. S. Dimitrijevic and N. Konjevic, J. Quant. Spectrosc. Radiat. Transfer **29**, 451 (1980).

The Rosseland mean opacity of a composite material at high temperatures

T. J. Orzechowski, M. D. Rosen, H. N. Kornblum, J. L. Porter*,
L. J. Suter, A. R. Thiessen, and R. Wallace

Lawrence Livermore National Laboratory
P. O. Box 808
Livermore, CA 94550

The Rosseland mean opacity can be used to describe radiation transport through high-opacity materials. This mean opacity is dominated by the minima in the frequency-dependent opacity. By mixing appropriate materials, we can fill in the low opacity regions of one material with the high opacity regions of another material, resulting in a material with a Rosseland mean opacity higher than either of the constituents. This composite material can be used to improve the energy balance in indirect-drive inertial confinement fusion. For a given laser energy, this can raise the temperature of the laser heated hohlraum, or for a given desired temperature, require less laser energy.

In the indirect drive approach (1) to inertial confinement fusion (2) (ICF) the radiation that drives the implosion of the fuel capsule is generated by the interaction of intense beams, either lasers (3) or particles (4), with the interior wall of a high-Z cavity, or hohlraum. This radiation is typically described by a blackbody spectrum with a temperature of about 250-eV. This high temperature radiation not only drives the fuel pellet compression, but also heats and ablates the hohlraum wall. The interaction of the radiation with the hohlraum wall is characterized by multiple absorption and reemission of the x-rays (5,6,7). The ratio of the reemitted flux to the incident flux is referred to as the albedo, α. The efficiency with which the radiation couples to the capsule depends on the albedo; increasing the albedo improves the drive efficiency. The incident flux is the sum of the reemitted flux plus the x-ray flux lost to the wall. This flux lost to the wall propagates through the wall in the form of a (diffusive) ablative heat wave (8). The rate of diffusion is (approximately) inversely proportional to the square root of the Rosseland mean opacity (9), which is used to describe radiation transport in optically thick materials when the matter and radiation are in thermodynamic equilibrium. It is defined as a weighted harmonic mean of the frequency dependent opacity. This mean opacity is dominated by the minima in the frequency dependent opacity. Increasing the Rosseland mean opacity reduces the radiation energy lost to the walls and hence for a given laser power (and x-ray conversion efficiency) increases the drive temperature and improves the coupling efficiency of the radiation to the fuel pellet.

Typically we use pure Au hohlraums heated to a temperature of ~250 eV. A frequency dependent opacity for Au at a density and temperature relevant to these experiments is shown in figure 1. This opacity was calculated using a very simple average atom model (10) for Au at 1.0 g/cm^3 and a temperature of 250 eV. There are significant windows in the opacity at energies around the peak of the blackbody spectrum. The gross structure of the opacity shown in figure 1 is dominated by the

---

* Present address Sandia National Laboratory, P. O. Box 5800, Albuquerque, N. M. 87185

bound-free (photoelectric) absorption coefficient: the sharp increases in opacity correspond to the photoionization of the various atomic shells (K, L, M...). Also shown in figure 1 is the weighting function $\partial B_\nu / \partial T$ used in the definition of the Rosseland mean opacity for a 250-eV blackbody distribution. As can be seen in the figure, the peak of the weighting function is fairly broad (~1 keV FWHM) and occurs at about 1 keV, which is near the minimum in the opacity between the N- and O-band absorption edges. In order to improve the efficiency of the hohlraum we need to blend-in materials whose high opacity regions compliment the low opacity regions of the original material. Figure 1 also shows the calculated frequency-dependent opacity for gadolinium at T=250 eV and a density of 1.0 g/cm^3. Gd was chosen because its regions of high opacity occurs around the same frequency as the holes in the Au opacity. For the frequency dependent opacities shown here, the Rosseland mean opacity for Au is 823 cm^2/g and for Gd it is 455 cm^2/g. In a properly designed material containing both atoms, the radiation samples the combined opacity of both materials resulting in a higher Rosseland mean opacity. Again, using the opacity model employed in generating figure 1, the Rosseland mean opacity of a 50:50 mixture of gold and gadolinium at a density of 1 g/cm2 and a temperature of 250 eV is 1390 cm^2/g. A more sophisticated opacity model, XSN (11), in which bound-bound transitions play a more central role in determining opacity, would give κ_R=1500 cm^2/g for Au, 1300 cm^2/g for Gd and 2500 cm^2/g for the 50:50 Au/Gd mixture at the same density and temperature.

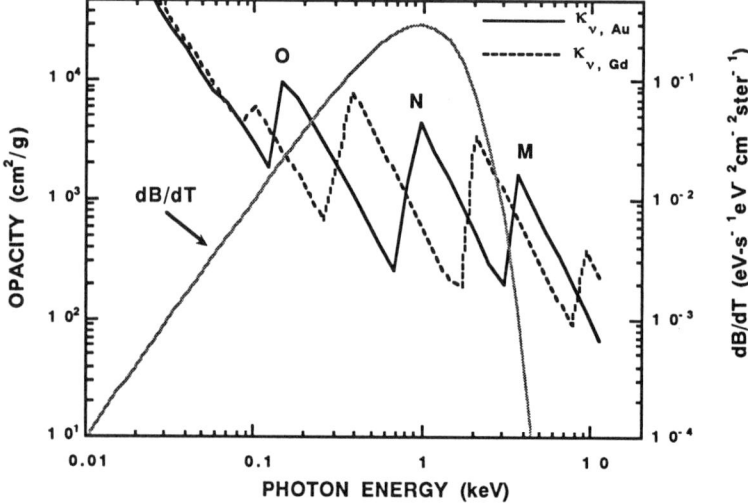

FIGURE 1. Frequency dependent opacity of Au and Gd. Also shown is the weighting function ($\partial B/\partial T$) corresponding to a 250-eV Planckian distribution

In order to determine the Rosseland mean opacity of a material, we measure the propagation time of a radiation heat wave (also referred to as a Marshak wave) through a well characterized sample of that material and compare the measurement to analytic (12,13,14) and numerical solutions. The non-linear diffusion equation that governs the Marshak wave behavior involves the specific energy density and the Rosseland mean opacity of the material. Using the XSN opacity model, the Rosseland mean opacity for Au in the temperature range of 100 to 300 eV is found to scale as $\kappa_R = \kappa_o \rho^{0.33} T^{-1}$, where κ_R is in units of cm²/g, and κ_o is 3500 when ρ is in g/cm³ and T is in heV (10^2 eV). The energy density of the material is approximated as $\varepsilon \sim \varepsilon_o T^{1.5}$. With these analytic models for κ_R and ε, the self-similar solution for the diffusion equation gives the energy lost to the wall and the position of the Marshak wave front as a position of time (6):

$$E_w \propto T^{3.0} t^{0.62} \kappa_o^{-0.4} \quad (1)$$

$$\rho X_M \propto T^{1.7} t^{0.55} \kappa_o^{-0.45} \quad (2)$$

These solutions were derived for a constant temperature. In a more general case the boundary temperature itself can vary in time, thus changing the temporal dependence of E_w and ρX_M. For example, in the case of a constant x-ray flux on the wall, $\dot{E}_w$ = const ($\sim t^0$) and from equation 1 we would have the temperature scaling as $T \sim t^{0.12}$. In this case the position of the Marshak wave would scale as $t^{0.78}$. This is the sort of behavior one would expect for a constant laser power and a constant x-ray conversion efficiency. In fact, the x-ray conversion efficiency increases slowly in time: we will assume that $\eta_{c.e.} \sim t^{0.2}$, although measurements (15) from Au disks indicate it could be significantly higher ($\eta_{c.e.} \sim t^{0.4}$). Using this temporal behavior for the x-ray flux (and hence wall loss), we find the temperature to scale as $t^{0.18}$ and the Marshak wave position to scale almost linearly with time:

$$\rho X_M \propto t^{0.9} \kappa_o^{-0.45} \quad (3)$$

This time dependence of the Marshak wave position is very close to that observed experimentally (13) for Au foils of varying thickness exposed to the same hohlraum drive used in these experiments ($\rho X_M \propto t$). Also, as we shall see below, this temporal dependence of the hohlraum temperature is close to that observed experimentally. Hence, this model is probably the best representation of the experiment described here.

To measure the propagation of the heat wave through a given material, we expose the sample to the near Planckian radiation distribution generated inside a standard Nova hohlraum (16). This hohlraum consists of a cylindrical cavity 2700-μm long and 1600 μm in diameter. The laser entrance holes (LEHs) in the ends of the hohlraum are 800 μm in diameter. The Nova laser beams, 5 from each end, enter the hohlraum through these LEHs and strike the inside of the hohlraum at 40° to the surface normal. The laser beams illuminate an annular region that is centered about 900 μm from the hohlraum midplane on each end of the hohlraum. For the experiments described here, the total laser energy is about 27 kJ and the laser pulse duration is 1 ns. The five beams on each end intersect in the LEH and are focused 1000 μm in front of the LEH. The laser beam illumination on the hohlraum wall is roughly a 400 μm x 600 μm ellipse. Thus the average laser intensity on the wall is about 10^{15} W/cm^2. The hohlraum wall temperature is monitored with an absolutely-calibrated multiple-channel soft x-ray spectrometer, DANTE (17). This diagnostic views a portion of the interior hohlraum wall through a small hole in the hohlraum. Care is taken that DANTE does not view a laser spot. In figure 2 we show the temporal profile of the total laser power and the hohlraum temperature. The laser pulse rises rather sharply (~100 ps) and exhibits a small oscillation (±5%) during the "flat-top" region of the pulse. The corresponding hohlraum temperature rises more slowly, reaching a temperature of about 200 eV at 300 ps. beyond this point the hohlraum temperature rises more slowly with time (~$t^{0.15}$) during the nominal constant laser power. This is characteristic of the Marshak wave behavior in a material exposed to a constant flux as opposed to a constant temperature. The test sample package covers a 600-μm high, 1200-μm long slot that is cut in the hohlraum wall and is centered about the hohlraum midplane. Various test materials cover portions of this slot with one section left uncovered to provide a fiducial signal at t=0 (the beginning of the drive pulse). The sample package usually contains a pure Au sample and a sample of the mixture. The fiducial is located

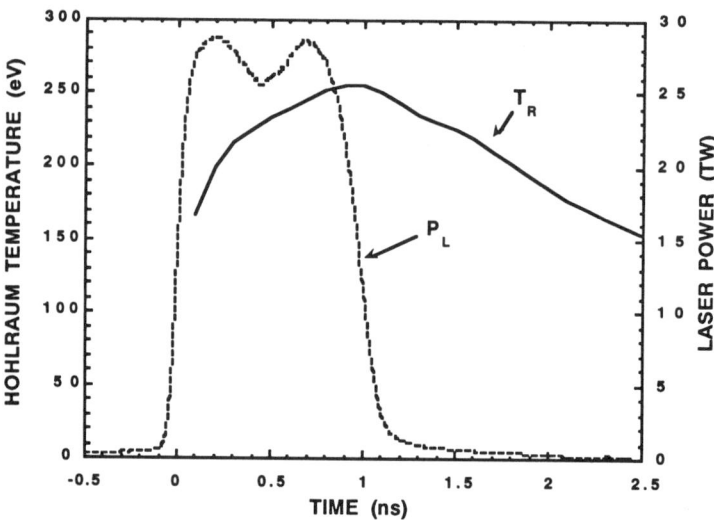

FIGURE 2. Typical 1-ns "square" laser pulse used to drive the hohlraum and measured hohlraum wall temperature.

in the center and the burnthrough foils are mounted on either side of it to mitigate any possible effects of hohlraum temperature nonuniformity in the axial direction (i. e. higher hohlraum temperature closer to the laser spots). The radiation inside the hohlraum drives the Marshak wave into the material. Sample thicknesses are usually on the order of one to a few microns so that the radiation propagates through the sample before the drive (i. e. the laser beams) turns off.

In the measurements described here, we investigate the transport of the thermal wave through pure Au foils and Au/Gd mixtures. Two methods of fabricating the mixture are to co-sputter the elements resulting in an amorphous material of Au-Gd, or to alternate layers of the constituent elements. In the latter method of fabrication, the areal density of each layer must be optically thin to the radiation so that the radiation samples both elements simultaneously and averages over the opacity of two elements. In this experiment, the Au/Gd samples are formed by depositing alternate layers of the two elements on a substrate that is later removed to provide a free standing sample. (A Au/Gd foil that was fabricated by co-sputtering the Au and Gd atoms onto a substrate was found to be unacceptable because the internal stresses generated by this fabrication technique resulted in an extremely fragile material.) Two different samples (of the multilayer variety) corresponding to different atomic fractions of Au and Gd were fabricated. One composite comprises 200 layer pairs of Au and Gd. The thickness of each Au layer is 75 Å and the thickness of each Gd layer is 75 Å. This sample is 33% Gd and 67% Au by atom. The overall thickness is 2.22 µm and the areal density of this sample is equivalent to 1.6 µm of Au (i. e. 3.15 mg/cm^2). The second sample comprises 146 layer pairs; the thickness of each Au layer is 35 Å, and the thickness of each Gd layer is 116 Å. In this case the sample is 67% Gd and 33% Au. The overall thickness of this second sample is 3.02 µm, again corresponding to an equivalent areal mass of Au of 3.15 mg/cm^2. In either of these samples, the individual thickness of each layer of either the Au or the Gd is much less than the range of a photon (100-1000 eV) in these materials. For example, the cold opacity of Au to x-rays between 100 and 1000 eV is between 0.5- and 1.5x10^4 cm^2/g. For solid density, then, the range of a photon is $(\kappa\rho)^{-1}$ or about 1000Å--much larger than the typical layer thickness (~10^2 Å) and much less than the typical sample thickness (~10^4 Å)

The thermal radiation corresponding to the hohlraum drive is monitored as a function of time as it burns through the different foils using a streaked x-ray imager (SXI). The SXI images the foil in one direction using a 20 µm wide imaging slit. The image is dispersed with a transmission grating oriented perpendicular to the imaging slit, and the energy at which we monitor the burnthrough is determined with an offset aperture located behind the transmission grating. All of the data discussed here corresponds to 225-eV radiation. The one-dimensional image is monitored with an x-ray streak camera.

In order to determine the ratio of the Rosseland mean opacity of the mixture to that of Au we used the results of self similar solutions (equation 3) and assume that the two foils are exposed to the same temperature. Independent measurements on the uniformity of the temperature in the region of the test patch indicated the hohlraum temperature to be uniform to within the accuracy of the measurement. The ratio of the Rosseland mean opacities, then depends on the ratio of the burnthrough times squared and the ratio of the areal masses of the foils to the 2.2 power. The foils were fabricated so that this latter quantity is nominally one, but the exact values were used to determining the ratio of opacities. Figure 3 shows the ratio of the Rosseland mean opacities of the Au/Gd foil to that of Au for the two different concentrations of Gd. The errors associated with the measurement correspond to uncertainties in the streak camera sweep speed (±15 ps) and in errors in determining the precise thickness of the foils (±250 Å). In addition, the measured concentration of Gd is accurate to about ±5%.

The solid curve in figure 3 shows the calculated Rosseland mean opacity of the Au/Gd mixture as a function of Gd concentration using the XSN opacity model and assuming a temperature of 250 eV and a density of 1.0 g/cm^2. These calculations indicate the maximum improvement in opacity corresponds to a 50:50 mixture of Au and Gd. The overall improvement in the opacity (over that of pure Au at the same temperature and density) is a factor of 1.7. This curve is normalized to the Rosseland mean opacity of Au at 1.0 g/cm^3 and a temperature of 250 eV (κ_R(Au)=1500 cm^2/g). The opacity of Gd at this temperature and density is 1300 cm^2/g. An independent series of experiments measured Marshak wave propagation through different thicknesses of Au foil (1-3 mm) and monitored the radiation at two different energies (225- and 550-eV). In these experiments the hohlraum temperatures were 250- to 265-eV and they validated the XSN opacity model to within 20% for Au at these temperatures. We indicate the results of those experiments by the datum at the pure Au end of the curve (fraction Gd=0).

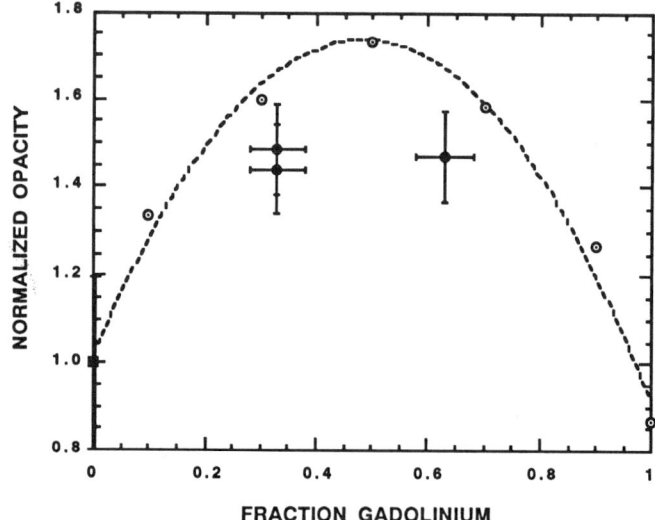

FIGURE 3. Rosseland mean opacity of Au/Gd composite normalized to that of pure Au at a temperature of 250 eV and a density of 1.0 g/cm^2. The open points correspond to the results of XSN calculations, and the solid line is a quadratic fit to those points. The solid points show the normalized opacity of the measurements as determined using equation 3.

The analytic models yield insight into the physical processes that govern the behavior of the Marshak wave propagation and illustrate its sensitivity to the various parameters that affect it (κ_R, T, etc.). However, the real situation is quite complicated. For example, the absorbed laser power may not be constant in time due to parametric instabilities such as SRS, or hohlraum conversion efficiency may behave differently from "disk" conversion efficiency. Furthermore, the opacity calculation shown in figure 3 represents one specific density and temperature; in reality, the experiment samples a range of densities and temperatures. In view of the relative simplicity of the model to the actual time dependent phenomenon, the model is remarkably accurate in that the measurements as interpreted by the model are only about 10 to 12% lower than the XSN calculation. Perhaps an even more sophisticated treatment of bound-bound transitions than XSN employs would account for the discrepancy. The most accurate way to determine the opacity of the composite is to simulate the experiment with a rad-hydro code. Initial LASNEX calculations monitoring a 250 eV photon energy channel yield a burnthrough time in a 2:1 Au/Gd mixture that is 1.3 times longer than that in an equivalent areal mass of pure Au. The measured burnthrough time (at 225 eV) is 1.2 times longer. Similarly, the simulation gives a burnthrough time that is 1.25 times longer than that in an equivalent areal mass of pure Au for a 1:2 Au/Gd mixture while the measured burnthrough time is 1.3 times longer. More detailed calculations are continuing. In addition, we are investigating different concentrations of Au/Gd to determine more precisely the effect of the Gd on the Au opacity.

We have demonstrated that by combining the appropriate elements we can produce a composite whose Rosseland mean opacity is higher than that of either of the constituents at a given temperature and density. The elements must be chosen so that the high opacity regions of one element overlap with the low opacity regions of the other. Because the composites have a higher reemission coefficient to the incident radiation, less energy is lost to the wall. For example in the scale-1 hohlraums used in these experiments the wall losses account for about 75% of the total energy lost (the remainder going out the LEHs; there are no capsule losses in these hohlraums). The observed 50% increase in the opacity of the composite could result in a 15% reduction in wall loss and in 12% less energy required to achieve the same hohlraum temperature. Alternatively, on Nova, the same amount of laser energy could lead to an increase in temperature of about 8 eV (about 12% more flux available to drive a capsule). The reduction in wall loss for the National Ignition Facility, the planned 1.8-MJ laser designed to achieve ignition in an ICF target, is even more dramatic. Here, for the point design ignition target (18), the energy lost to the hohlraum wall would be reduced by about 160 kJ if the pure Au hohlraum was replaced with a hohlraum made of the 37% Au/63% Gd composite. For an alternate ICF reactor based on a 10 MJ heavy ion driver (4) a reduction of 1.7 MJ in wall loss can be realized with the Au/Gd hohlraum wall.

ACKNOWLEDGMENTS

We would like to thank J. D. Kilkenny and B. Hammel for their support in conducting these experiments. We would also like to acknowledge the Nova operations staff and target fabrication group for making these experiments possible. This work was performed under the auspices of The U. S. DOE by the Lawrence Livermore National Laboratory under Contract No. W-7405-Eng-48

REFERENCES

1. Lindl, J. D., McCrory, R. L., and Campbell, E. M., *Physics Today*, **45** (9), pp. 32-40 (1992).

2. Nuckolls, J. H., Wood, L., Thiessen, A. R., and Zimmerman, G. B., *Nature*, **239**, pp. 139-142 (1972).

3. Lindl, J. D., *Physics of Plasmas*, **2**, pp. 3933-4024 (1995).

4. Lindl, J. D., Bangerter, R. O., Mark, J. W.-K., and Pan, Y. L., "Review of target studies for heavy ion fusion," *Heavy Ion Inertial Fusion*, edited by M. Reiser, T. Godlove, and R. O. Bangerter, AIP Conf. Proc. 152 (American Institute of Physics, New York, 1986)

5. Rosen, M. D., "Scaling Law for Radiation Temperature," Lawrence Livermore National Laboratory, Livermore, CA, UCRL-50055-79, 1979 (unpublished).

6. Rosen, M. D., "Marshak Waves: Constant Flux vs. Constant T--a (slight) Paradigm Shift (U)," Lawrence Livermore National Laboratory, Livermore, CA, UCRL-ID 119548, 1995 (unpublished).

7. Pakula, R., and Sigel, R., *Phys. Fluids*, **28**, pp. 232-244 (1985); **29**, p. 1340(E) (1986).

8. Marshak, R. E., *Phys. Fluids*, **1**, pp. 24-29 (1958).

9. Rosseland, S., *Monthly Notices of the Royal Astronomical Society*, **84**, No 7 (1924)

10. Larsen, J. T., "Hyades-a radiation hydrodynamics code for dense plasma studies," *Radiative Properties of Hot Dense Matter: Proceedings of the 4th International Workshop*, W. Goldstein, C. Hooper, J. Gauthier, J. Seely, R. Lee, eds. (World Scientific, Singapore, 1990)

11. Lokke, W. A., and Grasberger, W. H., "XSNQ-U, A non-LTE emission and absorption coefficient subroutine," Lawrence Livermore National Laboratory, Livermore, CA, UCRL-52276, 1977 (unpublished)

12. Sigel, R., et al., *Phys. Rev. Lett.*, **65**, pp. 587-590 (1990)

13. Porter, J. L., and Thiessen, A. R., "Summary of albedo experiments," Lawrence Livermore National Laboratory, Livermore, CA, CLY-92-059, 1992 (unpublished).

14. White, V. J. L., Foster, J. M., Hansom, J. C. V., and Rosen, P. A., *Phys. Rev E* **49**, pp. R4803-R4806 (1994)

15. Ze, F., Kania, D. R., Langer, S. H., Kornblum, H., Kauffman, R., Kilkenny, J., Campbell, E. M., and Tietbolt, G., *J. Appl. Phys.* **66**, pp. 1935-1939, (1989)

16. Kauffman, R. L., Suter, L., Darrow, C. B., Kilkenny, J. D., Kornblum, H. N., Montgomery, D. S., Phillion, D. W., Rosen, M. D., Thiessen, A. R., Wallace, R. J., and Ze, F., *Phys. Rev Lett.* **73**, pp. 2320-2323 (1994).

17. Kornblum, H. N., Kauffman, R. L., and Smith, J. A., *Rev. Sci. Instrum.* **57**, pp. 2179-2181 (1986).

18. Haan, S. W., et al., *Phys. Plasmas* **2** pp. 2480-2486 (1995)

Author Index

A

Abdallah, Jr., J., 131

B

Back, C. A., 123
Bailey, J. E., 245
Becker, K. H., 85
Beiersdorfer, P., 39
Boivin, R. L., 159
Borisov, A. B., 71
Bornath, Th., 215
Boyer, K., 71
Brickhouse, N. S., 31
Brown, G. V., 39
Busquet, M., 271

C

Carlson, A. L., 245
Chan, L. Y., 279
Chilla, J. L. A., 59
Clark, D., 59
Clark, R. E. H., 131
Côté, C. Y., 207
Crespo López-Urrutia, J., 39
Csanak, G., 131

D

Decaux, V., 39
Ditmire, T., 75
Donnelly, T. D., 75
Dunn, G. H., 187

F

Falcone, R. W., 75
Filuk, A. B., 245
Finkenthal, M., 11

G

Fontes, C. J., 131
Foord, M. E., 259
Fournier, K. B., 11

G

Gilbody, H. B., 169
Glenzer, S. H., 109, 123
Goldstein, W. H., 11
Griem, H. R., 159
Guethlein, G., 259

H

Hammer, J. H., 259
Hwang, W., 93

I

Ikhlef, A., 207

J

Jiang, Z., 207
Johnson, D. J., 245

K

Kahn, S. M., 39
Kawachi, T., 223
Kazantsev, S. A., 151
Kearney, K. J., 279
Kieffer, J. C., 207
Kilcrease, D. P., 131
Kim, Y.-K., 93
Kornblum, H. N., 287
Kremp, D., 215
Kruer, W. L., 259

L

Lake, P., 245
Lee, R. W., 123
Levinton, F. M., 143
Liedahl, D. A., 39
Lipschultz, B., 159
Lumma, D., 159

M

MacGowan, B. J., 123
Marconi, M. C., 59
Marmar, E. S., 11, 159
Maron, Y., 245
McCracken, G., 159
McGuire, E. J., 245
McPherson, A., 71
Mehlhorn, T. A., 245
Moreno, J. C., 123
Mostovych, A. N., 279
Murillo, M. S., 231

N

Nash, J. K., 123, 197

O

Orzechowski, T. J., 287

P

Pépin, H., 207
Perry, M. D., 75
Peyrusse, O., 207
Pointon, T. D., 245
Porter, J. L., 287
Post, D., 3
Powers, L. V., 123
Prenzel, R., 215

R

Reed, K. J., 39
Renk, T. J., 245
Rhodes, C. K., 71
Rice, J. E., 11
Rocca, J. J., 59
Rosen, M. D., 287
Rost, J. C., 159
Rowan, W. L., 21
Rudd, M. E., 93

S

Safronova, U. I., 11
Savin, D. W., 39
Schlanges, M., 215
Schultz, D. R., 197
Shepard, T. D., 123
Shlyaptsev, V. N., 59
Shull, J. M., 47
Stambulchik, E., 245
Stygar, W. A., 245
Suter, L. J., 287

T

Tarnovsky, V., 85
Terry, J. L., 11, 159
Thiessen, A. R., 287
Tomasel, F. G., 59

W

Wallace, R., 287
Weaver, J. L., 159
Welch, B. L., 159
Widmann, K., 39
Wiese, W. L., 177
Wilks, S. C., 259

Y

Young, P. E., 259

AIP Conference Proceedings

	Title	L.C. Number	ISBN
No. 140	Boron-Rich Solids (Albuquerque, NM, 1985)	86-70246	0-88318-339-0
No. 141	Gamma-Ray Bursts (Stanford, CA, 1984)	86-70761	0-88318-340-4
No. 142	Nuclear Structure at High Spin, Excitation, and Momentum Transfer (Indiana University, 1985)	86-70837	0-88318-341-2
No. 143	Mexican School of Particles and Fields (Oaxtepec, México, 1984)	86-81187	0-88318-342-0
No. 144	Magnetospheric Phenomena in Astrophysics (Los Alamos, NM, 1984)	86-71149	0-88318-343-9
No. 145	Polarized Beams at SSC & Polarized Antiprotons (Ann Arbor, MI & Bodega Bay, CA, 1985)	86-71343	0-88318-344-7
No. 146	Advances in Laser Science—I (Dallas, TX, 1985)	86-71536	0-88318-345-5
No. 147	Short Wavelength Coherent Radiation: Generation and Applications (Monterey, CA, 1986)	86-71674	0-88318-346-3
No. 148	Space Colonization: Technology and The Liberal Arts (Geneva, NY, 1985)	86-71675	0-88318-347-1
No. 149	Physics and Chemistry of Protective Coatings (Universal City, CA, 1985)	86-72019	0-88318-348-X
No. 150	Intersections Between Particle and Nuclear Physics (Lake Louise, Canada, 1986)	86-72018	0-88318-349-8
No. 151	Neural Networks for Computing (Snowbird, UT, 1986)	86-72481	0-88318-351-X
No. 152	Heavy Ion Inertial Fusion (Washington, DC, 1986)	86-73185	0-88318-352-8
No. 153	Physics of Particle Accelerators (SLAC Summer School, 1985) (Fermilab Summer School, 1984)	87-70103	0-88318-353-6
No. 154	Physics and Chemistry of Porous Media—II (Ridgefield, CT, 1986)	83-73640	0-88318-354-4
No. 155	The Galactic Center: Proceedings of the Symposium Honoring C. H. Townes (Berkeley, CA, 1986)	86-73186	0-88318-355-2
No. 156	Advanced Accelerator Concepts (Madison, WI, 1986)	87-70635	0-88318-358-0
No. 157	Stability of Amorphous Silicon Alloy Materials and Devices (Palo Alto, CA, 1987)	87-70990	0-88318-359-9

	Title	L.C. Number	ISBN
No. 158	Production and Neutralization of Negative Ions and Beams (Brookhaven, NY, 1986)	87-71695	0-88318-358-7
No. 159	Applications of Radio-Frequency Power to Plasma: Seventh Topical Conference (Kissimmee, FL, 1987)	87-71812	0-88318-359-5
No. 160	Advances in Laser Science—II (Seattle, WA, 1986)	87-71962	0-88318-360-9
No. 161	Electron Scattering in Nuclear and Particle Science: In Commemoration of the 35th Anniversary of the Lyman-Hanson-Scott Experiment (Urbana, IL, 1986)	87-72403	0-88318-361-7
No. 162	Few-Body Systems and Multiparticle Dynamics (Crystal City, VA, 1987)	87-72594	0-88318-362-5
No. 163	Pion–Nucleus Physics: Future Directions and New Facilities at LAMPF (Los Alamos, NM, 1987)	87-72961	0-88318-363-3
No. 164	Nuclei Far from Stability: Fifth International Conference (Rosseau Lake, ON, 1987)	87-73214	0-88318-364-1
No. 165	Thin Film Processing and Characterization of High-Temperature Superconductors (Anaheim, CA, 1987)	87-73420	0-88318-365-X
No. 166	Photovoltaic Safety (Denver, CO, 1988)	88-42854	0-88318-366-8
No. 167	Deposition and Growth: Limits for Microelectronics (Anaheim, CA, 1987)	88-71432	0-88318-367-6
No. 168	Atomic Processes in Plasmas (Santa Fe, NM, 1987)	88-71273	0-88318-368-4
No. 169	Modern Physics in America: A Michelson-Morley Centennial Symposium (Cleveland, OH, 1987)	88-71348	0-88318-369-2
No. 170	Nuclear Spectroscopy of Astrophysical Sources (Washington, DC, 1987)	88-71625	0-88318-370-6
No. 171	Vacuum Design of Advanced and Compact Synchrotron Light Sources (Upton, NY, 1988)	88-71824	0-88318-371-4
No. 172	Advances in Laser Science—III: Proceedings of the International Laser Science Conference (Atlantic City, NJ, 1987)	88-71879	0-88318-372-2
No. 173	Cooperative Networks in Physics Education (Oaxtepec, Mexico, 1987)	88-72091	0-88318-373-0

Title	L.C. Number	ISBN
No. 174 Radio Wave Scattering in the Interstellar Medium (San Diego, CA, 1988)	88-72092	0-88318-374-9
No. 175 Non-neutral Plasma Physics (Washington, DC, 1988)	88-72275	0-88318-375-7
No. 176 Intersections Between Particle and Nuclear Physics (Third International Conference) (Rockport, ME, 1988)	88-62535	0-88318-376-5
No. 177 Linear Accelerator and Beam Optics Codes (La Jolla, CA, 1988)	88-46074	0-88318-377-3
No. 178 Nuclear Arms Technologies in the 1990s (Washington, DC, 1988)	88-83262	0-88318-378-1
No. 179 The Michelson Era in American Science: 1870–1930 (Cleveland, OH, 1987)	88-83369	0-88318-379-X
No. 180 Frontiers in Science: International Symposium (Urbana, IL, 1987)	88-83526	0-88318-380-3
No. 181 Muon-Catalyzed Fusion (Sanibel Island, FL, 1988)	88-83636	0-88318-381-1
No. 182 High T_c Superconducting Thin Films, Devices, and Applications (Atlanta, GA, 1988)	88-03947	0-88318-382-X
No. 183 Cosmic Abundances of Matter (Minneapolis, MN, 1988)	89-80147	0-88318-383-8
No. 184 Physics of Particle Accelerators (Ithaca, NY, 1988)	89-83575	0-88318-384-6
No. 185 Glueballs, Hybrids, and Exotic Hadrons (Upton, NY, 1988)	89-83513	0-88318-385-4
No. 186 High-Energy Radiation Background in Space (Sanibel Island, FL, 1987)	89-83833	0-88318-386-2
No. 187 High-Energy Spin Physics (Minneapolis, MN, 1988)	89-83948	0-88318-387-0
No. 188 International Symposium on Electron Beam Ion Sources and their Applications (Upton, NY, 1988)	89-84343	0-88318-388-9
No. 189 Relativistic, Quantum Electrodynamic, and Weak Interaction Effects in Atoms (Santa Barbara, CA, 1988)	89-84431	0-88318-389-7
No. 190 Radio-frequency Power in Plasmas (Irvine, CA, 1989)	89-45805	0-88318-397-8
No. 191 Advances in Laser Science—IV (Atlanta, GA, 1988)	89-85595	0-88318-391-9

	Title	L.C. Number	ISBN
No. 192	Vacuum Mechatronics (First International Workshop) (Santa Barbara, CA, 1989)	89-45905	0-88318-394-3
No. 193	Advanced Accelerator Concepts (Lake Arrowhead, CA, 1989)	89-45914	0-88318-393-5
No. 194	Quantum Fluids and Solids—1989 (Gainesville, FL, 1989)	89-81079	0-88318-395-1
No. 195	Dense Z-Pinches (Laguna Beach, CA, 1989)	89-46212	0-88318-396-X
No. 196	Heavy Quark Physics (Ithaca, NY, 1989)	89-81583	0-88318-644-6
No. 197	Drops and Bubbles (Monterey, CA, 1988)	89-46360	0-88318-392-7
No. 198	Astrophysics in Antarctica (Newark, DE, 1989)	89-46421	0-88318-398-6
No. 199	Surface Conditioning of Vacuum Systems (Los Angeles, CA, 1989)	89-82542	0-88318-756-6
No. 200	High T_c Superconducting Thin Films: Processing, Characterization, and Applications (Boston, MA, 1989)	90-80006	0-88318-759-0
No. 201	QED Structure Functions (Ann Arbor, MI, 1989)	90-80229	0-88318-671-3
No. 202	NASA Workshop on Physics From a Lunar Base (Stanford, CA, 1989)	90-55073	0-88318-646-2
No. 203	Particle Astrophysics: The NASA Cosmic Ray Program for the 1990s and Beyond (Greenbelt, MD, 1989)	90-55077	0-88318-763-9
No. 204	Aspects of Electron-Molecule Scattering and Photoionization (New Haven, CT, 1989)	90-55175	0-88318-764-7
No. 205	The Physics of Electronic and Atomic Collisions (XVI International Conference) (New York, NY, 1989)	90-53183	0-88318-390-0
No. 206	Atomic Processes in Plasmas (Gaithersburg, MD, 1989)	90-55265	0-88318-769-8
No. 207	Astrophysics from the Moon (Annapolis, MD, 1990)	90-55582	0-88318-770-1
No. 208	Current Topics in Shock Waves (Bethlehem, PA, 1989)	90-55617	0-88318-776-0
No. 209	Computing for High Luminosity and High Intensity Facilities (Santa Fe, NM, 1990)	90-55634	0-88318-786-8
No. 210	Production and Neutralization of Negative Ions and Beams (Brookhaven, NY, 1990)	90-55316	0-88318-786-8

	Title	L.C. Number	ISBN
No. 211	High-Energy Astrophysics in the 21st Century (Taos, NM, 1989)	90-55644	0-88318-803-1
No. 212	Accelerator Instrumentation (Brookhaven, NY, 1989)	90-55838	0-88318-645-4
No. 213	Frontiers in Condensed Matter Theory (New York, NY, 1989)	90-6421	0-88318-771-X 0-88318-772-8 (pbk.)
No. 214	Beam Dynamics Issues of High-Luminosity Asymmetric Collider Rings (Berkeley, CA, 1990)	90-55857	0-88318-767-1
No. 215	X-Ray and Inner-Shell Processes (Knoxville, TN 1990)	90-84700	0-88318-790-6
No. 216	Spectral Line Shapes, Vol. 6 (Austin, TX 1990)	90-06278	0-88318-791-4
No. 217	Space Nuclear Power Systems (Albuquerque, NM 1991)	90-56220	0-88318-838-4
No. 218	Positron Beams for Solids and Surfaces (London, Canada 1990)	90-56407	0-88318-842-2
No. 219	Superconductivity and Its Applications (Buffalo, NY 1990)	91-55020	0-88318-835-X
No. 220	High Energy Gamma-Ray Astronomy (Ann Arbor, MI 1990)	91-70876	0-88318-812-0
No. 221	Particle Production Near Threshold (Nashville, IN 1990)	91-55134	0-88318-829-5
No. 222	After the First Three Minutes (College Park, MD 1990)	91-55214	0-88318-828-7
No. 223	Polarized Collider Workshop (University Park, PA 1990)	91-71303	0-88318-826-0
No. 224	LAMPF Workshop on (π, K) Physics (Los Alamos, NM 1990)	91-71304	0-88318-825-2
No. 225	Half Collision Resonance Phenomena in Molecules (Caracas, Venezuela 1990)	91-55210	0-88318-840-6
No. 226	The Living Cell in Four Dimensions (Gif sur Yvette, France 1990)	91-55209	0-88318-794-9
No. 227	Advanced Processing and Characterization Technologies (Clearwater, FL 1991)	91-55194	0-88318-910-0
No. 228	Anomalous Nuclear Effects in Deuterium/ Solid Systems (Provo, UT 1990)	91-55245	0-88318-833-3
No. 229	Accelerator Instrumentation (Batavia, IL 1990)	91-55347	0-88318-832-1
No. 230	Nonlinear Dynamics and Particle Acceleration (Tsukuba, Japan 1990)	91-55348	0-88318-824-4

	Title	L.C. Number	ISBN
No. 231	Boron-Rich Solids (Albuquerque, NM 1990)	91-53024	0-88318-793-4
No. 232	Gamma-Ray Line Astrophysics (Paris-Saclay, France 1990)	91-55492	0-88318-875-9
No. 233	Atomic Physics 12 (Ann Arbor, MI 1990)	91-55595	088318-811-2
No. 234	Amorphous Silicon Materials and Solar Cells (Denver, CO 1991)	91-55575	088318-831-7
No. 235	Physics and Chemistry of MCT and Novel IR Detector Materials (San Francisco, CA 1990)	91-55493	0-88318-931-3
No. 236	Vacuum Design of Synchrotron Light Sources (Argonne, IL 1990)	91-55527	0-88318-873-2
No. 237	Kent M. Terwilliger Memorial Symposium (Ann Arbor, MI 1989)	91-55576	0-88318-788-4
No. 238	Capture Gamma-Ray Spectroscopy (Pacific Grove, CA 1990)	91-57923	0-88318-830-9
No. 239	Advances in Biomolecular Simulations (Obernai, France 1991)	91-58106	0-88318-940-2
No. 240	Joint Soviet-American Workshop on the Physics of Semiconductor Lasers (Leningrad, USSR 1991)	91-58537	0-88318-936-4
No. 241	Scanned Probe Microscopy (Santa Barbara, CA 1991)	91-76758	0-88318-816-3
No. 242	Strong, Weak, and Electromagnetic Interactions in Nuclei, Atoms, and Astrophysics: A Workshop in Honor of Stewart D. Bloom's Retirement (Livermore, CA 1991)	91-76876	0-88318-943-7
No. 243	Intersections Between Particle and Nuclear Physics (Tucson, AZ 1991)	91-77580	0-88318-950-X
No. 244	Radio Frequency Power in Plasmas (Charleston, SC 1991)	91-77853	0-88318-937-2
No. 245	Basic Space Science (Bangalore, India 1991)	91-78379	0-88318-951-8
No. 246	Space Nuclear Power Systems (Albuquerque, NM 1992)	91-58793	1-56396-027-3 1-56396-026-5 (pbk.)
No. 247	Global Warming: Physics and Facts (Washington, DC 1991)	91-78423	0-88318-932-1
No. 248	Computer-Aided Statistical Physics (Taipei, Taiwan 1991)	91-78378	0-88318-942-9
No. 249	The Physics of Particle Accelerators (Upton, NY 1989, 1990)	92-52843	0-88318-789-2

Title	L.C. Number	ISBN
No. 250 Towards a Unified Picture of Nuclear Dynamics (Nikko, Japan 1991)	92-70143	0-88318-951-8
No. 251 Superconductivity and its Applications (Buffalo, NY 1991)	92-52726	1-56396-016-8
No. 252 Accelerator Instrumentation (Newport News, VA 1991)	92-70356	0-88318-934-8
No. 253 High-Brightness Beams for Advanced Accelerator Applications (College Park, MD 1991)	92-52705	0-88318-947-X
No. 254 Testing the AGN Paradigm (College Park, MD 1991)	92-52780	1-56396-009-5
No. 255 Advanced Beam Dynamics Workshop on Effects of Errors in Accelerators, Their Diagnosis and Corrections (Corpus Christi, TX 1991)	92-52842	1-56396-006-0
No. 256 Slow Dynamics in Condensed Matter (Fukuoka, Japan 1991)	92-53120	0-88318-938-0
No. 257 Atomic Processes in Plasmas (Portland, ME 1991)	91-08105	0-88318-939-9
No. 258 Synchrotron Radiation and Dynamic Phenomena (Grenoble, France 1991)	92-53790	1-56396-008-7
No. 259 Future Directions in Nuclear Physics with 4π Gamma Detection Systems of the New Generation (Strasbourg, France 1991)	92-53222	0-88318-952-6
No. 260 Computational Quantum Physics (Nashville, TN 1991)	92-71777	0-88318-933-X
No. 261 Rare and Exclusive B&K Decays and Novel Flavor Factories (Santa Monica, CA 1991)	92-71873	1-56396-055-9
No. 262 Molecular Electronics—Science and Technology (St. Thomas, Virgin Islands 1991)	92-72210	1-56396-041-9
No. 263 Stress-Induced Phenomena in Metallization: First International Workshop (Ithaca, NY 1991)	92-72292	1-56396-082-6
No. 264 Particle Acceleration in Cosmic Plasmas (Newark, DE 1991)	92-73316	0-88318-948-8
No. 265 Gamma-Ray Bursts (Huntsville, AL 1991)	92-73456	1-56396-018-4
No. 266 Group Theory in Physics (Cocoyoc, Morelos, Mexico 1991)	92-73457	1-56396-101-6
No. 267 Electromechanical Coupling of the Solar Atmosphere (Capri, Italy 1991)	92-82717	1-56396-110-5

	Title	L.C. Number	ISBN
No. 268	Photovoltaic Advanced Research & Development Project (Denver, CO 1992)	92-74159	1-56396-056-7
No. 269	CEBAF 1992 Summer Workshop (Newport News, VA 1992)	92-75403	1-56396-067-2
No. 270	Time Reversal—The Arthur Rich Memorial Symposium (Ann Arbor, MI 1991)	92-83852	1-56396-105-9
No. 271	Tenth Symposium Space Nuclear Power and Propulsion (Vols. I–III) (Albuquerque, NM 1993)	92-75162	1-56396-137-7 (set)
No. 272	Proceedings of the XXVI International Conference on High Energy Physics (Vols. I and II) (Dallas, TX 1992)	93-70412	1-56396-127-X (set)
No. 273	Superconductivity and Its Applications (Buffalo, NY 1992)	93-70502	1-56396-189-X
No. 274	VIth International Conference on the Physics of Highly Charged Ions (Manhattan, KS 1992)	93-70577	1-56396-102-4
No. 275	Atomic Physics 13 (Munich, Germany 1992)	93-70826	1-56396-057-5
No. 276	Very High Energy Cosmic-Ray Interactions: VIIth International Symposium (Ann Arbor, MI 1992)	93-71342	1-56396-038-9
No. 277	The World at Risk: Natural Hazards and Climate Change (Cambridge, MA 1992)	93-71333	1-56396-066-4
No. 278	Back to the Galaxy (College Park, MD 1992)	93-71543	1-56396-227-6
No. 279	Advanced Accelerator Concepts (Port Jefferson, NY 1992)	93-71773	1-56396-191-1
No. 280	Compton Gamma-Ray Observatory (St. Louis, MO 1992)	93-71830	1-56396-104-0
No. 281	Accelerator Instrumentation Fourth Annual Workshop (Berkeley, CA 1992)	93-072110	1-56396-190-3
No. 282	Quantum 1/f Noise & Other Low Frequency Fluctuations in Electronic Devices (St. Louis, MO 1992)	93-072366	1-56396-252-7
No. 283	Earth and Space Science Information Systems (Pasadena, CA 1992)	93-072360	1-56396-094-X
No. 284	US-Japan Workshop on Ion Temperature Gradient-Driven Turbulent Transport (Austin, TX 1993)	93-72460	1-56396-221-7

	Title	L.C. Number	ISBN
No. 285	Noise in Physical Systems and 1/f Fluctuations (St. Louis, MO 1993)	93-72575	1-56396-270-5
No. 286	Ordering Disorder: Prospect and Retrospect in Condensed Matter Physics: Proceedings of the Indo-U.S. Workshop (Hyderabad, India 1993)	93-072549	1-56396-255-1
No. 287	Production and Neutralization of Negative Ions and Beams: Sixth International Symposium (Upton, NY 1992)	93-72821	1-56396-103-2
No. 288	Laser Ablation: Mechanismas and Applications-II: Second International Conference (Knoxville, TN 1993)	93-73040	1-56396-226-8
No. 289	Radio Frequency Power in Plasmas: Tenth Topical Conference (Boston, MA 1993)	93-72964	1-56396-264-0
No. 290	Laser Spectroscopy: XIth International Conference (Hot Springs, VA 1993)	93-73050	1-56396-262-4
No. 291	Prairie View Summer Science Academy (Prairie View, TX 1992)	93-73081	1-56396-133-4
No. 292	Stability of Particle Motion in Storage Rings (Upton, NY 1992)	93-73534	1-56396-225-X
No. 293	Polarized Ion Sources and Polarized Gas Targets (Madison, WI 1993)	93-74102	1-56396-220-9
No. 294	High-Energy Solar Phenomena: A New Era of Spacecraft Measurements (Waterville Valley, NH 1993)	93-74147	1-56396-291-8
No. 295	The Physics of Electronic and Atomic Collisions: XVIII International Conference (Aarhus, Denmark, 1993)	93-74103	1-56396-290-X
No. 296	The Chaos Paradigm: Developments an Applications in Engineering and Science (Mystic, CT 1993)	93-74146	1-56396-254-3
No. 297	Computational Accelerator Physics (Los Alamos, NM 1993)	93-74205	1-56396-222-5
No. 298	Ultrafast Reaction Dynamics and Solvent Effects (Royaumont, France 1993)	93-074354	1-56396-280-2
No. 299	Dense Z-Pinches: Third International Conference (London, 1993)	93-074569	1-56396-297-7
No. 300	Discovery of Weak Neutral Currents: The Weak Interaction Before and After (Santa Monica, CA 1993)	94-70515	1-56396-306-X

	Title	L.C. Number	ISBN
No. 301	Eleventh Symposium Space Nuclear Power and Propulsion (3 Vols.) (Albuquerque, NM 1994)	92-75162	1-56396-305-1 (Set) 156396-301-9 (pbk. set)
No. 302	Lepton and Photon Interactions/ XVI International Symposium (Ithaca, NY 1993)	94-70079	1-56396-106-7
No. 303	Slow Positron Beam Techniques for Solids and Surfaces Fifth International Workshop (Jackson Hole, WY 1992)	94-71036	1-56396-267-5
No. 304	The Second Compton Symposium (College Park, MD 1993)	94-70742	1-56396-261-6
No. 305	Stress-Induced Phenomena in Metallization Second International Workshop (Austin, TX 1993)	94-70650	1-56396-251-9
No. 306	12th NREL Photovoltaic Program Review (Denver, CO 1993)	94-70748	1-56396-315-9
No. 307	Gamma-Ray Bursts Second Workshop (Huntsville, AL 1993)	94-71317	1-56396-336-1
No. 308	The Evolution of X-Ray Binaries (College Park, MD 1993)	94-76853	1-56396-329-9
No. 309	High-Pressure Science and Technology—1993 (Colorado Springs, CO 1993)	93-72821	1-56396-219-5 (Set)
No. 310	Analysis of Interplanetary Dust (Houston, TX 1993)	94-71292	1-56396-341-8
No. 311	Physics of High Energy Particles in Toroidal Systems (Irvine, CA 1993)	94-72098	1-56396-364-7
No. 312	Molecules and Grains in Space (Mont Sainte-Odile, France 1993)	94-72615	1-56396-355-8
No. 313	The Soft X-Ray Cosmos ROSAT Science Symposium (College Park, MD 1993)	94-72499	1-56396-327-2
No. 314	Advances in Plasma Physics Thomas H. Stix Symposium (Princeton, NJ 1992)	94-72721	1-56396-372-8
No. 315	Orbit Correction and Analysis in Circular Accelerators (Upton, NY 1993)	94-72257	1-56396-373-6
No. 316	Thirteenth International Conference on Thermoelectrics (Kansas City, Missouri 1994)	95-75634	1-56396-444-9

	Title	L.C. Number	ISBN
No. 317	Fifth Mexican School of Particles and Fields (Guanajuato, Mexico 1992)	94-72720	1-56396-378-7
No. 318	Laser Interaction and Related Plasma Phenomena 11th International Workshop (Monterey, CA 1993)	94-78097	1-56396-324-8
No. 319	Beam Instrumentation Workshop (Santa Fe, NM 1993)	94-78279	1-56396-389-2
No. 320	Basic Space Science (Lagos, Nigeria 1993)	94-79350	1-56396-328-0
No. 321	The First NREL Conference on Thermophotovoltaic Generation of Electricity (Copper Mountain, CO 1994)	94-72792	1-56396-353-1
No. 322	Atomic Processes in Plasmas Ninth APS Topical Conference (San Antonio, TX)	94-72923	1-56396-411-2
No. 323	Atomic Physics 14 Fourteenth International Conference on Atomic Physics (Boulder, CO 1994)	94-73219	1-56396-348-5
No. 324	Twelfth Symposium on Space Nuclear Power and Propulsion (Albuquerque, NM 1995)	94-73603	1-56396-427-9
No. 325	Conference on NASA Centers for Commercial Development of Space (Albuquerque, NM 1995)	94-73604	1-56396-431-7
No. 326	Accelerator Physics at the Superconducting Super Collider (Dallas, TX 1992-1993)	94-73609	1-56396-354-X
No. 327	Nuclei in the Cosmos III Third International Symposium on Nuclear Astrophysics (Assergi, Italy 1994)	95-75492	1-56396-436-8
No. 328	Spectral Line Shapes, Volume 8 12th ICSLS (Toronto, Canada 1994)	94-74309	1-56396-326-4
No. 329	Resonance Ionization Spectroscopy 1994 Seventh International Symposium (Bernkastel-Kues, Germany 1994)	95-75077	1-56396-437-6
No. 330	E.C.C.C. 1 Computational Chemistry F.E.C.S. Conference (Nancy, France 1994)	95-75843	1-56396-457-0
No. 331	Non-Neutral Plasma Physics II (Berkeley, CA 1994)	95-79630	1-56396-441-4
No. 332	X-Ray Lasers 1994 Fourth International Colloquium (Williamsburg, VA 1994)	95-76067	1-56396-375-2

	Title	L.C. Number	ISBN
No. 333	Beam Instrumentation Workshop (Vancouver, B. C., Canada 1994)	95-79635	1-56396-352-3
No. 334	Few-Body Problems in Physics (Williamsburg, VA 1994)	95-76481	1-56396-325-6
No. 335	Advanced Accelerator Concepts (Fontana, WI 1994)	95-78225	1-56396-476-7 (Set) 1-56396-474-0 (Book) 1-56396-475-9 (CD-Rom)
No. 336	Dark Matter (College Park, MD 1994)	95-76538	1-56396-438-4
No. 337	Pulsed RF Sources for Linear Colliders (Montauk, NY 1994)	95-76814	1-56396-408-2
No. 338	Intersections Between Particle and Nuclear Physics 5th Conference (St. Petersburg, FL 1994)	95-77076	1-56396-335-3
No. 339	Polarization Phenomena in Nuclear Physics Eighth International Symposium (Bloomington, IN 1994)	95-77216	1-56396-482-1
No. 340	Strangeness in Hadronic Matter (Tucson, AZ 1995)	95-77477	1-56396-489-9
No. 341	Volatiles in the Earth and Solar System (Pasadena, CA 1994)	95-77911	1-56396-409-0
No. 342	CAM-94 Physics Meeting (Cacun, Mexico 1994)	95-77851	1-56396-491-0
No. 343	High Energy Spin Physics Eleventh International Symposium (Bloomington, IN 1994)	95-78431	1-56396-374-4
No. 344	Nonlinear Dynamics in Particle Accelerators: Theory and Experiments (Arcidosso, Italy 1994)	95-78135	1-56396-446-5
No. 345	International Conference on Plasma Physics ICPP 1994 (Foz do Iguaçu, Brazil 1994)	95-78438	1-56396-496-1
No. 346	International Conference on Accelerator-Driven Transmutation Technologies and Applications (Las Vegas, NV 1994)	95-78691	1-56396-505-4
No. 347	Atomic Collisions: A Symposium in Honor of Christopher Bottcher (1945-1993) (Oak Ridge, TN 1994)	95-78689	1-56396-322-1
No. 348	Unveiling the Cosmic Infrared Background (College Park, MD, 1995)	95-83477	1-56396-508-9

Title	L.C. Number	ISBN
No. 349 Workshop on the Tau/Charm Factory (Argonne, IL, 1995)	95-81467	1-56396-523-2
No. 350 International Symposium on Vector Boson Self-Interactions (Los Angeles, CA 1995)	95-79865	1-56396-520-8
No. 351 The Physics of Beams Andrew Sessler Symposium (Los Angeles, CA 1993)	95-80479	1-56396-376-0
No. 352 Physics Potential and Development of $\mu^+\mu^-$ Colliders: Second Workshop (Sausalito, CA 1994)	95-81413	1-56396-506-2
No. 353 13th NREL Photovoltaic Program Review (Lakewood, CO 1995)	95-80662	1-56396-510-0
No. 354 Organic Coatings (Paris, France, 1995)	96-83019	1-56396-535-6
No. 355 Eleventh Topical Conference on Radio Frequency Power in Plasmas (Palm Springs, CA 1995)	95-80867	1-56396-536-4
No. 356 The Future of Accelerator Physics (Austin, TX 1994)	96-83292	1-56396-541-0
No. 357 10th Topical Workshop on Proton-Antiproton Collider Physics (Batavia, IL 1995)	95-83078	1-56396-543-7
No. 358 The Second NREL Conference on Thermophotovoltaic Generation of Electricity	95-83335	1-56396-509-7
No. 359 Workshops and Particles and Fields and Phenomenology of Fundamental Interactions (Puebla, Mexico 1995)	96-85996	1-56396-548-8
No. 360 The Physics of Electronic and Atomic Collisions XIX International Conference (Whistler, Canada, 1995)	95-83671	1-56396-440-6
No. 361 Space Technology and Applications International Forum (Albuquerque, NM 1996)	95-83440	1-56396-568-2
No. 362 Two-Center Effects in Ion-Atom Collisions (Lincoln, NE 1994)	96-83379	1-56396-342-6
No. 363 Phenomena in Ionized Gases XXII ICPIG (Hoboken, NJ, 1995)	96-83294	1-56396-550-X
No. 364 Fast Elementary Processes in Chemical and Biological Systems (Villeneuve d'Ascq, France, 1995)	96-83624	1-56396-564-X

	Title	L.C. Number	ISBN
No. 365	Latin-American School of Physics XXX ELAF Group Theory and Its Applications (México City, México, 1995)	96-83489	1-56396-567-4
No. 366	High Velocity Neutron Stars and Gamma-Ray Bursts (La Jolla, CA 1995)	96-84067	1-56396-593-3
No. 367	Micro Bunches Workshop (Upton, NY, 1995)	96-83482	1-56396-555-0
No. 368	Acoustic Particle Velocity Sensors: Design, Performance and Applications (Mystic, CT, 1995)	96-83548	1-56396-549-6
No. 369	Laser Interaction and Related Plasma Phenomena (Osaka, Japan 1995)	96-85009	1-56396-445-7
No. 370	Shock Compression of Condensed Matter-1995 (Seattle, WA 1995)	96-84595	1-56396-566-6
No. 371	Sixth Quantum 1/f Noise and Other Low Frequency Fluctuations in Electronic Devices Symposium (St. Louis, MO, 1994)	96-84200	1-56396-410-4
No. 372	Beam Dynamics and Technology Issues for + - Colliders 9th Advanced ICFA Beam Dynamics Workshop (Montauk, NY, 1995)	96-84189	1-56396-554-2
No. 373	Stress-Induced Phenomena in Metallization (Palo Alto, CA 1995)	96-84949	1-56396-439-2
No. 374	High Energy Solar Physics (Greenbelt, MD 1995)	96-84513	1-56396-542-9
No. 375	Chaotic, Fractal, and Nonlinear Signal Processing (Mystic, CT 1995)	96-85356	1-56396-443-0
No. 376	Chaos and the Changing Nature of Science and Medicine: An Introduction (Mobile, AL 1995)	96-85220	1-56396-442-2
No. 377	Space Charge Dominated Beams and Applications of High Brightness Beams (Bloomington, IN 1995)	96-85165	1-56396-625-7
No. 378	Surfaces, Vacuum, and Their Applications (Cancun, Mexico 1994)	96-85594	1-56396-418-X
No. 381	Atomic Processes in Plasmas (San Francisco, CA 1996)	96-86304	1-56396-552-6